I0036847

# Structure, Chemical Analysis, Biosynthesis, Metabolism, Molecular Engineering and Biological Functions of Phytoalexins

Special Issue Editor

Philippe Jeandet

MDPI • Basel • Beijing • Wuhan • Barcelona • Belgrade

**MDPI**

*Special Issue Editor*
Philippe Jeandet
University of Reims
France

*Editorial Office*
MDPI AG
St. Alban-Anlage 66
Basel, Switzerland

This edition is a reprint of the Special Issue published online in the open access journal *Molecules* (ISSN 1420-3049) from 2016–2018 (available at: http://www.mdpi.com/journal/molecules/special_issues/phytoalexins).

For citation purposes, cite each article independently as indicated on the article page online and as indicated below:

Lastname, F.M.; Lastname, F.M. Article title. *Journal Name*. **Year**. Article number, page range.

**First Edition 2018**

**ISBN 978-3-03842-755-1 (Pbk)**
**ISBN 978-3-03842-756-8 (PDF)**

# Table of Contents

About the Special Issue Editor . . . . . . . . . . . . . . . . . . . . . . . . . . . . . . . . . . . . . v

Preface to "Structure, Chemical Analysis, Biosynthesis, Metabolism, Molecular Engineering
and Biological Functions of Phytoalexins" . . . . . . . . . . . . . . . . . . . . . . . . . . vii

**Philippe Jeandet**
Structure, Chemical Analysis, Biosynthesis, Metabolism, Molecular Engineering, and
Biological Functions of Phytoalexins
doi: 10.3390/molecules23010061 . . . . . . . . . . . . . . . . . . . . . . . . . . 1

**Agnieszka Woźniak, Kinga Drzewiecka, Jacek Kęsy, Łukasz Marczak, Dorota Narożna,**
**Marcin Grobela, Rafał Motała, Jan Bocianowski and Iwona Morkunas**
The Influence of Lead on Generation of Signalling Molecules and Accumulation of
Flavonoids in Pea Seedlings in Response to Pea Aphid Infestation
doi: 10.3390/molecules22091404 . . . . . . . . . . . . . . . . . . . . . . . . . . 5

**Kelli Farrell, Md Asraful Jahan and Nik Kovinich**
Distinct Mechanisms of Biotic and Chemical Elicitors Enable Additive Elicitation of the
Anticancer Phytoalexin Glyceollin I
doi: 10.3390/molecules22081261 . . . . . . . . . . . . . . . . . . . . . . . . . . 35

**Sameh Selim, Jean Sanssené, Stéphanie Rossard and Josiane Courtois**
Systemic Induction of the Defensin and Phytoalexin Pisatin Pathways in Pea (*Pisum sativum*)
against *Aphanomyces euteiches* by Acetylated and Nonacetylated Oligogalacturonides
doi: 10.3390/molecules22061017 . . . . . . . . . . . . . . . . . . . . . . . . . . 48

**Leo-Paul Tisserant, Aziz Aziz, Nathalie Jullian, Philippe Jeandet, Christophe Clément,**
**Eric Courot and Michèle Boitel-Conti**
Enhanced Stilbene Production and Excretion in *Vitis vinifera* cv Pinot Noir Hairy
Root Cultures
doi: 10.3390/molecules21121703 . . . . . . . . . . . . . . . . . . . . . . . . . . 65

**Elías Hurtado-Gaitán, Susana Sellés-Marchart, Ascensión Martnez-Márquez,**
**Antonio Samper-Herrero and Roque Bru-Martnez**
A Focused Multiple Reaction Monitoring (MRM) Quantitative Method for Bioactive
Grapevine Stilbenes by Ultra-High-Performance Liquid Chromatography Coupled to
Triple-Quadrupole Mass Spectrometry (UHPLC-QqQ)
doi: 10.3390/molecules22030418 . . . . . . . . . . . . . . . . . . . . . . . . . . 82

**Seung Hwan Hwang, Ji Hun Paek and Soon Sung Lim**
Simultaneous Ultra Performance Liquid Chromatography Determination and Antioxidant
Activity of Linarin, Luteolin, Chlorogenic Acid and Apigenin in Different Parts of
Compositae Species
doi: 10.3390/molecules21111609 . . . . . . . . . . . . . . . . . . . . . . . . . . 97

**M. Soledade C. Pedras, Abbas Abdoli and Vijay K. Sarma-Mamillapalle**
Inhibitors of the Detoxifying Enzyme of the Phytoalexin Brassinin Based on Quinoline and
Isoquinoline Scaffolds
doi: 10.3390/molecules22081345 . . . . . . . . . . . . . . . . . . . . . . . . . . 111

**Young-Sun Moon, Leesun Kim, Hyang Sook Chun and Sung-Eun Lee**
4-Hydroxy-7-methyl-3-phenylcoumarin Suppresses Aflatoxin Biosynthesis via Downregulation of *aflK* Expressing Versicolorin B Synthase in *Aspergillus flavus*
doi: 10.3390/molecules22050712 . . . . . . . . . . . . . . . . . . . . . . . . . . . . **126**

**Wee Xian Lee, Dayang Fredalina Basri and Ahmad Rohi Ghazali**
Bactericidal Effect of Pterostilbene Alone and in Combination with Gentamicin against Human Pathogenic Bacteria
doi: 10.3390/molecules22030463 . . . . . . . . . . . . . . . . . . . . . . . . . . . . **135**

**Martina Chripkova, Frantisek Zigo and Jan Mojzis**
Antiproliferative Effect of Indole Phytoalexins
doi: 10.3390/molecules21121626 . . . . . . . . . . . . . . . . . . . . . . . . . . . . **147**

**Laetitia Nivelle, Jane Hubert, Eric Courot, Nicolas Borie, Jean-Hugues Renault, Jean-Marc Nuzillard, Dominique Harakat, Christophe Clément, Laurent Martiny, Dominique Delmas, Philippe Jeandet and Michel Tarpin**
Cytotoxicity of Labruscol, a New Resveratrol Dimer Produced by Grapevine Cell Suspensions, on Human Skin Melanoma Cancer Cell Line HT-144
doi: 10.3390/molecules22111940 . . . . . . . . . . . . . . . . . . . . . . . . . . . . **162**

**Virginie Aires, Dominique Delmas, Fatima Djouadi, Jean Bastin, Mustapha Cherkaoui-Malki and Norbert Latruffe**
Resveratrol-Induced Changes in MicroRNA Expression in Primary Human Fibroblasts Harboring Carnitine-Palmitoyl Transferase-2 Gene Mutation, Leading to Fatty Acid Oxidation Deficiency
doi: 10.3390/molecules23010007 . . . . . . . . . . . . . . . . . . . . . . . . . . . . **173**

**Paola Jara, Johana Spies, Constanza Cárcamo, Yennyfer Arancibia, Gabriela Vargas, Carolina Martin, Mónica Salas, Carola Otth and Angara Zambrano**
The Effect of Resveratrol on Cell Viability in the Burkitts Lymphoma Cell Line Ramos
doi: 10.3390/molecules23010014 . . . . . . . . . . . . . . . . . . . . . . . . . . . . **184**

# About the Special Issue Editor

**Philippe Jeandet** earned his doctorates in plant physiology and biochemistry in 1991 and 1996, respectively, from the University of Bourgogne (France). He started his research activities on resveratrol, a phytoalexin from the Vitaceae. He received an associate professor position at the University of Bourgogne. In 1997, Philippe Jeandet accepted a position as a professor and chairman of the laboratory of oenology and applied chemistry at the University of Reims. His research activities focused on physico-chemistry applied to wine and microbiology. He has been the director and adjunct director of the research unit "vine and wine of Champagne" and adjunct to the director of research and technology in the Champagne-Ardennes area. He is now leader of a research group on resveratrol. He has published over 270 papers in refereed journals or books, edited two books and four Special Issues and presented 260 communications to numerous symposia and congresses.

# Preface to "Structure, Chemical Analysis, Biosynthesis, Metabolism, Molecular Engineering and Biological Functions of Phytoalexins"

Ever since the concept of phytoalexins was proposed by Müller and Borger in 1940, these compounds have attracted considerable attention due to the central role they play in the defense mechanisms of various plants. Besides displaying antifungal activity in numerous plant–pathogen interactions, phytoalexins have been implicated in human health and disease as antioxidant, anticancer and cardioprotective agents as well as being supposed to act positively in neurodegenerative illnesses. More than 25 years after the work of Siemann and Creasy which established a relationship between the concentration of the phytoalexin resveratrol in wine and the beneficial effects of wine consumption on health, the relevant literature on phytoalexins and their role in health and disease has increased tremendously. Knowledge on phytoalexins relies on fields as diverse as organic synthesis, analytical chemistry, plant molecular pathology, biocontrol, biochemistry and various aspects of biomedicine and biotechnology. It is almost impossible to review all of these aspects and, therefore, an attempt is made here to illustrate some of them with a particular emphasis on the induction mechanisms of phytoalexin biosynthesis, methods for their analysis in complex matrices, fungal metabolism and phytoalexin bioactivity. Very diverse phytoalexins are described: the stilbene phytoalexins from grapevine; the pterocarpan phytoalexins pisatin and glyceollin I from pea and soybean; the indole phytoalexins brassinin and camalexin; the phenylpropanoid-derived phytoalexins coumarins; chlorogenic acid and the isoflavonoid phytoalexins luteolin, linarin and apigenin from the Compositae species. This book is divided into 13 chapters which are described in more detail in the editorial.

I hope this book will expose the need for and promise of phytoalexin research to the scientific community and encourage new colleagues to enter into this exciting and ever-growing field of research! This book will thus serve as a resource for teachers, researchers and students concerned with the study of phytoalexins.

I am grateful to the contributors of this book, who are all leading experts in their respective research areas as well as to the colleagues who took on the crucial task of evaluating all the submitted articles. I also wish to thank our publisher, MDPI, Derek J. McPhee, editor-in-chief of Molecules, the editorial staff of this journal, and Jade Lu, section managing editor, for their encouragement and expert guidance, which enabled the publication of this book.

This book is dedicated to the memoriam of Roger Bessis

**Philippe Jeandet**
*Special Issue Editor*

![molecules logo] *molecules*                    MDPI

*Editorial*

# Structure, Chemical Analysis, Biosynthesis, Metabolism, Molecular Engineering, and Biological Functions of Phytoalexins

**Philippe Jeandet**

Research Unit "Induced Resistance and Plant Bioprotection" EA 4707, SFR Condorcet FR CNRS 3417, Faculty of Sciences, University of Reims Champagne-Ardenne, PO Box 1039, 51687 Reims CEDEX 2, France; philippe.jeandet@univ-reims.fr

Received: 23 December 2017; Accepted: 26 December 2017; Published: 28 December 2017

Plants in their natural environment are facing large numbers of pathogenic microorganisms, mainly fungi and bacteria. To cope with these stresses, plants have evolved a variety of resistance mechanisms that can constitutively be expressed or induced. Phytoalexins, which are low-molecular-weight antimicrobial compounds produced by plants as a response to biotic and abiotic stresses, take part in this intricate defence system. This special issue is the continuation of that published in 2015 entitled "Phytoalexins: Current Progress and Future Prospects" through http://www.mdpi.com/journal/molecules/special_issues/phytoalexins-progress.

Phytoalexins display a wide range of properties as antifungal compounds in various plants or preventing actions against human diseases as antioxidant, anticancer and cardioprotective agents as well as being supposed to act positively in neurodegenerative diseases such as Alzheimer's and Parkinson diseases. These compounds have been the subject of numerous studies over the last two decades. Thirteen research and review articles have been published in this special issue: four articles concern the biosynthesis of phytoalexins as a response to biotic and/or abiotic elicitors capable of inducing their production in plants [1–4], two articles describe methods for phytoalexin analysis in complex matrices [5,6], one article reports on phytoalexin metabolism by fungi [7], and six articles focus on the biological activity of phytoalexins [8–13].

By definition, phytoalexins are non-constitutive compounds produced by plants solely as a response to potentially pathogenic microorganisms or a large number of biotic and chemical elicitors. In the work of Woźniak et al. [1], the ability of a chemical elicitor, lead employed at various doses, or a biotic factor, pea aphid infestation, to act on the signaling pathways (salicylic acid, SA and abscisic acid, ABA production) of pea seedlings was studied. Regulation of the level of these two signaling molecules was also assessed upon cross interactions between the abiotic factor (lead) and the biotic factor (aphid infestation). The elicitor-mediated increases of the SA and ABA pathways in pea resulted in a strong induction of the biosynthesis of the phytoalexin pisatin. In the article of Farrell et al. [2], two distinct elicitors were also used to enhance the production of glyceollin I, a phytoalexin from soybean. Combination of the chemical elicitor, silver nitrate (AgNO$_3$) with the wall glucan elicitor (WGE) from the pathogen *Phytophthora sojae* was shown to have an additive effect on the induction of glyceollin production in soybean, reaching up to 745 µg/g tissue. Both elicitors act by distinct mechanisms. WGE upregulates the genes working on the isoflavonoid and glyceollin pathways while AgNO$_3$ increases hydrolysis of 6″-O-malonyldaidzein to form daidzein, an intermediate in the glyceollin pathway.

Oligogalacturonides (OGs) are well known potent stimulators of the plant immune system. In the work of Selim et al. [3], the eliciting activity of two OG fractions of varying polymerization degrees, one non-acetylated and one 30% acetylated, was determined in pea against *Aphanomyces* root rot. Significant root infection reductions were observed in both cases. The OG-mediated increased resistance of pea to *Aphanomyces* root rot was linked namely to an upregulation of the genes involved in the phytoalexin pisatin pathway (phenylalanine ammonia lyase, chalcone synthase, and isoflavone reductase).

Study of the biological properties of phytoalexins is hampered by their limited supply and the impossibility to recover them in sufficient amounts by conventional plant extraction procedures or chemical synthesis. The use of biotechnological systems could thus represent powerful methods for the production at large-scale of these compounds. In the article of Tisserant et al. [4], hairy root cultures of grapevine obtained after transformation with *Rhizobacterium rhizogenes* were used for obtaining high-purity stilbene phytoalexins. A significant accumulation of resveratrol, piceid, and ε- and δ-viniferins was observed both in the fresh tissues and the extracellular medium as a response to a combination of two elicitors, methyljasmonate and cyclodextrins.

As phytoalexins are naturally occurring compounds of a very diverse nature, highly specific qualitative and quantitative analytical techniques are thus needed for their accurate determination in complex matrices such as biological fluids, plant extracts, or plant cell cultures. In the work of Hurtado-Gaitán et al. [5], a method coupling ultra-high-performance liquid chromatography to triple-quadrupole mass spectrometry operated in the multiple reaction monitoring mode was developed for the detection and the quantitation of five stilbene phytoalexins from grapevine (trans-resveratrol, trans-piceid, trans-piceatannol, trans-pterostilbene, and trans-ε-viniferin). The applicability of the technique was verified in various matrices including cell culture extracts and red wine, and the method was also used to follow the enzymatic conversion of trans-resveratrol to trans-piceatannol in the presence of NADPH substrates and grape protein extracts. In the work of Hwan Hwang et al. [6], four phytoalexins found in Compositae species, three isoflavonoid-type phytoalexins (linarin, luteolin, and apigenin) and one phenylpropanoid-related phytoalexin (chlorogenic acid) were analyzed and quantized by ultra-performance liquid chromatography. The technique employed led to a good resolution of the four phytoalexins with a reasonable analysis run time (14 min). Upstream applications of this method resulted in the determination of the antioxidant activity of the four phytoalexins.

The ability of a pathogenic microorganism to detoxify the phytoalexins to which it is exposed is an essential component of the cross talk between plants and pathogens. In the article of Pedras et al. [7], research on phytoalexin detoxification inhibitors, the so-called PALDOXINS, was developed. Work has focused on inhibitors of brassinin oxidase which is an inducible fungal enzyme from the plant pathogen, *Leptosphaeria maculans*, catalyzing the detoxification of the phytoalexin brassinin to indole-3-carboxaldehyde and S-methyl dithiocarbamate. It is suggested that quinoline-derived compounds, especially 3-ethyl-6-phenylquinoline, display the highest inhibiting activity of the brassinin oxidase.

Beside their antifungal properties in plants, phytoalexins show preventing activities against human diseases as antioxidant, anticancer, cardioprotective, antibacterial, and antifungal agents. Six articles report here on the biological implication of phytoalexins in human diseases [8–13].

In the work of Moon et al. [8], the antifungal activity of natural and synthetic coumarins, which are phenylpropanoid-derivated phytoalexins, was evaluated against *Aspergillus flavus*. This mold is responsible for the production of various aflatoxins, considered to be the most important carcinogenetic agents of natural origin. Among 26 tested coumarins, five compounds displayed potent antifungal and antiaflatoxigenic activities against *A. flavus*. The 4-hydroxy-7-methyl-3-phenyl coumarin especially showed a 50% inhibition of the fungal growth at a concentration of 100 µg/mL. Most interestingly, coumarins displayed remarkable inhibition effects at 10 µg/mL on the production of aflatoxins B1 and B2, being the activity of the 4-hydroxy-7-methyl-3-phenyl coumarin correlated with the downregulation of several genes (*aflD*, *aflQ*, *aflR*, and *aflK*) working on the biosynthetic route to aflatoxin.

Phytoalexins exert some inhibiting activity against a range of bacteria or fungi implicated in human diseases such as skin infection, candidiasis, gonorrhea, and respiratory tract infections. In the work of Lee et al. [9], the antibacterial activity of pterostilbene, a dimethylated phytoalexin derived from resveratrol, in combination with the antibiotic gentamicin was evaluated against six strains of Gram-positive and -negative bacteria. Results evidenced a synergistic action between the phytoalexin and the antibiotic against *Staphylococcus aureus* ATCC 25923, *Escherichia coli* O157, and *Pseudomonas aeruginosa* 15,442. Growth of the tested bacteria was completely inhibited by the synergistic action of pterostilbene and gentamicin within 2–8 h treatment with half of their minimum inhibitory concentrations.

Numerous phytoalexins have been reported to exhibit significant anticancer, chemopreventive, and antiproliferative activities. In the article of Chripkova et al. [10], the antiproliferative effects of indole phytoalexins including brassinin, homobrassinin, camalexin, and their synthetic derivatives have been reviewed. Mechanisms of their anticancer actions include induction of apoptosis and cell cycle arrest, inhibition of neovascularization, and modulation of the signaling pathways associated with malignant transformation or cell survival. Search for synthetic derivatives of indole phytoalexins such as the 2-amino derivatives of spiroindoline phytoalexins displaying high anticancer features was also described. In the work of Nivelle et al. [11], a new dimer of resveratrol called labruscol has been purified and identified from grapevine cell suspensions of *Vitis labrusca* L. cultivated in a 14 L bioreactor. The antiproliferative activity of labruscol was demonstrated, this compound exerting almost 100% of cell viability inhibition of the human skin melanoma cancer cell line HT-144 at a dose of 100 µM within 72 h of treatment. Moreover, at the very low concentration of 1.2 µM, labruscol showed a 40% inhibition of cancer cell invasion, an activity not displayed by resveratrol. It thus seems that labruscol possesses complementary properties of resveratrol in particular regarding cell invasion, suggesting its utilization in combination with resveratrol to improve its antiproliferative capacities. In the work of Aires et al. [12], the identification of microRNAs was described following treatment of human primary fibroblasts with resveratrol. This study focuses on the relation between resveratrol treatment and deficiency in Carnitine-Palmitoyl Transferase-2 (CTP2), a mitochondrial enzyme involved in long-chain fatty acids entry into the mitochondria for their β-oxidation and energy production. It has indeed already been shown that resveratrol treatment restores normal fatty acid oxidation rates in patients harboring CPT2-gene mutation. Data resulted in the identification of several microRNAs displaying altered expression levels in fibroblasts either in the presence or absence of CPT2 or in the presence or absence of resveratrol stimulation. In addition, putative target transcripts of the microRNAs were described, suggesting that their gene products are important for the detrimental effects of CPT2 and the beneficial effects of resveratrol.

Although resveratrol has been shown to prevent the proliferation of malignant cells, the molecular mechanisms mediating resveratrol specific effects on lymphoma cells remain unknown. To answer this question, Jara et al. investigated cell survival and gene expression in the Burkitt's lymphoma cell line Ramos upon treatment with resveratrol [13]. The data obtained suggest that resveratrol displays significant anti-proliferative and pro-apoptotic activities on those cells, modulating the expression of several genes implied in the apoptotic process as well as inducing the DNA damage response and DNA repairing. From a mechanistic point of view, the data clearly correlated the decrease in malignant cell survival with the activation of apoptotic markers such as caspase 3 and fragmented poly(ADP-ribose) polymerase 1 in a dose-dependent manner. Moreover, expression of the pro-apoptotic genes *Noxa* and *Puma* was increased in a time-dependent fashion after 1 h and 3 h of resveratrol treatment, but no effect was observed on the expression of the *Fas* gene. Additionally, resveratrol induced significant increases in proteins necessary for the initiation of the DNA repair pathway.

All these articles thus highlight the central role played by phytoalexins in plant–microbe interactions as well as in human diseases. This special issue is accessible through the following link: http://www.mdpi.com/journal/molecules/special_issues/phytoalexins.

**Acknowledgments:** The guest editor thanks all of the authors for their contributions to this special issue, all the reviewers for their work in evaluating the manuscripts, and Derek J. McPhee, the editor-in-chief of *Molecules* as well as the editorial staff of this journal, especially Jade Lu, Managing Editor, for their kind help in making this special issue. This special issue is dedicated to the memoriam of Roger Bessis.

**Conflicts of Interest:** The author declare no conflict of interest.

## References

1.  Woźniak, A.; Drzewiecka, K.; Kęsy, J.; Marczak, L.; Narożna, D.; Grobela, M.; Motała, R.; Bocianowski, J.; Morkunas, I. The influence of lead on generation of signaling molecules and accumulation of flavonoids in pea seedlings in response to pea aphid infestation. *Molecules* **2017**, *22*, 1404. [CrossRef] [PubMed]
2.  Farrell, K.; Jahan, M.A.; Kovinich, N. Distinct mechanisms of biotic and chemical elicitors enable additive elicitation of the anticancer phytoalexin glyceollin I. *Molecules* **2017**, *22*, 1261. [CrossRef] [PubMed]
3.  Selim, S.; Sanssené, J.; Rossard, R.; Courtois, J. Systemic induction of the defensin and phytoalexin pisatin pathways in pea (*Pisum sativum*) against *Aphanomyces euteiches* by acetylated and nonacetylated oligogalacturonides. *Molecules* **2017**, *22*, 1017. [CrossRef] [PubMed]
4.  Tisserant, L.-P.; Aziz, A.; Jullian, N.; Jeandet, P.; Clément, C.; Courot, E.; Boitel-Conti, M. Enhanced stilbene production and excretion in *Vitis vinifera* cv Pinot Noir hairy root cultures. *Molecules* **2016**, *21*, 1703. [CrossRef] [PubMed]
5.  Hurtado-Gaitán, E.; Sellés-Marchart, S.; Martínez-Márquez, A.; Samper-Herrero, A.; Bru-Martínez, R.-A. Focused multiple reaction monitoring (MRM) quantitative method for bioactive grapevine stilbenes by ultra-high-performance liquid chromatography coupled to triple-quadrupole mass spectrometry (UHPLC-QqQ). *Molecules* **2017**, *22*, 418. [CrossRef] [PubMed]
6.  Hwan Hwang, S.; Hun Paek, J.; Sung Lim, S. Simultaneous ultra-performance liquid chromatography determination and antioxidant activity of linarin, luteolin, chlorogenic acid and apigenin in different parts of Compositae species. *Molecules* **2016**, *21*, 1609. [CrossRef] [PubMed]
7.  Pedras, M.-S.-C.; Abdoli, A.; Sarma-Mamillapalle, V.-K. Inhibitors of the detoxifying enzyme of the phytoalexin brassinin based on quinoline and isoquinoline scaffolds. *Molecules* **2017**, *22*, 1345. [CrossRef] [PubMed]
8.  Moon, Y.-S.; Kim, L.; Sook Chun, H.; Lee, S.-E. 4-Hydroxy-7-methyl-3-phenylcoumarin suppresses aflatoxin biosynthesis via downregulation of aflK expressing versicolorin B synthase in *Aspergillus flavus*. *Molecules* **2017**, *22*, 712. [CrossRef] [PubMed]
9.  Lee, W.X.; Basri, D.-F.; Ghazali, A.-R. Bactericidal effect of pterostilbene alone and in combination with gentamicin against human pathogenic bacteria. *Molecules* **2017**, *22*, 463. [CrossRef] [PubMed]
10. Chripkova, M.; Zigo, F.; Mojzis, J. Antiproliferative effect of indole phytoalexins. *Molecules* **2016**, *21*, 1626. [CrossRef] [PubMed]
11. Nivelle, L.; Hubert, J.; Courot, E.; Borie, N.; Renault, J.-H.; Nuzillard, J.-M.; Harakat, D.; Clément, C.; Martiny, L.; Delmas, D.; et al. Cytotoxicity of labruscol, a new resveratrol dimer produced by grapevine cell suspensions, on human skin melanoma cancer cell line HT-144. *Molecules* **2017**, *22*, 1940. [CrossRef] [PubMed]
12. Aires, V.; Delmas, D.; Djouadi, F.; Bastin, J.; Cherkaoui Malki, M.; Latruffe, N. Resveratrol-induced changes in microRNA expression in human primary fibroblasts harboring carnitine-palmitoyl transferase-2 (CPT2) gene mutation, leading to fatty acid oxidation deficiency. *Molecules* **2018**, *23*, 7. [CrossRef] [PubMed]
13. Jara, P.; Spies, J.; Carcamo, C.; Arancibia, Y.; Vargas, G.; Martin, C.; Salas, M.; Otth, C.; Zambrano, A. The effect of resveratrol on cell viability in the Burkitt's lymphoma cell line Ramos. *Molecules* **2018**, *23*, 14. [CrossRef] [PubMed]

*molecules*

MDPI

*Article*

# The Influence of Lead on Generation of Signalling Molecules and Accumulation of Flavonoids in Pea Seedlings in Response to Pea Aphid Infestation

Agnieszka Woźniak [1], Kinga Drzewiecka [2], Jacek Kęsy [3], Łukasz Marczak [4], Dorota Narożna [5], Marcin Grobela [6], Rafał Motała [6], Jan Bocianowski [7] and Iwona Morkunas [1,*]

[1] Department of Plant Physiology, Poznań University of Life Sciences, Wołyńska 35, 60-637 Poznań, Poland; agnieszkam.wozniak@gmail.com
[2] Department of Chemistry, Poznań University of Life Sciences, Wojska Polskiego 75, 60-625 Poznań, Poland; kinga.drzewiecka@gmail.com
[3] Chair of Plant Physiology and Biotechnology, Nicolaus Copernicus University, Gagarina 9, 87-100 Toruń, Poland; kesy@umk.pl
[4] Institute of Bioorganic Chemistry, Polish Academy of Sciences, Noskowskiego 12/14, 61-704 Poznań, Poland; lukasmar@ibch.poznan.pl
[5] Department of Biochemistry and Biotechnology, Poznań University of Life Sciences, Dojazd 11, 60-632 Poznań, Poland; dorna@o2.pl
[6] Department of Ecology and Environmental Protection, Laboratory of Environmental Analyses, the Institute of Plant Protection National Research Institute, Węgorka 20, 60-101 Poznań, Poland; grobela@iorpib.poznan.pl (M.G.); r.motala@iorpib.poznan.pl (R.M.)
[7] Department of Mathematical and Statistical Methods, Poznań University of Life Sciences, Wojska Polskiego 28, 60-637 Poznań, Poland; jboc@up.poznan.pl
* Correspondence: iwona.morkunas@gmail.com or iwona.morkunas@mail.up.poznan.pl; Tel.: +48-61-846-6040; Fax: +48-61-848-7179

Received: 30 June 2017; Accepted: 21 August 2017; Published: 24 August 2017

**Abstract:** The aim of this study was to investigate the effect of an abiotic factor, i.e., lead at various concentrations (low causing a hormesis effect and causing high toxicity effects), on the generation of signalling molecules in pea (*Pisum sativum* L. cv. Cysterski) seedlings and then during infestation by the pea aphid (*Acyrthosiphon pisum* Harris). The second objective was to verify whether the presence of lead in pea seedling organs and induction of signalling pathways dependent on the concentration of this metal trigger defense responses to *A. pisum*. Therefore, the profile of flavonoids and expression levels of genes encoding enzymes of the flavonoid biosynthesis pathway (phenylalanine ammonialyase and chalcone synthase) were determined. A significant accumulation of total salicylic acid (TSA) and abscisic acid (ABA) was recorded in the roots and leaves of pea seedlings growing on lead-supplemented medium and next during infestation by aphids. Increased generation of these phytohormones strongly enhanced the biosynthesis of flavonoids, including a phytoalexin, pisatin. This research provides insights into the cross-talk between the abiotic (lead) and biotic factor (aphid infestation) on the level of the generation of signalling molecules and their role in the induction of flavonoid biosynthesis.

**Keywords:** lead; *Acyrthosiphon pisum*; signalling molecules; flavonoids; pisatin; flavonoid biosynthesis enzymes; defense responses; *Pisum sativum*

## 1. Introduction

Under natural conditions we may frequently observe the effect of many stress factors acting simultaneously or sequentially. Plants demonstrate a great ability to adapt their metabolism to rapid changes in the environment [1] and therefore they have developed a wide range of mechanisms to cope with abiotic and biotic stresses. It has been established that plant defenses against these stresses may imply common and/or complementary pathways of signal perception, signal transduction and metabolism [2,3]. Plants exposed to these stresses respond on multiple levels. To date molecular mechanisms involved in plant defenses against the above-mentioned stress factors were revealed independently and singly, thus further research is required to identify convergence points between abiotic and biotic stress signalling pathways. Fuijta et al. [4] reported that signalling molecules, transcription factors and kinases may be important common players that are involved in the crosstalk between stress signalling pathways. It has been suggested that phytohormone and reactive oxygen species (ROS)/ reactive nitrogen species (RNS) signalling pathways play key roles in the cross-talk between biotic and abiotic stress signalling. These cross-talk signalling pathways regulate metabolic processes in the context of plant defense.

Since in the course of the coevolution of plants and biotic factors, including herbivores, undergo mutual adaptation, abiotic factors also significantly affect that process. Poschenrieder et al. [5] reported that during their coevolution with plants, pathogens and herbivores compete in an environment where efficient metal ion acquisition and ion homeostasis are essential for survival. Nevertheless, to date no studies have been conducted on plant-insect interactions involving the stress response signalling system in plants in relation to heavy metal concentration in the environment. Only plant signalling in response to herbivory, including aphids, has been well documented [6–8]. Invertebrates, especially insects, are good models to study heavy metal toxicity and can be bioindicators of environmental pollution. Insects, including aphids, play a definite role in the trophic chain and as food for other organisms they may constitute an important path for the bioaccumulation of heavy metals.

Studies in this work emphasise the important role of salicylic acid (SA) and abscisic acid (ABA) in defense responses associated with the accumulation of flavonoids in edible pea exposed to varying lead concentrations, i.e., at a low concentration inducing the metabolic status of the plants, potentially leading to the hormesis effect, and at a high concentration causing a toxic effect, as well as during infestation of *A. pisum*. This is the first report revealing the effect of lead as an abiotic factor and phytophages (*A. pisum*) as a biotic factor on the biosynthesis of pisatin, a phytoalexin characteristic of *Pisum sativum* L., which may serve a significant role in its defense strategy. Pisatin is believed to play a key role during abiotic and biotic stress responses. Jeandet et al. [9] reported that phytoalexins are biocidal compounds synthesised by and accumulated in plants as a response to biotic and abiotic stresses, which play important roles in their defense systems. Significantly enhanced production of phytoalexins was also observed in response to the elicitation of signalling molecules such as SA, methyl jasmonate and methyl-β-cyclodextrins in plants [10]. The induction of phytoalexin biosynthesis was demonstrated in many plant species in response to insects [11–20]. Dual-choice tests involving varied phytoalexin contents carried out by Hart [21] revealed that an isoflavonoid phytoalexin(s) had feeding-deterrent properties towards insects. Additionally, it has been revealed that several isoflavonoid phytoalexins, including coumestrol and genistein, deterred insect feeding [22,23]. The anti-nutritional effects of flavonoids on insects have also been confirmed by other research results [24–26]. Moreover, an isoflavone genistein and a flavone luteolin were shown to have an impact on the prolonged period of stylet probing, reduced salivation and passive ingestion of the pea aphid, *A. pisum* [27]. Simmonds [12,28] reported that flavonoids modulate the feeding and oviposition behaviour of insects. The aphicidal effect of flavonoids against aphids was manifested by mortality of nymphs and apterous adults [29]. It was suggested that flavonoids may be used as a bio-insecticide within the framework of integrated pest management (IPM) programmes. On the other hand, Diaz Napal and Palacios [30] demonstrated that flavonoids can also be phagostimulants when applied at a low concentration.

Moreover, the accumulation of phytoalexins was also demonstrated in plant responses to heavy metals [31–37]. The concentration of heavy metals, including lead, has been increasing in the environment as a result of progressive industrialisation. Ashraf et al. [38] reported that recent rates of soil contamination with various heavy metals leading to their introduction to agro-ecosystems and their transfer to human beings through the food chain are alarming and observed on a global scale. It has been documented that in terrestrial ecosystems soil is the primary source of heavy metal transfer to agricultural produce [39]. A proportion of these metals also enters plant systems from the external atmosphere surrounding the plants [40], thus affecting productivity and crop quality. Surface waters may also be contaminated with lead due to the use of nitrogen fertilisers containing this metal [41]. Pourrut et al. [42] reported that among heavy metals lead is the second most harmful pollutant, second only to arsenic, according to the new European REACH regulations.

Edible pea, a crop object of our research, is used on a broad scale due to the high protein content in its seeds. Proper understanding of resistance mechanisms in crop plants is the foundation of integrated pest management. Additionally, insects playing a distinct role in the trophic chain and as food for other organisms may be an important element in the bioaccumulation of heavy metals.

The first objective was to investigate the effect of lead on the generation of signalling molecules such as phytohormones, e.g., SA and ABA, and next to determine how cross-interactions of both stress factors, i.e., lead and *A. pisum*, regulate the level of these signalling molecules and affect flavonoid biosynthesis. The second objective was to determine the level of flavonoids, especially a phytoalexin, pisatin, in response to the impact of the above-mentioned stressors. Flavonoids are a remarkable group of plant metabolites that are important elements of the defence system of legumes in interactions with biotic stress factors [14]. No other class of secondary products has been credited with so many and such diverse key functions in plants. Additionally, within the second research objective the level of expression was determined for genes encoding enzymes of the flavonoid biosynthesis pathway, i.e., phenylalanine ammonialyase (PAL), an enzyme initiating phenylpropanoid metabolism, and chalcone synthase (CHS), which catalyses the first committed step in the flavonoid biosynthetic pathway. It is known that flavonoids may be found either in a free state or conjugated as esters or glycosides. Biological activity in an interaction with biotic stressors was revealed by free flavonoid aglycones, released from glycosides with the use of glucosidases [43]. For this reason, in this study we analysed changes in the activity of β-glucosidase, an enzyme which hydrolyses flavonoid glucosides. In turn, PAL is an enzyme that catalyses a reaction converting L-phenylalanine to ammonia and *trans*-cinnamic acid. PAL is the first enzyme of the phenylpropanoid pathway, via which polyphenol compounds, such as flavonoids, are biosynthesised in plants. Additionally, this enzyme initiates one of the pathways of SA biosynthesis in plants [44]. Therefore, phenylalanine as a substrate is transformed by PAL to cinnamic acid, which then can be converted to *o*-coumaric acid, and in subsequent reactions to SA. Cinnamic acid may also be converted to benzoates, and then with the participation of the BA2H enzyme (benzoic acid-2-hydroxylase) to SA [45]. Moreover, the third objective was to determine the effect of lead at varying concentrations (i.e., at a low concentration inducing the metabolic status of the plants, potentially leading to the hormesis effect, and at a high concentration causing a toxic effect on the growth of pea seedlings. At the same time, we investigated lead content in roots and leaves of pea seedlings growing at varied lead concentrations in the medium and during cross-interactions of lead and infestation of a phytophage with the piercing-sucking mouthpart, i.e., *A. pisum*, as well as lead content in bodies of the insect *A. pisum*.

We assume in this study that at a low concentration of lead in the substrate this metal will be accumulated mainly in roots of pea seedlings, while at higher concentrations some of the accumulated lead will be transported to leaves. In the case of the application of these two different lead concentrations we may expect differences in the intensity of generation of signalling molecules such as SA and ABA, in the sequence/period of their generation, changes in the transduction of signals from roots to leaves and in triggering of defense responses, i.e., flavonoid accumulation, including pisatin. In view of the above we expect that at a low concentration lead deposited mainly in roots will induce synthesis of signalling molecules and the signal will be transmitted to leaves, which contributes to an enhanced stress defense potential of pea seedlings. In turn, in the case of the applied toxic concentration of lead in the substrate a certain pool of this metal will be transported from roots to leaves, thus physiological and biochemical changes in leaves of pea seedlings will firstly be a consequence of the direct effect of lead.

## 2. Results

### 2.1. The Effect of Lead and A. pisum on SA Accumulation in Pea Seedlings

Already at the beginning of the experiment, i.e., 4 days after the administration of lead, before aphid transfer to pea seedlings, a significant accumulation was recorded for total salicylic acid (TSA) (Figure 1a,b), i.e., the sum of free (SA) (Figure 1c,d) and glucoside-bound salicylic acid (SAG) (Figure 1e,f) in the roots and leaves of pea seedlings growing on the Hoagland medium with 0.075 and 0.5 mM $Pb(NO_3)_2$. Statistical analysis confirmed significance of differences in these results (Table S1a,b) in Supplementary Materials). Additionally, at 0 h of the experimental SA concentration in the roots of pea seedlings cultured at 0.075 and 0.5 mM $Pb(NO_3)_2$ was two and 500 times higher than in the control. In turn, at the application of 0.075 and 0.5 mM $Pb(NO_3)_2$ the level of SA in leaves of pea seedlings was 2- and 20-fold greater than in the control. The level of TSA in these organs of pea seedlings cultured at the high lead concentration (0.5 mM $Pb(NO_3)_2$) was significantly higher than in the organs of seedlings growing on the medium with 0.075 $Pb(NO_3)_2$ or in the control (pea seedlings cultured with no addition of lead and not colonised by pea aphids). The highest TSA level was recorded in 72-h roots (10,555.97 ng $g^{-1}$ FW) and leaves (773 ng $g^{-1}$ FW) of pea seedlings growing at the higher tested lead concentration of 0.5 mM $Pb(NO_3)_2$ and colonised by pea aphids *A. pisum*. The high accumulation of TSA was also found in the roots of pea seedlings growing on the Hoagland medium with 0.5 mM $Pb(NO_3)_2$ and not colonised by *A. pisum*. Moreover, pea aphid feeding significantly enhanced the accumulation of TSA in leaves, both in those of pea seedlings growing on the medium with a low lead level (0.075 mM $Pb^{2+}$+aphids variant) and a high lead concentration (0.5 mM $Pb^{2+}$+aphids variant), or without it (+aphids variant). In leaves of pea seedlings growing on the Hoagland medium with 0.075 mM $Pb(NO_3)_2$ and colonised by *A. pisum* the concentration of TSA increased versus infestation time, but it was significantly lower than in the case of leaves of seedlings cultured with the high lead concentration.

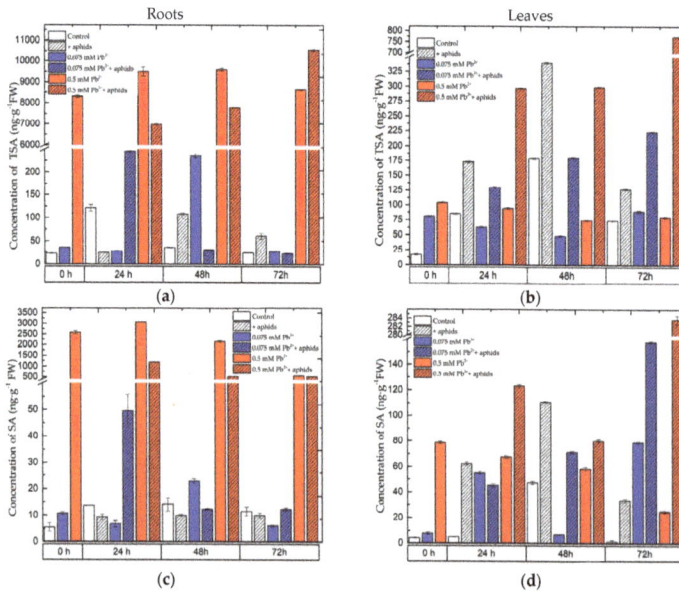

**Figure 1.** *Cont.*

8

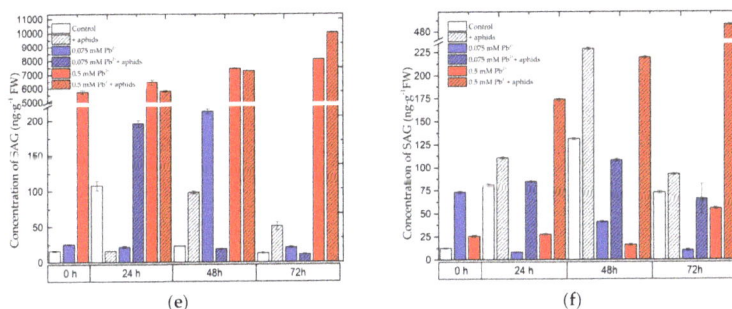

(e)    (f)

**Figure 1.** The effect of lead and *A. pisum* on accumulation of total salicylic acid (TSA) (**a,b**), salicylic acid (SA) (**c,d**) and salicylic acid glucoside (SAG) (**e,f**) in roots (**a,c,e**) and leaves (**b,d,f**) of pea seedlings. The data were obtained in three independent experiments and statistically analysed using ANOVA (*p*-values at $\alpha = 0.05$). Hypotheses on the equality of means were verified by the two-sample *t*-test. To account for multiple testing, we used the Bonferroni correction (statistically significant differences are shown in Table S1).

In 24-h roots of pea seedlings from the 0.075 mM $Pb(NO_3)_2$+aphids variant the level of free salicylic acid (SA) was observed to increase in comparison to the control, +aphids and +0.075 mM $Pb^{2+}$ variants. In turn, in 48- and 72-h leaves the accumulation of SA in the 0.075 mM $Pb^{2+}$+aphids variant was greater than in the control and 0.075 mM $Pb(NO_3)_2$ variants. Besides, the level of TSA in the roots was much higher than in the leaves. The A. pisum infestation alone caused a significant increase in the SA level, as the concentration of SA in these leaves was higher than in the control leaves.

## 2.2. The Effect of Lead and A. pisum on ABA Accumulation in Pea Seedlings

The level of abscisic acid (ABA) was markedly higher in the leaves than in the roots (Figure 2a,b). At the application of the toxic lead concentration in the substrate (0.5 mM $Pb(NO_3)_2$) a very strong accumulation of ABA was observed already at the beginning of the experiment (0 h), i.e., after 4 days from the administration of lead at 0.5 mM $Pb(NO_3)_2$ . The greatest accumulation of ABA was recorded in 24-h roots (15 ng g$^{-1}$ FW) and leaves (99.82 ng g$^{-1}$ FW) of pea seedlings growing at the higher tested lead concentration of 0.5 mM $Pb(NO_3)_2$. Additionally, the high ABA level in these leaves (0.5 mM $Pb^{2+}$ variant) was maintained at all time points of the experiment. An increase in ABA levels was also observed from 24 to 72 hpi in the leaves of pea seedlings growing on the medium with 0.5 mM $Pb(NO_3)_2$ and colonised by *A. pisum*, but only at 72 hpi it was higher than in the 0.5 mM $Pb^{2+}$ variant and in the other variants. In turn, an increasing ABA level was also recorded in leaves of pea seedlings growing on the medium with the low lead level (0.075 mM $Pb^{2+}$ variant) and infested by *A. pisum* (0.075 mM $Pb^{2+}$+aphids variant), as the concentration of ABA in these leaves was 9.99 and 14.53 ng g$^{-1}$ FW, respectively. In contrast, in the control it was 6.47 ng g$^{-1}$ FW. Statistical analysis showed highly significant differences in these results (Table S1a,b). Also aphid feeding alone caused ABA accumulation in 72-h leaves. In addition, the highest ABA accumulation was recorded at 24 h of the experiment in the roots of the 0.5 mM $Pb^{2+}$ variant. Up to 48 hpi ABA accumulation was also demonstrated in the roots of the 0.5 mM $Pb^{2+}$+aphids variant. In turn, in the roots of pea seedlings growing on the medium with 0.075 mM $Pb(NO_3)_2$ an increase in ABA levels was recorded in comparison to the control, but only at 72 h of the experiment both in the 0.075 mM $Pb^{2+}$ and 0.075 mM $Pb^{2+}$+aphids variants. Additionally, ABA accumulation occurred also as a result of *A. pisum* feeding mainly at 24 hpi (+aphids variant).

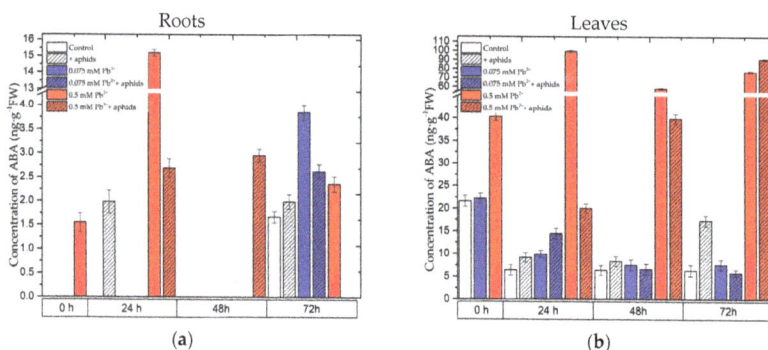

**Figure 2.** The effect of lead and *A. pisum* on accumulation of abscisic acid (ABA) in roots (**a**) and leaves (**b**) of pea seedlings. The data were obtained in three independent experiments and statistically analysed using ANOVA (*p*-values at $\alpha = 0.05$). Hypotheses on the equality of means were verified by the two-sample *t*-test. To account for multiple testing, we used the Bonferroni correction (statistically significant differences are shown in Table S1).

## 2.3. The Effect of Lead and A. pisum on Accumulation of Flavonoids in Pea Seedlings

The flavonoid profile in roots and leaves of pea seedlings revealed the presence of pisatin, 2′OH-genistein hexoside, Glc-Glc-Glc rhamnose, Glc-Glc-Glc kaempferol, Glc-Glc-Glc rhamnose isomer 1, Glc-Glc-Glc rhamnose isomer 2, 2′OH-genistein tetrahexoside, quercetin hexoside, Glc-Glc-Glc kaempferol and Glc-Glc-Glc-Rha quercetin (Figures 3 and 4, Figure S1).

### 2.3.1. The Effect of Lead and A. pisum on Accumulation of Pisatin in Pea Seedlings

It should be mentioned here that at the beginning of the experiment (i.e., 4 days after the administration of lead) the level of pisatin in the roots of the 0.5 mM $Pb^{2+}$ variant was 19 times higher than in the control, while in the roots of the 0.075 mM $Pb^{2+}$ variant its was two times higher than in the control (Figure 3a,b). Statistical analysis showed differences in these results to be highly significant (Table S1a,b). In turn, in leaves of the 0.5 mM $Pb^{2+}$ variant at 0 h of the experiment the concentration of pisatin was 2-fold greater than in the other variants. Moreover, at all the time points of the experiment a significant pisatin accumulation was demonstrated in the organs of pea seedlings after lead administration at the high concentration. While a very strong pisatin accumulation was noted in the roots of the 0.5 mM $Pb^{2+}$ and 0.5 mM $Pb^{2+}$+aphids variants, in those of pea seedlings growing on the medium with 0.5 mM $Pb(NO_3)_2$ and not colonised by pea aphids (0.5 mM $Pb^{2+}$ variant) the level of pisatin was higher than it was in the variant with 0.5 mM $Pb(NO_3)_2$ and pea aphids (0.5 mM $Pb^{2+}$+aphids variant). In turn, in the roots of pea seedlings growing on the medium with 0.075 mM $Pb(NO_3)_2$ (0.075 mM $Pb^{2+}$ and 0.075 mM $Pb^{2+}$+aphids variants), the level of pisatin at 24 h and 48 h of the experiment was 2- and over 2.5-fold greater in relation to the control, respectively. Additionally, it should be mentioned here that pisatin level in the leaves of pea seedlings was lower than in roots. The highest pisatin accumulation was observed at 48 h of the experiment both in the roots and leaves of pea seedlings. However, in the leaves of the 0.5 mM $Pb^{2+}$+aphids variant the level of this metabolite was highest. It should be stressed that the cross-talk of these two stressors, i.e., lead at 0.5 mM $Pb(NO_3)_2$ and pea aphid feeding, caused the strongest accumulation of pisatin at all time points after infestation. Additionally, from 48 hpi an increase was observed in pisatin levels in leaves of seedlings cultured at a low concentration of lead in the medium (0.075 mM $Pb^{2+}$ variant). The cross-talk of lead at 0.075 mM $Pb(NO_3)_2$ and pea aphid infestation caused an accumulation of pisatin in leaves at 72 hpi, but this accumulation was significantly lower than in the leaves of the 0.5 mM $Pb^{2+}$ and 0.5 mM $Pb^{2+}$+aphids variants.

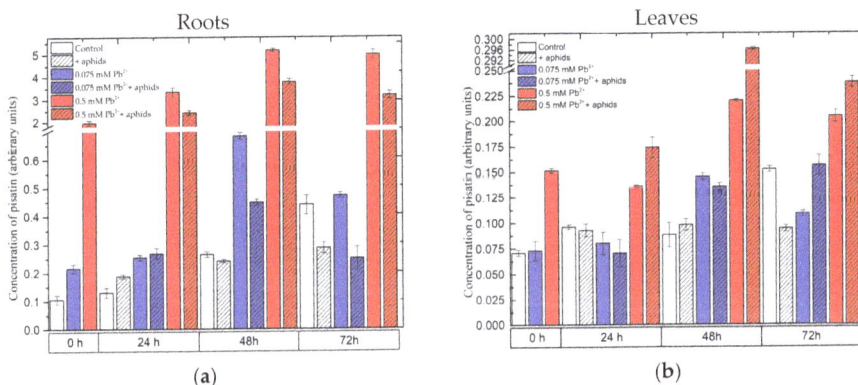

**Figure 3.** The effect of lead and *A. pisum* on accumulation of pisatin in roots (**a**) and leaves (**b**) of pea seedlings. The data were obtained in three independent experiments and statistically analysed using ANOVA (*p*-values at $\alpha = 0.05$). Hypotheses on the equality of means were verified by the two-sample *t*-test. To account for multiple testing, we used the Bonferroni correction (statistically significant differences are shown in Table S1).

2.3.2. The Effect of Lead and *A. pisum* on the Level of Isoflavonoid and Flavonoid Glycosides in Pea Seedlings

Four days after the administration of lead at a high concentration (0 h of the experiment), before transferring aphids to pea seedlings, generally a significant accumulation of isoflavonoid and flavonoid glycosides was recorded (Figure 4, Figure S1 in Supplementary Materials). At this time point in the leaves of pea seedlings growing on the medium with 0.5 mM $Pb(NO_3)_2$, the levels of 2'OH-genistein hexoside, Glc-Glc-Glc rhamnose isomer 1, Glc-Glc-Glc rhamnose isomer 2,2'OH-genistein tetrahexoside, quercetin hexoside, Glc-Glc-Glc kaempferol and Glc-Glc-Glc-Rha quercetin were higher than in the other experimental variants (control and 0.075 mM $Pb^{2+}$ variants). Statistical analysis showed highly significant differences in these results (Table S1a,b). In turn, an opposite trend was observed in the roots, i.e., a decrease in the content of the above-mentioned metabolites in relation to the control, with the exception of Glc-Glc-Glc rhamnose isomer 2, which was present only in the leaves (Figure 4, Figure S1). At subsequent time points both in the roots and leaves of pea seedlings, both non-infested and infested by *A. pisum*, growing on the medium with 0.5 mM $Pb(NO_3)_2$ (0.5 mM $Pb^{2+}$ and 0.5 mM $Pb^{2+}$+aphids variants), generally high accumulation was found for 2'OH-genistein hexoside, Glc-Glc-Glc rhamnose, Glc-Glc-Glc rhamnose isomer 1, Glc-Glc-Glc rhamnose isomer 2 (only in the leaves) and 2'OH-genistein tetrahexoside. However, concentrations of theses metabolites in tissues affected by both factors, i.e., abiotic (lead) and biotic (aphid) (0.5 mM $Pb^{2+}$+aphids variant), were lower than in those affected by lead alone (0.5 mM $Pb^{2+}$ variant). At 24 h of the experiment, during the influence of only lead applied at the low concentration (0.075 mM $Pb^{2+}$ variant), as well as cross-talk between lead and *A. pisum* (0.075 mM $Pb^{2+}$+aphids variant) and only *A. pisum* infestation (+aphids), levels of such metabolites as 2'OH-genistein hexoside, 2'OH-genistein tetrahexoside, Glc-Glc-Glc rhamnose, Glc-Glc-Glc rhamnose isomer 1, Glc-Glc Glc rhamnose isomer 2, quercetin hexoside, Glc-Glc-Glc kaempferol and Glc-Glc-Glc-Rha quercetin were observed to increase in leaves of pea seedlings. Besides, generally at 48 hpi in leaves of pea seedlings infested by *A. pisum* and growing on the medium with 0.075 mM $Pb(NO_3)_2$ the levels of 2'OH-genistein tetrahexoside, Glc-Glc-Glc rhamnose, Glc-Glc-Glc rhamnose isomer 1, Glc-Glc-Glc rhamnose isomer 2, Glc-Glc-Glc kaempferol and Glc-Glc-Glc-Rha quercetin decreased in relation to those in leaves of the 0.075 mM $Pb^{2+}$ variant. Moreover, at 72 h of the experiment, i.e., 7 days after lead application, levels of these metabolites in the leaves and roots of pea seedlings cultured on the medium with 0.075 mM $Pb(NO_3)_2$ (0.075 mM $Pb^{2+}$ variant) were markedly reduced in relation to the control.

### 2.3.3. Quantitative Analysis of Metabolites in the Roots and Leaves of Pea Seedlings

Comparative metabolomic analyses were performed between each group of samples (variants) both for roots and for leaves (Figure 5). PCA differentiated all the analysed experimental groups, whereas the group of samples from pea seedlings treated with the higher lead concentration was very clearly separated in the case of roots, while the additional treatment with aphids also gave a distinctly separated group. Such good separation was not the case for pea seedling leaves, although also in this case samples obtained from pea seedling leaves treated with the higher lead concentration are easily distinguished from the others. In contrast, the effect of aphid presence on the metabolome was not as evident as in the case of roots. In turn, in leaves significant changes were observed between the metabolome of such groups as 0.5 mM $Pb^{2+}$, 0.5 mM $Pb^{2+}$+aphids and 0.075 mM $Pb^{2+}$, 0.075 mM $Pb^{2+}$+aphids, +aphids. Moreover, it is interesting that quantitative analyses of metabolites for roots demonstrated that the control group, the group growing in the medium with the low lead concentration (0.075 mM $Pb^{2+}$ variant ) and the group only infested by pea aphids ( +aphids variant) were clustered close to each other. Only the 0.5 mM $Pb^{2+}$+aphids variant was located separately, although it was still near them.

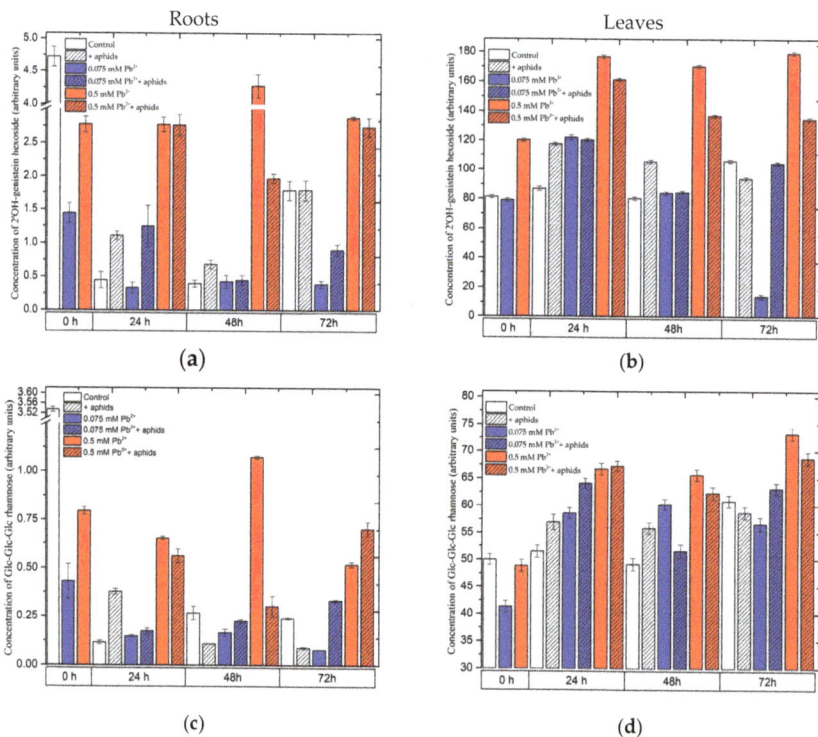

**Figure 4.** *Cont.*

(e)

(f)

1-pisatin, 2- 2'OH-genistein-tetrahexoside, 3- Glc-Glc-Glc-rhamnose isomer1, 4- Glc-Glc-Glc-rhamnose isomer2, 5- quercetin-hexoside,6- 2'OH-genistein-hexoside

(g)

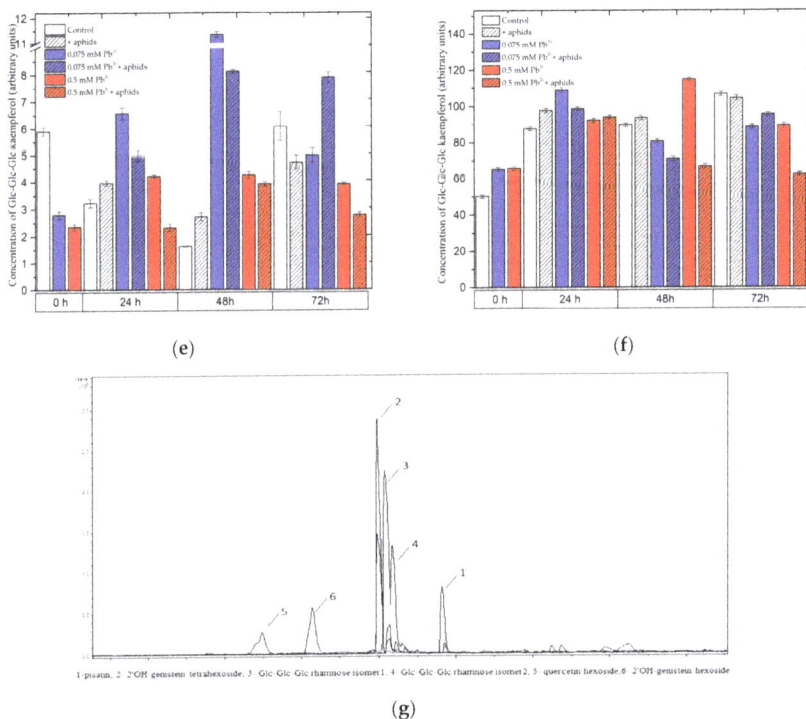

**Figure 4.** The effect of lead and *A. pisum* on the level of isoflavonoid and flavonoid glycosides 2'OH genistein hexoside (**a,b**), Glc-Glc-Glc rhamnose (**c,d**), Glc-Glc-Glc kaempferol (**e,f**), in roots (**a,c,e**) and leaves (**b,d,f**) of pea seedlings. An LC-MS extracted ion chromatogram showing profiles of phenolic componds found in leaves (**g**). Arbitrary unit means a relative unit of measurement to show the amount of substance (intensity), the reference measurement is dependent on individual measurement on MS (ion counts on MS detector). The data were obtained in three independent experiments and statistically analysed using ANOVA ($p$-values at $\alpha = 0.05$). Hypotheses on the equality of means were verified by the two-sample $t$-test. To account for multiple testing, we used the Bonferroni correction (statistically significant differences are shown in Table S1).

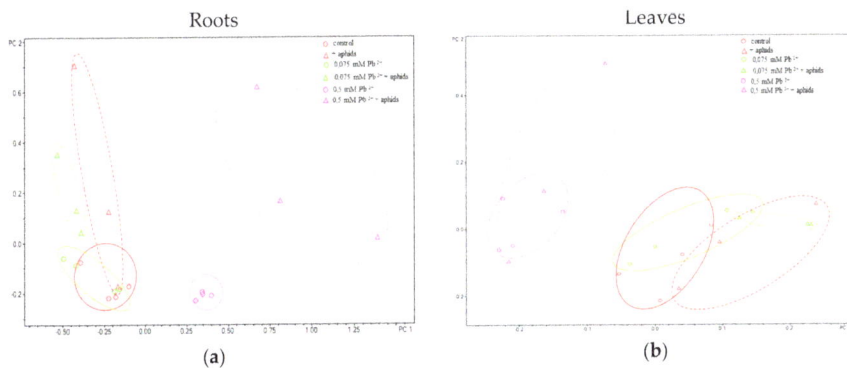

Roots

Leaves

(a)

(b)

**Figure 5.** *Cont.*

13

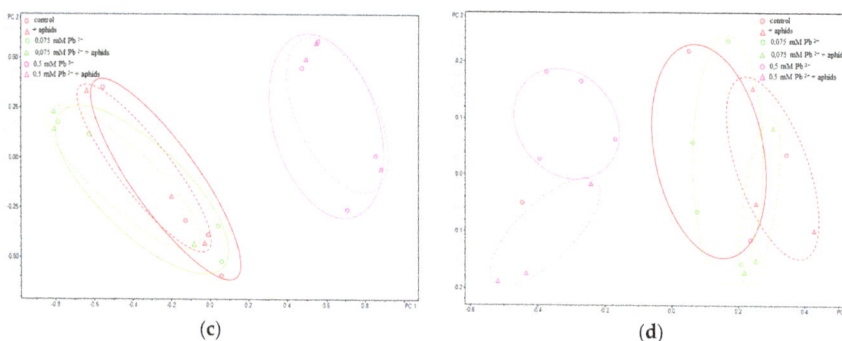

(c)

(d)

**Figure 5.** Quantitative analysis of metabolites in roots (**a,c**) and leaves (**b,d**) of pea seedlings. The Principal Components Analysis (PCA) of the positive ion (**a,b**) and negative ion (**c,d**) MS shows distinct and close clustered groups of these ions for experimental variants, i.e., control, +aphids, 0.075 mM $Pb^{2+}$, 0.075 mM $Pb^{2+}$+aphids, 0.5 mM $Pb^{2+}$ and 0.5 mM $Pb^{2+}$+aphids

Unfortunately, it was impossible to identify all the differentiating components of the sample due to the difficulty in interpreting the MS2 spectra, although spectral analysis showed that partially differentiation of the samples corresponded to flavonoid derivatives. For this reason, the MS2 spectra of flavonoids were analysed both for positive and negative ions and the quantity of identified compounds was measured using the targeted analysis approach by determining the areas of the corresponding chromatographic peaks.

### 2.4. The Effect of Lead and A. pisum on Expression Levels of Phenylalanine Ammonialyase and Chalcone Synthase Genes in Pea Seedlings

Semi-quantitative RT-PCR analysis revealed that pea aphid feeding alone (+aphids), lead administration at the high and low concentrations in the medium (0.5 mM $Pb^{2+}$ and 0.075 mM $Pb^{2+}$) as well as the cross-talk of lead and *A. pisum* infestation (0.5 mM $Pb^{2+}$+aphids and 0.075 mM $Pb^{2+}$+ aphids) upregulated mRNA levels for PAL and CHS in relation to the control (Figure 6). However, during exposure to lead at the high concentration and *A. pisum* (0.5 mM $Pb^{2+}$+aphids variant), the expression level of genes encoding PAL in the 0.5 mM $Pb^{2+}$+aphids leaves was lower than in the 0.5 mM $Pb^{2+}$ leaves, while it was generally higher than in leaves infested by aphids only (+aphids variant). Besides, the low concentration of lead (0.075 $Pb(NO_3)_2$) caused also upregulation of mRNA levels of the PAL and CHS genes. At the application of the low lead concentration the highest relative mRNA level for PAL was observed at 48 hpi in the leaves of seedlings growing on the medium with 0.075 $Pb(NO_3)_2$ and colonised by pea aphids (0.075 mM $Pb^{2+}$+aphids variant). However, these stress factors had a much stronger effect on the upregulation of CHS rather than PAL. The very high upregulation of mRNA for the CHS genes was observed as a result of the impact of lead or the cross-talk of lead and *A. pisum*, especially at the toxic concentration of lead. Also, the low lead level and *A. pisum* infestation (0.075 mM $Pb^{2+}$+aphids variant) raised the mRNA level for CHS, but it was much less markedly than in the case of lead or the cross-talk of lead at the toxic concentration and *A. pisum*.

### 2.5. The Effect of Lead and A. pisum on β-Glucosidase Activity in Pea Seedlings

In the roots of pea seedlings growing on the Hoagland medium supplemented with lead, already after 4 days from the administration of lead an increase in β-glucosidase activity was observed, being higher at the high lead concentration in comparison to that at the low lead concentration (Figure 7, Table S1a,b). A very high stimulation of β-glucosidase activity was recorded in roots of pea seedlings of the 0.5 mM $Pb^{2+}$ and 0.5 mM $Pb^{2+}$+aphids variants. At 24 hpi during the cross-talk of lead (0.5 mM $Pb(NO_3)_2$) and *A. pisum*, the activity of this enzyme in roots was highest among the experimental variants tested and this high level was maintained at subsequent time points after infestation. Moreover, it is of interest that at 72 hpi this activity dropped in roots of the 0.5 mM $Pb^{2+}$+aphids variant in relation to the activity of β-glucosidase in roots of the 0.5 mM $Pb^{2+}$ variant. A similar trend was observed in leaves at 72 hpi. In turn, the activity of the enzyme in leaves of seedlings cultured on the medium with lead was lower than in the control at 0 h of the experiment. At 48 and 72 h under the influence of lead alone applied at the low concentration and the cross-talk of lead and *A. pisum* (0.075 mM $Pb^{2+}$ and 0.075 mM $Pb^{2+}$+aphids variants), a reduction in β-glucosidase activity was observed in relation to the control. Additionally, aphid infestation alone (+aphids variant) stimulated an increase in β-glucosidase activity, as at 24 and 72 hpi the activity of this enzyme was higher than in the control. It should be added that the lowest activity of the enzyme was recorded in the leaves of pea seedlings growing on the medium with lead applied at the high concentration (0.5 mM $Pb^{2+}$ variant) and in leaves of the 0.5 mM $Pb^{2+}$+aphids variant.

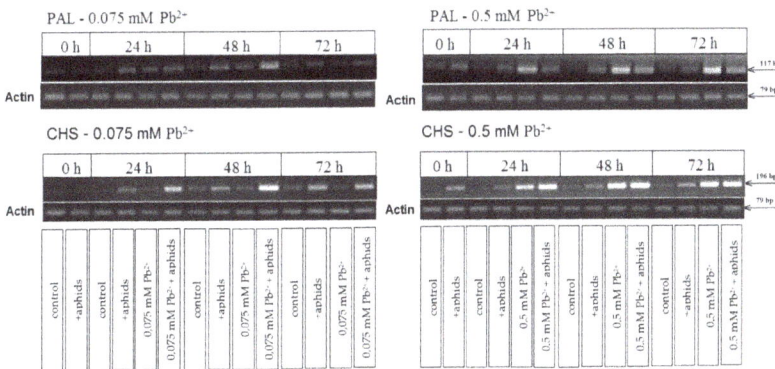

**Figure 6.** The effect of lead and *A. pisum* on expression levels of phenylalanine ammonialyase (PAL) and chalcone synthase (CHS) genes in leaves of pea seedlings. The data were obtained in three independent experiments and statistically analysed using ANOVA (*p*-values at α = 0.05). Hypotheses on the equality of means were verified by the two-sample *t*-test. To account for multiple testing, we used the Bonferroni correction (statistically significant differences are shown in Table S1).

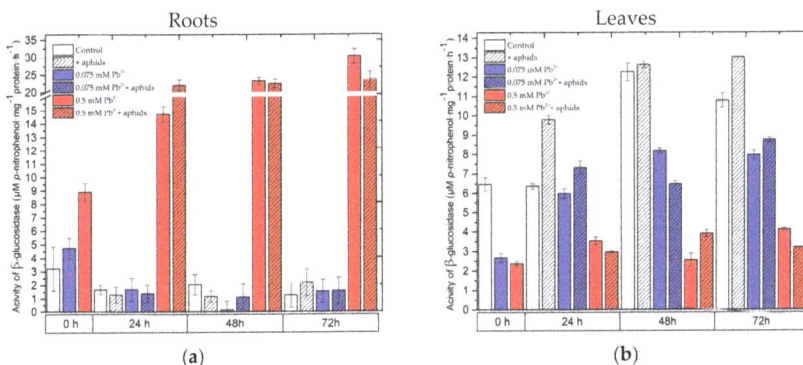

**Figure 7.** The effect of lead and *A. pisum* on β-glucosidase activity in roots (**a**) and leaves (**b**) of pea seedlings. The data were obtained in three independent experiments and statistically analysed using ANOVA (*p*-values at α = 0.05). Hypotheses on the equality of means were verified by the two-sample *t*-test. To account for multiple testing, we used the Bonferroni correction (statistically significant differences are shown in Table S1).

## 2.6. The Effect of Lead and A. pisum on Activity of Phenylalanine Ammonialyase in Pea Seedlings

The activity of phenylalanine ammonialyase (PAL) differed in the roots and the leaves of pea seedlings (Figure 8). At all time points of the experiments a decrease in PAL activity was observed in the roots of pea seedlings exposed to stress factors in relation to the control, with the exception of the activity of this enzyme in roots exposed to lead at the low concentration at 72 h (0.075 mM $Pb^{2+}$ variant). It should be noted that the strong reduction in enzyme activity was recorded in roots of pea seedlings exposed to lead alone at the higher concentration (0.5 mM $Pb^{2+}$ variant) and during the cross-talk of lead and the pea aphid (0.5 mM $Pb^{2+}$+aphids variant). In turn, during the impact of these stress factors, i.e., aphid infestation (+aphids), lead at the low concentration (0.075 mM $Pb^{2+}$) and the cross-talk of lead and aphids (0.075 mM $Pb^{2+}$+aphids), the reduction in PAL activity in roots of seedlings exposed to these stress factors was not marked when compared to the control. Moreover, in leaves of seedlings growing for 4 days on the medium with lead (0 h of the experiment), PAL activity increased both at the low and at the high lead concentration, although it was higher in leaves growing on the medium with 0.5 mM $Pb(NO_3)_2$ than in leaves growing on the medium with 0.075 mM $Pb(NO_3)_2$. At subsequent time points a higher PAL activity was detected in leaves infested by aphids (+aphids variant) and in leaves of seedlings growing on the medium supplemented with lead at the high concentration (0.5 mM $Pb^{2+}$ variant) as well as during the cross-talk of lead and A. pisum (0.5 mM $Pb^{2+}$+aphids variant) in relation to the control. A very high PAL activity was observed at 24 and 72 h of the experiment in the seedlings growing on the medium with 0.5 mM $Pb(NO_3)_2$ and infested by pea aphids (0.5 mM $Pb^{2+}$+aphids variant). For lead application at the low concentration, at 48 h and 72 h of the experiment in leaves of pea seedlings PAL activity was found to decrease in relation to the control. Statistical analysis showed significant differences in these results (Table S1a,b). Additionally, it should be mentioned that after 4 days from the administration of the high lead concentration (0 h of the experiment), a 2-fold decrease in enzyme activity was recorded.

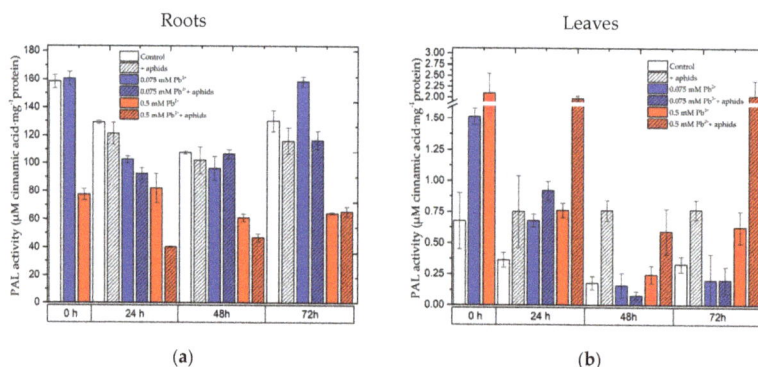

**Figure 8.** The effect of lead and A. *pisum* on activity of phenylalanine ammonialyase in roots (**a**) and leaves (**b**) of pea seedlings. The data were obtained in three independent experiments and statistically analysed using ANOVA (*p*-values at $\alpha = 0.05$). Hypotheses on the equality of means were verified by the two-sample *t*-test. To account for multiple testing, we used the Bonferroni correction (statistically significant differences are shown in Table S1).

## 2.7. The Effect of Lead and A. pisum on the Growth of Pea Seedlings

The administration of lead to the medium caused symptoms of phytotoxicity, which were varied and depended on the concentration of lead in the medium. Symptoms of phytotoxicity were visible the earliest on the roots due to their direct contact with lead ions (results not shown). Calculations of the Index of Tolerance (IT) showed that among the tested lead concentrations (0.05 and 0.075 mM $Pb(NO_3)_2$) inhibition of root elongation in relation to the control was found at the concentration of

0.5 mM $Pb(NO_3)_2$. Additionally, feeding of *A. pisum* on pea seedlings cultured with an addition of lead at the above-mentioned concentrations only slightly reduced elongation of pea seedlings (Figure 9).

We need to stress here that it was the lower of the tested lead concentration, i.e., 0.075 mM $Pb(NO_3)_2$. The effect of lead alone at the above concentration, as well as the simultaneous effect of lead (0.075 mM) and *A. pisum* generally did not inhibit the increment in length of pea seedlings, as the values of these growth indexes were only slightly higher than in the control. Stimulation of shoot and root length of pea seedlings by the low lead concentration (0.075 mM $Pb(NO_3)_2$) was visible at the beginning of the experiment (after 4 days from the administration of lead before transferring the aphids to pea seedlings) and at the subsequent time points, i.e., 24 h and 48 h for shoots. In contrast, a significant inhibition of root and shoot growth of pea seedlings was found after 4 days from the administration of lead at the high concentration. The roots in the 0.5 mM $Pb^{2+}$ variant were two times shorter than those of the control and the roots of 0.075 mM $Pb^{2+}$ variants. At subsequent time points of the experiment (24, 48 and 72 hpi) limitation of root elongation growth was also found in the 0.5 mM $Pb^{2+}$ and 0.5 mM $Pb^{2+}$+aphids variants. Therefore, the applied toxic concentration (0.5 mM $Pb(NO_3)_2$) caused the strong inhibition of growth in pea seedlings, while the applied low concentration (0.075 mM $Pb(NO_3)_2$) caused stimulation of growth in pea seedling shoots up to 48 h, with these differences being statistically significant (Table S1a,b).

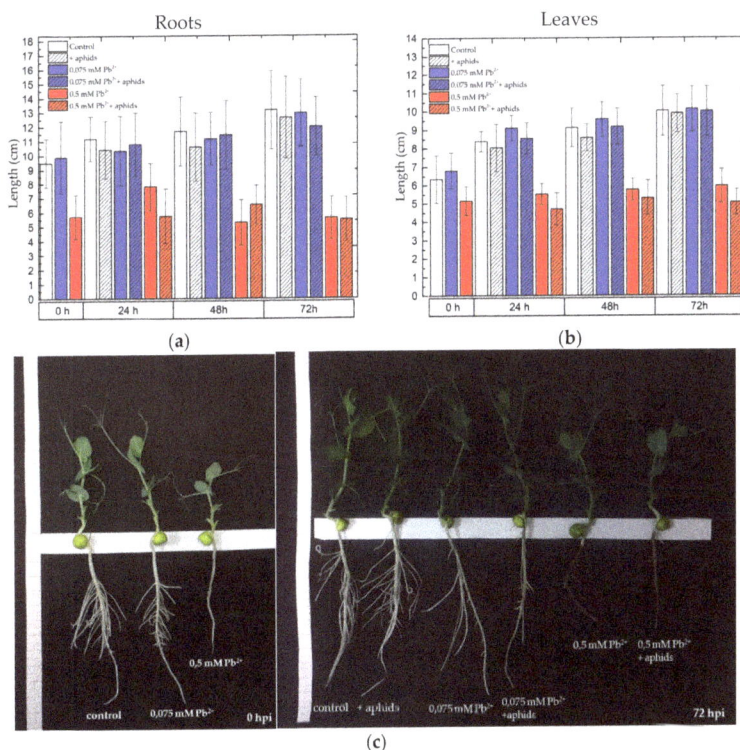

**Figure 9.** The effect of lead and *A. pisum* on growth of roots (**a**) and shoots (**b**) of pea seedlings. Pea seedlings after 4 days from the administration of lead and prior to transferring aphids onto pea seedlings (at 0 h) and 72 h after pea aphid infestation; (**c**). The data were obtained in ten independent experiments and statistically analysed using ANOVA (*p*-values at $\alpha = 0.05$). Hypotheses on the equality of means were verified by the two-sample *t*-test. To account for multiple testing, we used the Bonferroni correction (statistically significant differences are shown in Table S1).

*2.8. The Content of Lead in the Roots and Leaves of Pea Seedlings and in Bodies of Pea Aphid*

Analysis of lead content in pea seedlings revealed that this heavy metal was accumulated in large quantities in the roots of pea seedlings. Already at 0 h of the experiments, i.e., after 4 days from the administration of lead, accumulation of this element was observed both in the roots of seedlings growing on the medium with 0.5 mM $Pb(NO_3)_2$ and 0.075 mM $Pb(NO_3)_2$ (Figure 10, Table S1). However, lead concentration in the tissue of roots cultured on the medium with the high lead concentration was over 3.5 times higher than in the tissues cultured on the medium with its low concentration. At subsequent time points of the experiment, lead content in roots growing on the medium with 0.5 mM $Pb(NO_3)_2$ increased as compared to that in roots at 0 h. Thus, the following values were recorded: at 0 h 18,569.94 mg·$kg^{-1}$, at 24 h 24,571.18 mg·$kg^{-1}$, at 48 h 24,911.73 mg·$kg^{-1}$ and at 72 h 27,750.66 mg·$kg^{-1}$, respectively. In roots of pea seedlings growing on the medium with lead at the high and the low concentrations and infested by *A. pisum* (0.075 mM $Pb^{2+}$ + aphids and 0.5 mM $Pb^{2+}$ + aphids variants), the content of lead increased slightly, but only at 24 hpi, while at the next time point it was lower than in the 0.075 mM $Pb^{2+}$ and 0.5 mM $Pb^{2+}$ variants. Furthermore, at the administration of the lower lead concentration to the medium its contents in the seedling roots were as follows: at 0 h 5027.8 mg·$kg^{-1}$, 24 h 4343.16 mg·$kg^{-1}$, 48 h 5589.32 mg·$kg^{-1}$ and 72 h 5202.08 mg·$kg^{-1}$. Administration of lead at 0.5 mM $Pb(NO_3)_2$ caused an increased lead content in the seedling leaves. Lead contents in leaves of the 0.5 mM $Pb^{2+}$ variant were as follows: at 0 h 19.64158 mg·$kg^{-1}$, 24 h 30.04383 mg·$kg^{-1}$, 48 h 23.7783 mg·$kg^{-1}$ and 72 h 37.78 mg·$kg^{-1}$. In turn, lead contents in leaves of the 0.075 mM $Pb^{2+}$ variant at 0 h was 1.89381 mg·$kg^{-1}$, at 24 h 1.72839 mg·$kg^{-1}$, 48 h 2.49095 mg·$kg^{-1}$ and 72 h 5.05971 mg·$kg^{-1}$, respectively. Additionally, pea aphid infestation caused increased lead contents in leaves of pea seedlings at all time points after infestation. The highest content was recorded at 72 hpi in pea seedlings growing on the medium with 0.5 mM $Pb(NO_3)_2$ and infested by *A. pisum*, amounting to 67.28 mg·$kg^{-1}$. Additionally, as it was mentioned above, pea aphid feeding stimulated lead uptake and increased lead contents also in leaves of pea seedlings in the 0.075 mM $Pb^{2+}$ + aphids variant, i.e., at 24 hpi it was 2.84 times, at 48 h 2.87 times and at 72 h 2.64 times in relation to lead contents in leaves of the 0.075 mM $Pb^{2+}$ variant, respectively. Moreover, it should be mentioned here that at the beginning of the experiment (i.e., 4 days after lead administration), lead concentration in leaves of pea seedlings growing on the medium with 0.5 mM $Pb(NO_3)_2$ was 10 times higher than in leaves of seedlings growing on the medium with 0.075 mM $Pb(NO_3)_2$.

Lead content in bodies of pea aphids increased with infestation time, i.e., in the period from 24 to 72 hpi (Figure 10c, Table S2). The highest lead content was recorded at 72 hpi in bodies of pea aphids, ranging from 175.1867 mg·$kg^{-1}$ to 431.62322 mg·$kg^{-1}$. In turn, in bodies of pea aphids feeding on pea seedlings growing on the medium with 0.075 mM $Pb(NO_3)_2$ the content of lead was markedly lower than in the bodies of pea aphids feeding on pea seedlings growing on the medium with 0.5 mM $Pb(NO_3)_2$, with the levels ranging from 24.0655 mg·$kg^{-1}$ to 30.8712 mg·$kg^{-1}$, respectively.

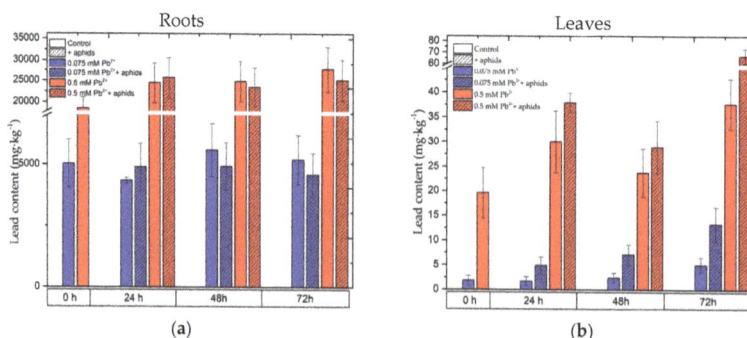

(a)        (b)

**Figure 10.** *Cont.*

(c)

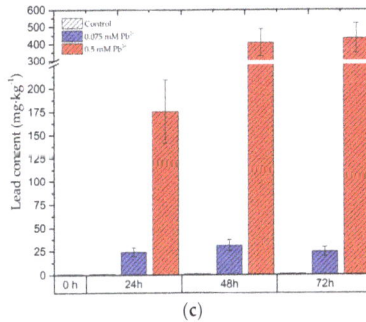

**Figure 10.** Lead contents in roots (**a**) and leaves (**b**) of pea seedlings and in bodies of pea aphids (**c**). The data were obtained in three independent experiments and statistically analysed using ANOVA (*p*-values at α = 0.05). Hypotheses on the equality of means were verified by the two-sample *t*-test. To account for multiple testing, we used the Bonferroni correction (statistically significant differences are shown in Table S1, S2).

## 3. Discussion

An important goal from the point of view of ecology is to study interactions between biotic and abiotic stresses at different levels of biological organisation [46]. Exposure of plants to one stress alters their metabolic status, which can influence the intensity or ability to adaptively respond to a subsequent stress. The results obtained within this study show for the first time the relationship between different concentrations of lead in the medium in terms of their effect on the growth of pea seedlings, pea aphid infestation and defensive response of plants manifested in the generation of signalling molecules. Moreover, it is a novel finding to show an enhanced generation of phytohormones such as SA and ABA in roots and leaves of pea seedlings and induction of biosynthesis of secondary metabolites, i.e., flavonoids, in the context of the effect of lead and the cross-talk of lead and a phytophage with the piercing-sucking mouthpart, i.e., *A. pisum.* Presently more evidence needs to be provided for the simultaneous impact of environmental factors on the defense strategy of the plants, i.e., to show the relationship of an abiotic stress factor (heavy metal) and plant defenses induced by a biotic stress factor (aphids). Also, the results within this research work provide new insights concerning the phenomenon of hormesis: the abiotic factor (lead)—plant (signalling molecules and flavonoids, including pisatin)—herbivore (the insect with a piercing-sucking mouthpart).

We demonstrated that after administration of lead at a high concentration in the substrate at 0.5 mM Pb(NO$_3$)$_2$, a high concentration of this element was observed first in the roots and then in the leaves of pea seedlings. Also, lead was downloaded by pea aphids, especially in large amounts when aphids were feeding on pea seedlings growing on the medium with 0.5 mM Pb(NO$_3$)$_2$. In turn, at the low concentration of 0.075 mM Pb(NO$_3$)$_2$, this element was mainly accumulated in the roots, while only small amounts appeared in the leaves and next in bodies of *A. pisum*, which was in line with our assumption. At a low concentration of lead deposited mainly in roots the consequence was an enhanced metabolic status of pea seedlings, which elicited defense mechanisms of the leaves to the biotic stress factor. Therefore, after 4 days from the administration of lead, in 0 h-leaves of the 0.075 mM Pb(NO$_3$)$_2$ variant an increased level of signalling molecules such as SA was noted. Statistical analysis showed significant differences between levels of SA or SAG in leaves of the 0.075 mM Pb(NO$_3$)$_2$ variant and the control variant. However, the content of an isoflavonoid phytoalexin, pisatin, in leaves with a low lead level at this point time was not increased in relation to the control leaves, while an increase was observed only for some glucosides and the PAL activity, an enzyme initiating phenylpropanoid metabolism. The increase in pisatin levels in the above-mentioned leaves was found at 48 and 72 hpi as a result of lead or lead and *A. pisum* interactions; however, it was significantly lower than in the case of the impact of lead with the high concentration. It should be emphasized that the different concentrations of the abiotic factor in the medium caused the different defensive response on multiple levels. In addition, direct contact of the stress factor (lead) with the roots caused a stronger defense response of the roots in comparison to the leaves. An interesting observation was noted after pea aphid feeding on pea seedlings cultured on the medium with lead at both concentrations. It was shown that *A. pisum* infestation modulated molecular and metabolic responses in the pea seedlings both with the high and low lead concentration. These changes included both the roots and the leaves, which indicates systemic signalling from the leaves to the roots, triggered by *A. pisum* feeding on the pea seedlings. The analysis of generation intensity for TSA (SA and SAG) and ABA in the roots and the leaves of pea seedlings growing on the medium with lead and during the treatment of both lead and *A. pisum* shows that SA is a molecule intensively stimulated both by lead alone and the cross-talk of lead and *A. pisum*. Also, a strong SA accumulation was found in the roots and the leaves during *A. pisum* feeding infestation alone, while higher levels of TSA were detected in the roots rather than the leaves. In turn, an opposite result was obtained for ABA, as a higher level of this molecule was found in the leaves than in the roots. Moreover, in the leaves the very strong TSA accumulation was a result of the amplification of the signal coming from lead and *A. pisum*, with the response being stronger at the high lead concentration than at the low concentration. In the roots a similar effect for SA was observed, but it was visible only at 72 hpi. Moreover, it should be stressed here that for ABA a slightly different trend was recorded. Lead treatment stimulated ABA accumulation most strongly. The amplification of the signal from lead and *A. pisum* was only observed at 72 hpi in the leaves with a high lead content and at 24 hpi in the leaves with a low lead content, respectively.

In the published literature the important role of SA and ABA was demonstrated both in responses to biotic and abiotic stresses, as well as in the development of plants ([47–51] and references in this article). These phytohormones play a crucial role in adaptive responses to stresses. SA as a signal molecule induces plant defense responses to biotic stress factors, including insect pests ([52,53] and references in this article). Besides, SA has been assigned roles in improved plant tolerance to stress via SA-mediated control of major metabolic processes in plants [51]. SA treatments induced different physiological and biochemical processes (e.g., enhanced the level of oxidative stress, the expression of genes involved in flavonoid metabolism and the amount of non-enzymatic antioxidants) in the leaves and roots of plants; however, the response of these organs varied [10]. Moreover, salicylates are molecules that may activate defense genes in plant response to stress. Upon intensification of these signals, an increase in the synthesis of allelochemicals and defence proteins was triggered [54], providing protection to the plants. It was also suggested that SA antagonises the oxidative damaging effects of lead both directly by activating the antioxidative enzymatic system and indirectly by decreasing lead uptake from soil [55]. Additionally, Chen et al. [56] demonstrated that SA alleviates lead-induced membrane disruptions and plays a positive role in rice seedlings protecting them against toxicity. Additionally, it has been revealed that lead causes an increase in endogenous free SA levels in all organs of *Zygophyllum fabago* [57]. The application of SA on wheat induced plant resistance to aphids [58]. El-Khawas [59] reported that priming plants with SA induced resistance of crop plants against herbivores, showing it to be successful in *Pisum sativum* against the leafminer *Liriomyza trifolii*. Therefore, an increase in the SA level was often observed in plant response to abiotic and biotic stresses, including aphid infestation [8,60]. Additionally, there are reports concerning the significance of SA in the minimisation of impacts of heavy metals, including lead [56,61–63]. It is believed that phytohormones such as SA and ABA can be promising compounds for the reduction of plant sensitivity to stresses, which may be important for agriculture [10,64]. The results of our experiments have shown that lead treatment and aphid attack seem to affect SA and pisatin accumulation synergistically. Lead treatment caused stimulation of PAL activity in leaves of seedlings after 4 days (0 h). Additionally, the cross-talk of these two stressors, i.e., lead at 0.5 mM $Pb(NO_3)_2$ and *A. pisum* significantly induced the activity of the enzyme in leaves of the 0.5 mM $Pb(NO_3)_2$ + aphids variant. What is more, amplification of the signal induced by the cross-talk of lead and *A. pisum* attack strongly upregulated the expression of PAL and CHS genes in the leaves of pea seedlings; it was stronger for CHS than for PAL. On the other hand, a synergistic positive role with jasmonate-induced defenses against herbivores and an antagonistic role with salicylate-based resistance to some pathogens was observed in the case of plant responses to water and salt stress by Thaler and Bostock [46]. Besides, the role of ABA as a central regulator of abiotic stress response in plants has been proven in numerous studies [65–67]. ABA concentration in plant tissues is known to increase when plants are exposed to heavy metals, suggesting an involvement of this phytohormone in the induction of protective mechanisms against heavy metal toxicity [68–70]. Stroiński et al. [71] reported that ABA was required in the transduction of the cadmium (Cd) signal to potato roots. For example, cadmium treatment leads to increased endogenous ABA levels in roots of *Typha latifolia* and *Phragmites australis* [72], in potato tubers [73] as well as rice plants [74]. The same effect was verified in several other studies. When mercury (Hg), Cd and copper (Cu) solutions were applied to wheat seeds during germination, ABA levels increased [75]. In cucumbers, seed germination decreased and ABA content increased under $Cu^{2+}$ and zinc ($Zn^{2+}$) stress [76]. Similarly, increased amounts of ABA were detected in germinating chickpea (*Cicer arietinum*) seeds under lead (Pb) toxicity conditions [77], as well as crowberries (*Empetrum nigrum*) exposed to Cu and nickel (Ni) [78]. Strong SA and ABA generation found in *P. sativum* L. cv. Cysterski seedlings during the influence of lead and the cross-talk of lead and *A. pisum*, especially at a high lead concentration, may indicate the involvement of these molecules in defensive response. However, as it was mentioned above, lead and aphid attack seem to affect SA and pisatin accumulation synergistically. In the case of ABA, this dependence was demonstrated only at 72 hpi. In the published literature some reports have shown that ABA was involved in the modulation of the phenylpropanoid pathway in berry skins [79,80]. Exogenous applications of ABA to red grape berries caused an increase in the flavonoid content [79–82]. In turn, in stressed plants ABA treatment regulated the expression of genes encoding enzymes of the flavonoid biosynthesis pathway [83]. In the case of our experiments, apart from an enhanced generation of signalling molecules under stress conditions, upregulation of phenylpropanoid metabolism genes, a strong accumulation of flavonoids and a strong stimulation of the activity of β-glucosidase were also recorded. The upregulation of CHS mRNA and pisatin accumulation in leaves of pea seedlings infested by the pea aphid at varying population sizes was observed in our previous studies [20]. We demonstrated also the participation of another signal molecule such as nitric oxide (NO) in pisatin accumulation and upregulation of the relative mRNA levels for PAL in leaves pretreated with NO donors (GSNO and SNP), both infested and non-infested tissues [84]. Moreover, there are reports concerning a transcriptional and posttranscriptional control of PAL expression under heavy metals stress [36]. The Pb induced PAL mRNA levels in legume plants exposed to heavy metals stress.

Additionally, our results show also a significant accumulation of flavonoid and isoflavonoid glycosides in response to lead alone and to the cross-talk of lead and the phytophage. Higher levels of these metabolites were recorded in the roots than in the leaves. As it was already mentioned above, the level of glycosides was also higher in the organs of pea seedlings growing with a high lead concentration in comparison to seedlings cultured with a low lead concentration. Generally, the dominant trend was associated with reduced levels of the metabolites in tissues affected by lead and aphids in comparison to those affected by lead treatment. In contrast, the reduction of flavonoid and isoflavonoid glycosides was observed mainly at the administration of the toxic dose of lead in the medium and *A. pisum* infestation in relation to their levels in leaves of the 0.5 mM $Pb^{2+}$ variant. Additionally, the reduction of these metabolites was stronger in the roots than in the leaves. The reduction of glycosides accompanied a high activity of β-glucosidase in pea seedlings growing with lead, or the variant with lead and infested by aphids, or only that with the attack by aphids. An increase in the activity of this enzyme may be related with the release of metabolites toxic for aphids or participation in the reconstruction of cell wall damage and the strengthening of its structure [85]. In turn, in our experiments reduction of the enzyme activity at 72 hpi in both leaves and roots may be due to the effect of *A. pisum* effectors that may block defense mechanisms of the host plant, including hydrolysis of glucosides to free aglycones. Many of the isoflavonoids and flavonoids have reactive free hydroxyl groups which are biologically active [86]. A diverse group of phytoalexins, particularly their chemical diversity, the main biosynthetic pathways and their regulatory mechanisms, phytoalexin gene transfer in plants and their role as antibiotic agents, were presented by Jeandet [87]. Also, information has also been provided concerning modulation of phytoalexin levels through engineering of plant hormones, defence-related markers or elicitors [88]. Our earlier research results [20] revealed also subcellular location of flavonoids using confocal microscopy in leaves of pea seedlings infested by aphids [20]. Furthermore, Adrian et al. [89] for the first time showed that aluminum chloride can act as a potent elicitor of resveratrol synthesis in grapevine leaves. Strong evidence was provided that a metallic salt can act as a direct inducer of phytoalexin response in grapevines.

Additionally, very interesting results of these studies showed that the toxic concentration of lead strongly inhibited growth of pea seedlings (root length in the 0.5 mM $Pb^{2+}$ variant was about 2 times shorter than in the control and 0.075 mM $Pb^{2+}$ variants), while the low concentration of 0.075 mM $Pb(NO_3)_2$ stimulated slightly growth, especially the shoots of seedlings, thus indicating the hormesis effect. Extremely interesting are results of quantitative analysis of metabolites in the roots and leaves of pea seedlings. Here the Principal Component Analysis (PCA) of the positive and negative ions MS revealed that groups ions in the control, + aphids, 0.075 mM $Pb^{2+}$, 0.075 mM $Pb^{2+}$ + aphids variants are clustered close. There may be a link between growth and the metabolome of plants from these variants. In turn, groups of the 0.5 mM $Pb^{2+}$ and 0.5 mM $Pb^{2+}$ + aphids variants were distinct in relation to the other.

It is known that at elevated concentrations heavy metals negatively affect morphology, physiology and biochemistry of plants [90–94]. The inhibition growth of lupin roots was observed in response to heavy metals (Pb, Cd and Cu), which was accompanied by an increased synthesis and accumulation of a 16 kDa polypeptide [95]. In turn, Rucińska et al. [96] reported that at higher lead concentrations the formation of both free radicals and reactive oxygen species is beyond the capacity of the antioxidant system, which in turn may contribute to reduced root growth. Lead can induce several morphological, physiological and biochemical dysfunctions in plants, such as a decrease in seed germination, plant growth, chlorophyll production, etc., while it also causes lipid peroxidation, oxidative stress and DNA damage [97]. A decrease was also recorded in dry weight of both roots and shoots [98,99]. Lead can reduce 50% of growth when applied at a concentration of 1000 mg/L in *Eichhornia crassipes* (water hyacinth). Also, an increase in lead concentration caused a decrease in chlorophyll content and an increase in the activity of antioxidative enzymes [100]. The results of the studies showed that Pb is a highly toxic element that can be accumulated in the cell wall, the cell membrane, vacuoles, mitochondria and peroxisomes. Małecka et al. [101]. Krzesłowska et al. [102] reported that plant cell wall (CW) remodeling, in particular formation of CW thickenings (CWTs) abundant in low-methylesterified pectins (pectin epitope JIM5 - JIM5-P) able to bind metal ions, including Pb, certainly increases the CW capacity for toxic metal ion binding and compartmentalisation. Rabęda et al. [103] reported that binding trace metals by JIM5-P may lead to a reduction of cell elongation and plant growth inhibition. Moreover, results of our research highlighted lead-induced hormetic growth response. Hormesis and the paradoxical effect of Pb at a low concentration on growth of wheat seedlings was revealed by Erofeeva et al. [104], but it was noted for Cd. In turn, a hormetic response of *Lonicera japonica* induced by Cd at low concentrations ($\leq$10 mg·kg$^{-1}$) closely related to the increase in net photosynthesis was demonstrated by Jia et al. [105]. Numerous vivid discussions concerning the importance of hormesis responses in toxicology and risk assessment may be followed in scientific papers and in the media, particularly during the last 15 years. It is assumed that hormesis is an adaptive response to stress. Recently, in a further attempt to convert the classical descriptive term of hormesis into a mechanistically based concept, it was proposed to consider hormesis as an adaptive response that is "providing a quantitative estimate of biological plasticity" [5]. In the published literature some reports indicate that herbivorous insects feeding on contaminated plants may suffer from developmental disorders manifested e.g., in an alteration of some life history traits, such as changes in morphology. When specimens of the cabbage aphid *Brevicoryne brassicae* L. were reared on lead contaminated host plants, they were smaller and showed a considerable degree of morphological asymmetry in comparison to aphids on non-contaminated plants [106,107]. Moreover, Poschenrieder et al. [2] reported that heavy metals in plants may affect herbivores, e.g., by acting as an antifeedant or a systemic plant pesticide, or by fortifying their antioxidant defenses.

In summary, this study supplies additional new information on the responses of pea seedlings to cross-talk of a heavy metal, lead, and aphid attacks. We have shown a relationship between SA and ABA-mediated signalling and their impact on the accumulation of flavonoids, including an isoflavonoid phytoalexin, pisatin. Also, we have demonstrated changes in the expression of key enzymes of the phenylpropanoid pathway (PAL and CHS) and the activity of β-glucosidase, an enzyme hydrolysing flavonoid glucosides to free aglycones, in the context of the cross-talk of lead and aphid attack. Moreover, our results demonstrate that a response of the plant to one stress with varying severity influences its response to other stresses.

In the future it is necessary to provide more evidence for the cross-tolerance phenomenon. Moreover, it is interesting to examine the interactions between hormone-signaling pathways and their impact on defence responses of plants and the ecological pattern of association of stresses.

## 4. Materials and Methods

### 4.1. Plant Material and Growth Conditions

Pea (*Pisum sativum* L. cv. Cysterski) seeds of the S-elite class were used in the experiments, these were obtained from the Plant Breeding Company at Tulce near Poznan in Poland. Surface-sterilisation of seeds was performed as described by Mai et al. [8] and Morkunas et al. [20]. After 6 h of imbibition the seeds were transferred onto filter paper (in Petri dishes) and immersed in a small amount of water in order to support further absorption. After a subsequent 66 h the seed coats were removed from the germinating seeds. Next the germinating seeds (35 pieces) were transferred to hydroponic grow boxes containing Hoagland medium (the control). Hydroponic boxes were covered with dark foil to mimic soil conditions. The experiments were conducted on seedlings of pea (*P. sativum*). Pea seedlings during the first four days were cultured in hydroponic cultures on the Hoagland medium in a phytotrone at 22–23 °C, 65% relative humidity, and light intensity of 130–150 μM photons m$^{-2}$s$^{-1}$ with the 14/10 h (light/dark) photoperiod. Next, on the fifth day the medium was replaced in all the hydroponic variants and lead was added to the medium at 0.075 Pb(NO$_3$)$_2$ and 0.5 mM Pb(NO$_3$)$_2$. After 4 days from the administration of lead, pea aphids were transferred onto pea seedlings. The following experimental variants were used: the control pea seedlings cultured with no addition of lead and not colonised by pea aphids (*Acyrthosiphon pisum*), pea seedlings growing on the Hoagland medium with different concentrations of lead ions, i.e., 0.075 mM Pb(NO$_3$)$_2$ and 0.5 mM Pb(NO$_3$)$_2$ and pea seedlings growing on the Hoagland medium with different concentrations of lead ions and colonised by pea aphids *A. pisum*, pea seedlings growing on the Hoagland medium colonised by pea aphids *A. pisum*. Samples for analyses were collected 4 days after lead administration and prior to transferring aphids onto pea seedlings (at 0 h), and next after 24, 48, 72 h of the action of the two stress factors, i.e., lead and *A. pisum*. Hydroponic cultures were aerated with an aeration system. Pea seedlings, both the control and growing in the presence of lead ions and seedlings growing in the presence of lead ions for colonisation by *A. pisum*, were cultured in glass aquariums (30 cm × 22 cm × 28 cm) and protected with gauze. Main experimental treatments with the insects comprised leaves of pea seedlings, while analyses were also performed for roots, in order to obtain comprehensive information concerning dependencies of root-leaves after the addition of lead to the medium at different concentrations. The experiments were conducted only using adult insects.

## 4.2. Aphids and Infestation Experiment

*Acyrthosiphon pisum* (Harris), originally cultured and supplied by the Department of Entomology, the Poznań University of Life Sciences, Poland, was reared on *Pisum sativum* L. in a growth chamber under conditions as specified above. On day 11 of culture pea seedlings were infested with 20 apterous adult females of *A. pisum*. The aphid populations were monitored throughout all the experiments [108]. The control pea seedlings were cultured with no addition of lead and not colonised by pea aphids (*Acyrthosiphon pisum*).

## 4.3. Detection of Salicylic Acid (SA)

Salicylic acid was extracted and quantified following the HPLC method, as previously described by Yalpani et al. [109] and modified by Mai et al. [8]. Frozen roots and leaves were ground in liquid nitrogen to a homogenous powder, from which approximately 0.50 g was taken for analysis. Salicylic acid was extracted twice with methanol (90% followed by a straight solvent), strongly stirred and centrifuged at 12,000× $g$ for 10 min at 4 °C. The resultant supernatant was divided into two equal batches and the solvent was evaporated to dryness under a stream of nitrogen. Dry extract was dissolved in 5% trichloroacetic acid (TCA), and after centrifugation free salicylic acid (SA) was extracted three times with the extractive organic mixture of ethyl acetate:cyclopentane:isopropanol (100:99:1; $v/v/v$). In order to determine free and glucoside bound salicylic acid (TSA), 40 units of β-glucosidase in Na-acetate buffer (100 mM, pH 5.2) were added to the second part of the dry extract and incubated for 90 min at 37 °C. Lyophilised β-glucosidase from almonds (Sigma-Aldrich) was used; according to the manufacturer's specification one unit liberates 1.0 μmole of glucose from salicin per min at pH 5.0 and 37 °C. The reaction was terminated by the addition of 5% TCA and salicylic acid was extracted as described above. After solvent evaporation, the dry residue was dissolved in a mobile phase (200 mM K-acetate buffer, pH 5.0 and 0.5 mM ethylenedinitrilotetraacetic acid, EDTA) and analyzed by HPLC coupled with fluorometric detection in a Waters Company chromatograph (Milford, MA, USA) composed of a 2699 Separation Module Alliance and 2475 Multi- Fluorescence Detector. Chromatographic separation was obtained in a Spherisorb ODS2 WATERS Company column (4.6 mm × 10 mm, 3 μm). Detection parameters were λ = 295 nm for excitation and 405 nm for emission. The content of glucoside bound salicylic acid (SAG) was calculated as the difference between total and free salicylic acid (TSA-SA) and the results were expressed as nanograms per gram of fresh weight material (ng g$^{-1}$ FW).

## 4.4. Detection of Abscisic Acid (ABA)

For ABA measurement samples consisting of approx. 1 g frozen tissue were homogenised in liquid nitrogen and extracted twice with 20 mL 80% ($v/v$) methanol containing 20 mg/L butylated hydroxytoluene. Subsequently 100 ng [$^2$H$_6$]ABA were added. The methanol fraction was removed under reduced pressure and the aqueous phase was acidified to pH 2.0 with 12 M HCl and centrifuged at 10,000× $g$ for 15 min to remove chlorophyll. The supernatant was partitioned three times against ethyl acetate and dried under vacuum. The dry residue was dissolved in 5 mL 1 M formic acid (FA) and loaded on a Discovery® DSC−18 SPE cartridge (Supelco Inc., Bellefonte, PA, USA) pre-conditioned with 4 mL MeOH and allowed to equilibrate with 4.0 mL of 1 M FA. The column was then washed with 4 mL 1 M FA, 20% of methanol in 1M FA and finally target phytohormones were eluted with 4 mL of 80% methanol in water. The eluate was evaporated and further purified by HPLC using a SUPELCOSIL ABZ + PLUS column (250 mm × 4.5 mm, 5 μm article size; Supelco). The samples were reconstituted in 130 μL of 20 % methanol and chromatographed with a linear gradient of 20–80% methanol in 0.1 M formic acid for 20 min, flow rate 1.0 mL/min at 22 °C. The fractions collected at 12.5 ± 0.5 min were evaporated to dryness, methylated with diazomethane, dissolved in 30 μL of methanol and analyzed by GC/MS–SIM (Auto-System XL coupled to a TurboMass, Perkin-Elmer, Walthman, MA, USA) using a MDN-5 column (30 m × 0.25 mm, 0.25 μm phase thickness, Supelco).

The GC temperature program was 60 °C for 1 min, 60–250 °C at 10 °C/min, flow rate 1.5 mL/min, injection port was 280 °C and electron potential 70 eV. The retention times of ABA and $[^2H_6]$ ABA were 14.07 and 14.3 min, respectively. GC/MS–SIM was performed by monitoring $m/z$ 190 for endogenous ABA and 194 for $[^2H_6]$ ABA, 130 for endogenous IAA and 132 $[^2H_2]$ IAA according to the method described by Vine et al. [110].

### 4.5. Analysis of Flavonoids

#### 4.5.1. Isolation of Phenolic Compounds

Plant material, previously frozen at −80 °C was homogenised in 80% methanol (20 mL·g$^{-1}$·FW) and sonicated for 3 min in a VirTis VirSonic 60 sonicator [111]. The suspension was filtered through a Büchner funnel and concentrated under vacuum at 40 °C. Plant extract samples for LC analyses were prepared from 0.5 g FW pea tissue. The samples were purified and concentrated by solid-phase extraction on cartridges containing a cation exchanger and RP C−18 silica gel (Alltech, Carnforth, UK) used in tandem, according to the method proposed by Stobiecki et al. [112].

#### 4.5.2. Liquid Chromatography–Mass Spectrometry (LC/UV/ESI/MS/MS)

Plant extract samples were analysed using a Waters UPLC Acquity system coupled with a micrOToF-Q mass spectrometer (Bruker Daltonics, Bremen, Germany). An Agilent Poroshell RP-C18 column (100 mm × 2.1 mm; 2.7 µm) was used. During LC analyses elution was performed using two solvent mixtures: A (95% $H_2O$, 4.5% acetonitrile, 0.5% acetic acid; $v/v/v$) and B (95% acetonitrile, 4.5% $H_2O$, 0.5% acetic acid; $v/v/v$). Elution steps were as follows: 0–5 min 10%–30% B, 5–12 min isocratic at 30% B, 12–13 min linear gradient up to 95% of B and 13–15 min isocratic at 95% of B. Pisatin and flavonols were identified by comparing their retention times and mass spectra with the data from respective standards. The micrOToF-Q mass spectrometer consisted of an ESI source operating at a voltage of ±4.5 kV, nebulisation with nitrogen at 1.2 bar and dry gas flow of 8.0 L/min at 220 °C. The instrument was operated using the micrOToF Control program version 2.3 and data were analysed using the Bruker Data Analysis ver. 4 package. Targeted MS/MS experiments were performed using a collision energy ranging from 10 to 25 eV, depending on the molecular masses of the compounds. The instrument operated at a resolution of minimum 15,000 full widths at half maximum.

#### 4.5.3. Quantitative Analysis of Metabolites

For targeted quantitative analysis, the extracted ion chromatogram traces were used, with peaks plotted for exact monoisotopic masses of compounds. Such traces were prepared for each compound earlier and identified based on their MS2 spectra. For calibration purposes, *p*-hydroxybenzoic acid was added to each analysed sample as the internal standard at a final concentration of 125 µM (LC retention time and MS spectra did not interfere with those of the studied compounds). Total quantitative analysis was conducted based on all signals acquired in the positive MS mode using the Profile Analysis 2.1 software (Bruker Daltonics). Multivariate analyses were carried out by the unsupervised principal component analysis (PCA).

### 4.6. Total RNA Extraction and Semiquantitative RT-PCR Analysis

Pea seedling leaves (0.50 g) were frozen in liquid nitrogen and ground with a mortar and pestle in the presence of liquid nitrogen. For RT-PCR analyses of the target gene, total RNA was isolated from 45 mg tissue using the SV Total RNA Isolation System (Promega, Manheim, Germany) according to the recommendations of the manufacturer [84,113]. The RNA level in samples was assayed spectrophotometrically at 260 nm. The $A_{260}/A_{280}$ ratio varied from 1.8 to 2.0. The cDNA samples for RT-PCR experiments were synthesised from 1 µg of total RNA and oligo (dT)$_{18}$ primers using the High Capacity cDNA Reverse Transcription Kit (Applied Biosystems, Life Technologies Polska, Warszawa, Poland). One µL of the each cDNA was used as a template for the PCR reaction with specific PCR primers. Thermal cycling conditions in the PAL, CHS and actin genes expression assay consisted of an initial denaturation at 95 °C for 5 min, followed by 25 cycles at 95 °C for 30 s, 58 °C for 30 s and 72 °C for 45 s. The PCR products were analysed by agarose gel electrophoresis (1.5%). The specific primers were used for PCR reactions: PAL F 5′-GAGCAGCACAACCAAGATG-3′ and PAL R 5′-CTCCCTCTCAATTGACTTGG-3′, CHS F 5′-AGGTCTTCGCCCATATGTGAA-3′ and CHSR 5′-GTGCTTGTCCAACAAGACTGT-3′. The fragment of *Pisum sativum* actin (HQ231775) coding sequence was amplified as a reference gene using actin F 5′-GCATTGTAGGTCGTCCTCG-3′ and actin R 5′-TGTGCCTCATCACCAACATAT-3′ primers.

*4.7. Extraction and Assay of β 1,3-Glucosidase Activity*

The activity of β 1,3-glucosidase (EC 3.2.1.21) was determined spectrophotometrically (Lambda 15 UV-Vis spectrophotometer, Perkin Elmer, Norwalk, CT, USA) applying the method proposed by Nichols et al. [114] and modified by Morkunas et al. [115]. Leaves and roots of pea seedlings (500 mg) were ground at 4 °C in 0.05 M phosphate buffer of pH 7.0 and 1% polyvinylpyrrolidone (PVP). The enzyme activity was determined in the supernatant obtained after centrifugation at $15,000 \times g$ for 20 min. The mixture containing 0.2 mL phosphate buffer (0.05 M, pH 7.0), 0.2 mL extract and 0.2 mL 4-nitrophenyl-β-D-glucopyranoside as substrate ($2 \, mg \cdot mL^{-1}$) was incubated for one hour at 35 °C. Afterwards, 0.6 mL 0.2 M $Na_2CO_3$ was added. The formation of *p*-nitrophenol (*p*-NP) was followed at 400 nm. The activity was measured in three replications and expressed as µM *p*-nitrophenol $mg^{-1} \cdot protein \cdot h^{-1}$.

*4.8. Extraction and Assay of Phenylalanine Ammonialyase (PAL) Activity*

The activity of PAL (EC 4.3.1.24) was determined with a modified method of Cahill and McComb [116] as modified by Morkunas et al. [43]. The amount of 0.50 g of frozen leaves was homogenised at 4 °C with a mortar and pestle in 4 mL of 100 mM Tris–HCl buffer (pH 8.9) containing 5 mM mercaptoethanol, and 0.050 g PVP. Afterwards the homogenate was centrifuged at $12,000 \times g$ for 20 min at 4 °C. The supernatant was used for enzyme analyses. The reaction mixtures contained 0.50 mL of 20 mM borate buffer (pH 8.9), 0.50 mL of 10 mM l-phenylalanine, and 0.50 mL extract in a total volume of 1.5 mL. A sample without the substrate l-phenylalanine was used as a blank. The reaction proceeded for 24 h at 30 °C and was interrupted by the addition of 1.5 mL 2 N HCl. PAL activity was measured by the change of absorbance at 290 nm due to the formation of *trans*-cinnamic acid using a Perkin Elmer Lambda 15 UV-Vis spectrophotometer. The activity of PAL is expressed as µM *trans*-cinnamic acid expressed per mg protein per hour (µM *trans*-cinnamic acid· $mg^{-1}$ protein $h^{-1}$).

*4.9. Determination of Lead Content*

All reagents and standards were of at least analytical grade. All solutions were prepared using ultrapure water 18.2 MΩ·cm. Root and leaf samples of pea seedlings were rinsed with ultrapure water (CDRX-200, Polwater, Kraków, Poland) and air dried at room temperature. Air dried samples were then ground (Planetary Micro Mill Pulverisette 7 premium line, Fritsch, Idar-Oberstein, Germany) and oven dried at 70 °C before weighing. Subsamples (0.5 g) were digested in 10 mL 65% nitric acid (Suprapur, Merck, Darmstadt, Germany) using a closed vessel digestion microwave system (MARS 5, CEM Corp., Matthews, NC, USA). The digested solutions were diluted to 50 mL using ultrapure water. The aphid samples were washed and dried as described for the plant samples. 0.2 g subsamples of aphids were digested in 10 mL 65% nitric acid (Suprapur, Merck) at 80 °C until the solution became clear. The digested solutions were diluted to 25 mL using ultrapure water. The lead concentrations were determined by flame atomic absorption spectroscopy (FAAS) (SpectrAA 240FS, Varian, Palo Alto, CA, USA) and graphite furnace atomic absorption spectroscopy with the Zeeman correction (GFAAS) (SpectrAA 240Z, Varian). Certified reference materials were analysed to verify the accuracy and precision of the measurements. Observed concentrations for the certified INCT-MPH-2 mixed Polish herbs sample (Institute of Nuclear Chemistry and Technology, Poland) were 2.02 mg·$kg^{-1}$ Pb (certified

2.16 $\pm$ 0.23 mg·kg$^{-1}$) for plant samples and 1.99 mg·kg$^{-1}$ Pb for aphid samples. The Laboratory is accredited according to ISO/IEC 17025.

*4.10. Statistical Analysis*

All determinations were conducted within three independent experiments. Additionally, three biological replicates per experimental variant were made for a given experiment. Analysis of variance (ANOVA) was used to verify the significance of means from independent experiments within a given experimental variant. The elementary comparisons between particular levels of the analysed factor in different times (independently) were tested using the two-sample $t$-test for equal means for all the observed traits. To account for multiple testing, we used the Bonferroni correction. Moreover, comparisons related to the following plant material variants, i.e., the control vs. the 0.075 mM Pb$^{2+}$ variant; the control vs. 0.5 mM Pb$^{2+}$ variant; the control vs. the +aphids variant; the control vs. 0.075 mM Pb$^{2+}$+aphids variant; the control vs. 0.5 mM Pb$^{2+}$+aphids variant;. 0.075 mM Pb$^{2+}$ variant vs. 0.5 mM Pb$^{2+}$ variant; 0.075 mM Pb$^{2+}$ variant vs. 0.075 mM Pb$^{2+}$+aphids variant; 0.5 mM Pb$^{2+}$ variant vs. 0.5 mM Pb$^{2+}$+aphids variant; 0.075 mM Pb$^{2+}$+aphids variant vs. 0.5 mM Pb$^{2+}$+aphids variant; +aphids variant vs. 0.075 mM Pb$^{2+}$+aphids variant; +aphids variant vs. 0.5 mM Pb$^{2+}$+aphids variant. In turn, comparisons related to the following aphid variants, i.e., the control vs. 0.075 mM Pb$^{2+}$ variant; the control vs. 0.5 mM Pb$^{2+}$ variant; 0.075 mM Pb$^{2+}$ variant vs. 0.5 mM Pb$^{2+}$ variant, respectively. The figures present data obtained as means of triplicates for each variant along with standard errors of mean (SE). All the analyses were conducted using the GenStat v. 17 statistical software package.

## 5. Conclusions

In the presented study we revealed that lead at various concentrations (low causing the hormesis effect vs. high causing the toxic effect) and the cross-talk of lead and *A. pisum* induced generation of SA and ABA in pea seedlings. Increased generation of these phytohormones strongly enhanced the biosynthesis of flavonoids, including a phytoalexin, pisatin. Strong generation of SA and ABA found in pea seedlings as a result of the influence of lead alone, especially at a lead high concentration, and the cross-talk of lead and *A. pisum* may indicate the involvement of these molecules in defensive responses, with these responses being stronger at exposure to a toxic lead dose rather than at a low dose of lead.

**Supplementary Materials:** Supplementary materials are available online.

**Acknowledgments:** This work was supported in part by the Poznań University of Life Sciences young scientist fund for A.W., project no. 507.645.83 (in 2016) and 507.645.29 (in 2017). This concept and part of the research was the basis of a research project submitted to the National Polish Science Center (NCN, registration No.: 2016/21/B/NZ3/00636 and No.:2017/25/N/NZ9/00704).

**Author Contributions:** A.W., a Ph.D. student of I.M. designed the studies, wrote and prepared the manuscript, performed most experiments, and analysed and interpreted the data; I.M. created the concept and designed the studies (this concept was the basis of a research project for the Polish National Science Centre-NCN, registration No.: 2016/21/B/NZ3/00636 and No.: 2017/25/N/NZ9/00704), analysed literature and wrote the manuscript, analysed and interpreted the data, supervised the organisation of the study; Ł.M. contributed to measurements of flavonoid concentrations; D.N. was responsible for performed RT-PCR analyses; J.K. contributed to measurements of abscisic acid concentrations, K.D. contributed to measurements of salicylic acid concentrations; J.B. contributed to performing elementary comparisons between particular levels of analysed factors at different time points using the two-sample $t$-test for equal means for all observed traits; M.G. and R.M. contributed to measurements of lead contents in plant material and aphid bodies.

**Conflicts of Interest:** The authors declare no conflict of interests. The founding sponsors had no role in the design of the study; in the collection, analyses, or interpretation of data; in the writing of the manuscript, and in the decision to publish the results.

## Abbreviations

| | |
|---|---|
| ABA | abscisic acid |
| ANOVA | analysis of variance |
| Cd | cadmium |
| CHS | chalcone synthase |
| EDTA | ethylenedinitrilotetraacetic acid |
| ESI | electrospray ionization |
| FW | fresh weight |
| g | gram |
| Glc | glucose |
| h | hour |
| FA | formic acid |
| GC-MS | gas chromatography/mass spectrometry |
| hpi | hour post infestation |
| HPLC | high-performance liquid chromatography |
| IAA | indole-3yl-acetic acid |
| LC/UV/ESI/MS/MS | Liquid chromatography/ultraviolet detection/electrospray–mass spectrometry (tandem mass spectrometry) |
| MS | mass spectrometry |
| MS2 | second stage of mass spectrometry |
| PAL | phenylalanine ammonia-lyase |
| PCA | principal component analysis |
| Pb | lead |
| REACH | Registration, Evaluation, Authorisation and Restriction of Chemicals |
| RT-PCR | Reverse transcription polymerase chain reaction |
| SA | salicylic acid |
| SAG | salicylic acid glucoside |
| SIM | Selected ion monitoring |
| TCA | trichloroacetic acid |
| TSA | total salicylic acid |
| UV-Vis | ultraviolet-visible spectrophotometry |

## References

1. Maksymiec, W. Signaling responses in plants to heavy metal stress. *Acta Physiol. Plant.* **2007**, *29*, 177–187. [CrossRef]
2. Poschenrieder, Ch.; Tolrà, R.; Barceló, J. Can metal defend plants against biotic stress? *Trends Plant Sci.* **2006**, *11*, 288–295. [CrossRef] [PubMed]
3. Rejeb, K.B.; Abdelly, C.; Savouré, A. How reactive oxygen species and proline face stress together. *Plant Physiol. Biochem.* **2014**, *80*, 278–284. [CrossRef] [PubMed]
4. Fujita, M.; Fujita, Y.; Noutoshi, Y.; Takahashi, F.; Narusaka, Y.; Yamaguchi-Shinozaki, K.; Shinozaki, K. Crosstalk between abiotic and biotic stress responses: A current view from the points of convergence in the stress signaling networks. *Curr. Opin. Plant Biol.* **2006**, *9*, 436–442. [CrossRef] [PubMed]
5. Poschenrieder, Ch.; Cabot, C.; Martos, S.; Gallego, B.; Barceló, J. Do toxic ions induce hormesis in plants? *Plant Sci.* **2013**, *212*, 15–25. [CrossRef] [PubMed]
6. Arimura, G.; Maffei, M.E. Calcium and secondary CPK signaling in plants in response to herbivore attack. *Biochem. Biophys. Res. Commun.* **2010**, *400*, 455–460. [CrossRef] [PubMed]
7. Arimura, G.; Ozawa, R.; Maffei, M.E. Recent advances in plant early signaling in response to herbivory. *Int. J. Mol. Sci.* **2011**, *12*, 3723–3739. [CrossRef] [PubMed]
8. Mai, V.Ch.; Drzewiecka, K.; Jeleń, H.; Narożna, D.; Rucińka-Sobkowiak, R.; Kęsy, J.; Floryszak-Wieczorek, J.; Gabryś, B.; Morkunas, I. Differential induction of *Pisum sativum* defense signaling molecules in response to pea aphid infestation. *Plant Sci.* **2014**, *221–222*, 1–12. [CrossRef] [PubMed]

9. Jeandet, P.; Courot, E.; Clément, Ch.; Ricord, S.; Crouzet, J.; Aziz, A.; Cordelier, S. Molecular engineering of phytoalexins in plants: Benefits and limitations for food and agriculture. *J. Agric. Food Chem.* **2017**, *65*, 2643–2644. [CrossRef] [PubMed]

10. Gondor, O.K.; Pál, M.; Darkó, É.; Janda, T.; Szalai, G. Salicylic acid and sodium salicylate alleviate cadmium toxicity to different extents in maize (*Zea mays* L.). *PLoS ONE* **2016**, *11*, 1–18. [CrossRef] [PubMed]

11. Lattanzio, V.; Arpaia, S.; Cardinali, A.; di Venere, D.; Linsalata, V. Role of endogenous flavonoids in resistance mechanism of *Vigna* to aphids. *J. Agric. Food Chem.* **2000**, *48*, 5316–5320. [CrossRef] [PubMed]

12. Simmonds, M.S.J. Flavonoid–insect interactions: Recent advances in our knowledge. *Phytochem* **2003**, *6*, 21–30. [CrossRef]

13. Leitner, M.; Boland, W.; Mithöfer, A. Direct and indirect defences induced by piercing-sucking and chewing herbivores in *Medicago truncatula*. *New Phytol.* **2005**, *167*, 597–606. [CrossRef] [PubMed]

14. Gould, K.S.; Lister, C. Flavonoid functions in plants. In *Flavonoids: Chemistry, Biochemistry and Applications*; Andersen, Ø.M., Markham, K.R., Eds.; CRC Press: London, UK; New York, NY, USA, 2006; pp. 397–441, ISBN 0-8493-2021-6.

15. Bernards, M.A.; Båstrup-Spohr, L. Phenylpropanoid metabolism induced by wounding and insect herbivory. In *Induced Plant Resistance to Herbivory*; Schaller, A., Ed.; Springer: New York, NY, USA, 2008; pp. 189–211, ISBN 978-1-4020-8181-1.

16. Lev-Yadun, S.; Gould, K.S. Role of anthocyanins in plant defense. In *Anthocyanins Biosynthesis, Functions and Applications*; Gould, K.S., Davies, K.M., Winefield, C., Eds.; Springer: Berlin, Germany, 2009; pp. 21–48, ISBN 978-0-387-77334-6.

17. Mewis, I.; Khan, M.A.M.; Glawischnig, E.; Schreiner, M.; Ulrichs, Ch. Water stress and aphid feeding differentially influence metabolite composition in *Arabidopsis thaliana* (L.). *PLoS ONE* **2012**, *7*(11), e48661. [CrossRef] [PubMed]

18. Prince, D.C.; Drurey, C.; Zipfel, C.; Hogenhout, S.A. The leucine-rich repeat receptor-like kinase Brassinosteroid Insensitive1-Associated Kinase1 and the Cytochrome P450 PHYTOALEXIN DEFICIENT3 contribute to innate immunity to aphids in arabidopsis. *Plant Physiol.* **2014**, *164*, 2207–2219. [CrossRef] [PubMed]

19. Golan, K.; Sempruch, C.; Górska-Drabik, E.; Czerniewicz, P.; Łagowska, B.; Kot, I.; Kmieć, K.; Magierowicz, K.; Leszczyński, B. Accumulation of amino acids and phenolic compounds in biochemical plants responses to feeding of two different herbivorous arthropod pests. *Arthropod-Plant Interact.* **2017**, 1–8. [CrossRef]

20. Morkunas, I.; Woźniak, A.; Formela, M.; Mai, V.Ch.; Marczak, Ł.; Narożna, D.; Borowiak-Sobkowiak, B.; Kühn, Ch.; Grimm, B. Pea aphid infestation induces changes in flavonoids, antioxidative defence, soluble sugars and sugar transporter expression in leaves of pea seedlings. *Protoplasma* **2015**, *253*, 1063–1079. [CrossRef] [PubMed]

21. Hart, S.; Kogan, M.; Paxton, J.D. Effect of soybean phytoalexins on the herbivorus insects mexican bean bettle and soybean looper. *J. Chem. Ecol.* **1983**, *9*, 657–672. [CrossRef] [PubMed]

22. Russel, G.B.; Sutherland, O.R.W.; Hutchins, R.F.N.; Christmas, P.E. Vestitol: A phytoalexin with insect feeding-deterrent activity. *J. Chem. Ecol.* **1978**, *4*, 571–579. [CrossRef]

23. Sutherland, O.; Russell, G.; Biggs, D.; Lane, G. Insect feeding deterrent activity of phytoalexin isoflavonoids. *Biochem. Syst. Ecol.* **1980**, *8*, 73–75. [CrossRef]

24. Morimoto, M.; Kumeda, S.; Komai, K. Insect antifeedant flavonoids from *Gnaphalium affin*. *J. Agric. Food Chem.* **2000**, *48*, 1888–1891. [CrossRef] [PubMed]

25. Onyilagha, J.C.; Lazorko, J.; Gruber, M.Y.; Soroka, J.J.; Erlandson, M.A. Effect of flavonoids on feeding preference and development of the crucifer pest *Mamestra configurata* Walker. *J. Chem. Ecol.* **2004**, *30*, 109–124. [CrossRef] [PubMed]

26. Zhou, D.S.; Wang, C.Z.; Loon, J.J.A. Chemosensory basis of behavioural plasticity in response to deterrent plant chemicals in the larva of the small cabbage white butterfly *Pieris rapae*. *J. Insect Physiol.* **2009**, *55*, 788–792. [CrossRef] [PubMed]

27. Goławska, S.; Łukasik, I. Antifeedant activity of luteolin and genistein against the pea aphid, *Acyrthosiphon pisum*. *J. Pest Sci.* **2012**, *85*, 443–450. [CrossRef] [PubMed]

28. Simmonds, M.S.J. Importance of flavonoids in insect-plant interactions: Feeding and oviposition. *Phytochemistry* **2001**, *56*, 245–252. [CrossRef]

29. Ateyyat, M.; Abu-Romman, S.; Abu-Darwish, M.; Ghabeish, I. Impact of flavonoids against woolly apple aphid, *Eriosoma lanigerum* (Hausmann) and its sole parasitoid, *Aphelinus mali* (Hald.). *J. Agric. Sci.* **2012**, *4*, 227–236. [CrossRef]

30. Diaz Napal, G.N.; Palacios, S.M. Bioinsecticidal effect of flavonoids pinocembrin and quercetin against *Spodoptera frugiperda*. *J. Pest Sci.* **2015**, *88*, 629–635. [CrossRef]

31. Diaz, J.; Bernal, A.; Pomar, F.; Merino, F. Induction of shikimate dehydrogenase and peroxidase in pepper (*Capsicum annuum* L.) seedling in response to copper stress and its relation to lignification. *Plant Sci.* **2001**, *161*, 179–188. [CrossRef]

32. Sakihama, Y.; Cohen, M.F.; Grace, S.C.; Yamasaki, H. Plant phenolic antioxidant and prooxidant activities: Phenolics-induced oxidative damage mediated by metals in plants. *Toxicology* **2002**, *177*, 67–80. [CrossRef]

33. Smeets, K.; Cuypers, A.; Lambrechts, A.; Semane, B.; Hoet, P.; Laere, A.V.; Vangorsveld, J. Induction of oxidative stress and antioxidative mechanisms in *Phaseolus vulgaris* after Cd application. *J. Plant Physiol. Biochem.* **2005**, *43*, 437–444. [CrossRef] [PubMed]

34. Jahangir, M.; Abdel-Farid, I.B.; Kim, H.K.; Choi, Y.H.; Verpoorte, R. Healthy and unhealthy plants: The effect of stress on the metabolism of Brassicaceae. *Environ. Exp. Bot.* **2009**, *67*, 23–33. [CrossRef]

35. Posmyk, M.M.; Kontek, R.; Janas, K.M. Antioxidant enzymes activity and phenolic compounds content in red cabbage seedlings expose to copper stress. *Ecotoxicol. Env. Saf.* **2009**, *72*, 596–602. [CrossRef] [PubMed]

36. Pawlak-Sprada, S.; Arasimowicz-Jelonek, M.; Podgórska, M.; Deckert, J. Activation of phenylpropanoid pathway in legume plants expose to heavy metals: Part I. Effects of cadmium and lead on phenylalanine ammonia-lyase gene expression, enzyme activity and lignin content. *Acta Biochim. Pol.* **2011**, *58*, 211–216. [PubMed]

37. Rastgoo, L.; Alemzadeh, A. Biochemical responses of Gouan (*Aeluropus littoralis*) to heavy metals stress. *Aust. J. Crop Sci.* **2011**, *5*, 375–383.

38. Ashraf, U.; Kanu, A.S.; Deng, Q.; Mo, Z.; Pan, S.; Tian, H.; Tang, X. Lead (Pb) toxicity; physio-biochemical mechanisms, grain yield, quality, and Pb distribution proportions in scented rice. *Front. Plant Sci.* **2017**, *8*, 259. [CrossRef] [PubMed]

39. Arshad, M.; Silvestre, J.; Pinelli, E.; Kallerhoff, J.; Kaemmerer, M.; Tarigo, A.; Shahid, M.; Guiresse, M.; Pradere, P.; Dumat, C. A field study of lead phytoextraction by various scented Pelargonium cultivars. *Chemosphere* **2008**, *71*, 2187–2192. [CrossRef] [PubMed]

40. Uzu, G.; Sobanska, S.; Sarret, G.; Munoz, M.; Dumat, C. Foliar lead uptake by lettuce exposed to atmospheric fallouts. *Environ. Sci. Technol.* **2010**, *44*, 1036–1042. [CrossRef] [PubMed]

41. Krzywy, I.; Krzywy, E.; Pastuszak-Gabinowska, M.; Brodkiewicz, A. Lead- is there something to be afraid of? *Ann. Acad. Med. Stetin.* **2010**, *56*, 118–128. [PubMed]

42. Pourrut, B.; Shahid, M.; Dumat, C.; Winterton, P.; Pinelli, E. Lead uptake, toxicity, and detoxification in plants. In *Reviews of Environmental Contamination and Toxicology*; Whitacre, D.M., Ed.; Springer: Berlin, Germany, 2011; Volume 213, pp. 113–136, ISBN 978-1-4419-9859-0.

43. Morkunas, I.; Marczak, Ł.; Stachowiak, J.; Stobiecki, M. Sucrose-induced lupine defense against *Fusarium oxysporum*: Sucrose-stimulated accumulation of isoflavonoids as a defense response of lupine to *Fusarium oxysporum*. *Plant Physiol. Biochem.* **2005**, *43*, 363–373. [CrossRef] [PubMed]

44. Chen, Z.; Zheng, Z.; Huang, J.; Lai, Z.; Fan, B. Biosynthesis of salicylic acid in plants. *Plant Signal. Behav.* **2009**, *4*, 493–496. [CrossRef] [PubMed]

45. Liu, Y.; Liu, H.; Pan, Q.; Yang, H.; Zhan, J.; Huang, W. The plasma membrane H$^+$-ATPase is related to the development of salicylic acid-induced thermotolerance in pea leaves. *Planta* **2009**, *229*, 1087–1098. [CrossRef] [PubMed]

46. Thaler, J.S.; Bostock, R.M. Interactions between abscisic acid-mediated responses and plant resistance to pathogens and insects. *Ecology* **2004**, *85*, 48–58. [CrossRef]

47. Metwally, A.; Finkermeier, I.; Georgi, M.; Dietz, K.J. Salicylic acid alleviates the cadmium toxicity in barley seedlings. *Plant Physiol.* **2003**, *132*, 272–281. [CrossRef] [PubMed]

48. Horvath, E.; Szalai, G.; Janda, T. Induction of abiotic stress tolerance by salicylic acid signalling. *J. Plant Growth Reg.* **2007**, *26*, 290–300. [CrossRef]

49. Vlot, A.C.; Dempsey, D.A.; Klessig, D.F. Salicylic acid, a multifaceted hormone to combat disease. *Annu. Rev. Phytopathol.* **2009**, *47*, 177–206. [CrossRef] [PubMed]

50. Leng, P.; Yuan, B.; Guo, Y.; Chen, P. The role of abscisic acid in fruit ripening and responses to abiotic stress. *J. Exp. Bot.* **2014**, *65*, 4577–4588. [CrossRef] [PubMed]

51. Khan, M.I.R.; Fatma, M.; Per, T.S.; Anjum, N.A.; Khan, N.A. Salicylic acid-induced abiotic stress tolerance and underlying mechanisms in plants. *Front. Plant Sci.* **2015**, *6*, 462. [CrossRef] [PubMed]

52. Pickett, J.A.; Poppy, G.M. Switching on plant genes by external chemical signals. *Trends Plant Sci.* **2001**, *6*, 137–139. [CrossRef]

53. Ton, J.; Flors, V.; Mauch-Mani, B. The multifacede role of ABA in disease resistance. *Trends Plant Sci.* **2009**, *14*, 310–317. [CrossRef] [PubMed]

54. Li, X.; Schuler, M.A.; Berenbaum, M.R. Jasmonate and salicylate induce expression of herbivore cytochrome P450 genes. *Nature* **2002**, *419*, 712–715. [CrossRef] [PubMed]

55. Jing, Y.; He, Z.; Yang, X. Role of soil rhizobacteria in phytoremediation of heavy metal contaminated soils. *J. Zhejiang Univ. Sci. B* **2007**, 192–207. [CrossRef] [PubMed]

56. Chen, J.; Zhu, C.; Li, L.P.; Sun, Z.Y. Effects of exogenous salicylic acid on growth and $H_2O_2$– metabolizing enzymes in rice seedlings under lead stress. *J. Environ. Sci.* **2007**, *19*, 44–49. [CrossRef]

57. López-Orenes, A.; Martinez-Pérez, A.; Calderón, A.A.; Ferrer, M.A. Pb-induced response in *Zygophyllum fabago* plants are organ-dependent and modulated by salicylic acid. *Plant Physiol. Biochem.* **2014**, *84*, 57–66. [CrossRef] [PubMed]

58. Mahmoud, E.M.F.; Mahfouz, H.M. Effects of salicylic acid elicitor against aphids on wheat and detection of infestation using infrared thermal imaging technique in Ismailia. *Pestic. Phytomed. (Belgrade)* **2015**, *30*, 91–97. [CrossRef]

59. El-Khawas, S.A. Priming *Pisum sativum* with salicylic acid against the leafminer *Liriomyza trifolii*. *Afr. J. Agric. Res.* **2012**, *34*, 4731–4737.

60. Janda, T.; Gondor, O.K.; Yordanova, R.; Szalai, G.; Pál, M. Salicylic acid and photosynthesis: Signalling and effects. *Acta Physiol. Plant.* **2014**, *36*, 2537–2546. [CrossRef]

61. Krantev, A.; Yordanova, R.; Janda, T.; Szalai, G.; Popova, L. Treatment with salicylic acid decreases the effect of cadmium on photosynthesis in maize plants. *J. Plant Physiol.* **2008**, *165*, 920–931. [CrossRef] [PubMed]

62. Zengin, F. Exogenous treatment with salicylic acid alleviating copper toxicity in bean seedlings. *Proc. Natl. Acad. Sci. India Sec. B Biol. Sci.* **2014**, *84*, 749–755. [CrossRef]

63. Zhang, Y.; Xu, S.; Yang, S.; Chen, Y. Salicylic acid alleviates cadmium-induced inhibition of growth and photosynthesis through upregulating antioxidant defense system in two melon cultivars (*Cucumis melo* L.). *Protoplasma* **2015**, *252*, 911–924. [CrossRef] [PubMed]

64. Horváth, E.; Pál, M.; Szalai, G.; Páldi, E.; Janda, T. Exogenous 4-hydroxybenzoic acid and salicylic acid modulate the effect of short-term drought and freezing stress on wheat plants. *Biol. Plant.* **2007**, *51*, 480–487. [CrossRef]

65. Bartels, D.; Sunkar, R. Drought and salt tolerance in Plants. *CRC Crit. Rev. Plant. Sci.* **2005**, *24*, 23–58. [CrossRef]

66. Tuteja, N. Abscisic acid and abiotic stress signaling. *Plant Signal. Behav.* **2007**, *2*, 135–138. [CrossRef] [PubMed]

67. Danquah, A.; de Zelicourt, A.; Colcombet, J.; Hirt, H. The role of ABA and MAPK signaling pathways in plant abiotic stress responses. *Biotechnol. Adv.* **2014**, *32*, 40–52. [CrossRef] [PubMed]

68. Rauser, W.E.; Dumbroff, E.B. Effects of excess cobalt, nickel and zinc on the water relations of *Phaseolus vulgaris*. *Environ. Exp. Bot.* **1981**, *21*, 249–255. [CrossRef]

69. Poschenrieder, C.; Gunsé, B.; Barceló, J. Influence of cadmium on water relations, stomatal resistance, and abscisic acid content in expanding bean leaves. *Plant Physiol.* **1989**, *90*, 1365–1371. [CrossRef] [PubMed]

70. Hollenbach, B.; Schreiber, L.; Hartung, W.; Dietz, K.J. Cadmium leads to stimulated expression of the lipid transfer protein genes in barley: Implications for the involvement of lipid transfer proteins in wax assembly. *Planta* **1997**, *203*, 9–19. [CrossRef] [PubMed]

71. Stroiński, A.; Giżewska, K.; Zielezińska, M. Abscisic acid is required in transduction of cadmium signal to potato roots. *Biol. Plant.* **2013**, *57*, 121–127. [CrossRef]

72. Fediuc, E.; Lips, H.; Erdei, L. *O*-acetylserine (thiol) lyase activity in Pragmites and Typha plants under cadmium and NaCl stress conditions and the involvement of ABA in the stress response. *J. Plant. Physiol.* **2005**, *162*, 865–872. [CrossRef] [PubMed]

73. Stroiński, A.; Chadzinikolau, T.; Giżewska, K.; Zielezińska, M. ABA or cadmium induced phytochelatin synthesis in potato tubers. *Biol. Plant.* **2010**, *54*, 117–120. [CrossRef]
74. Kim, Y.-H.; Khan, A.L.; Kim, D.-H.; Lee, S.-Y.; Kim, K.-M.; Waqas, M.; Jung, H.-Y.; Shin, J.-H.; Kim, J.-G.; Lee, I.-J. Silicon mitigates heavy metal stress by regulating P-type heavy metal ATPases, *Oryza sativa* low silicon genes, and endogenous phytohormones. *BMC Plant. Biol.* **2014**, *14*, 13. [CrossRef] [PubMed]
75. Munzuro, Ö.; Fikriye, K.Z.; Yahyagil, Z. The abscisic acid levels of wheat (*Triticum aestivum* L. cv. Çakmak 79) seeds that were germinated under heavy metal (Hg$^{++}$, Cd$^{++}$, Cu$^{++}$) stress. *J. Sci.* **2008**, *21*, 1–7.
76. Wang, Y.; Wang, Y.; Kai, W.; Zhao, B.; Chen, P.; Sun, L.; Ji, K.; Li, Q.; Dai, S.; Sun, Y.; et al. Transcriptional regulation of abscisic acid signal core components during cucumber seed germination and under Cu$^{2+}$, Zn$^{2+}$, NaCl and simulated acid rain stresses. *Plant Physiol. Biochem.* **2014**, *76*, 67–76. [CrossRef] [PubMed]
77. Atici, Ö.; Ağar, G.; Battal, P. Changes in phytohormone contents in chickpea seeds germinating under lead or zinc stress. *Biol. Plant.* **2005**, *49*, 215–222. [CrossRef]
78. Monni, S.; Uhlig, C.; Hansen, E.; Magel, E. Ecophysiological responses of *Empetrum nigrum* to heavy metal pollution. *Environ. Pollut.* **2001**, *112*, 121–129. [CrossRef]
79. Jeong, S.; Goto-Yamamoto, N.; Kobayashi, S.; Esaka, M. Effects of plant hormones and shading on the accumulation of anthocyanins and the expression of anthocyanin biosynthetic genes in grape berry skins. *Plant Sci.* **2004**, *167*, 247–252. [CrossRef]
80. Wheeler, S.; Loveys, B.; Ford, C.; Davies, C. The relationship between the expression of abscisic acid biosynthesis genes, accumulation of abscisic acid and the promotion of *Vitis vinifera* L. berry ripening by abscisic acid. *Aust. J. Grape Wine Res.* **2009**, *15*, 195–204. [CrossRef]
81. Giribaldi, M.; Geny, L.; Delrot, S.; Schubert, A. Proteomic analysis of the effects of ABA treatments on ripening *Vitis vinifera* berries. *J. Exp. Bot.* **2010**, *61*, 2447–2458. [CrossRef] [PubMed]
82. Koyama, K.; Sadamatsu, K.; Goto-Yamamoto, N. Abscisic acid stimulated ripening and gene expression in berry skins of the Cabernet Sauvignon grape. *Funct. Integr. Genom.* **2010**, *10*, 367–381. [CrossRef] [PubMed]
83. Cantín, C.M.; Fidelibus, M.W.; Crisosto, C.H. Application of abscisic acid (ABA) at veraison advanced red color development and maintained postharvest quality of 'Crimson Seedless' grapes. *Postharvest Biol. Technol.* **2007**, *46*, 237–241. [CrossRef]
84. Woźniak, A.; Formela, M.; Bilman, P.; Grześkiewicz, K.; Bednarski, W.; Marczak, Ł.; Narożna, D.; Dancewicz, K.; Mai, V.Ch.; Borowiak-Sobkowiak, B.; et al. The dynamics of the defense strategy of pea induced by exogenous nitric oxide in response to aphid infestation. *Int. J. Mol. Sci.* **2017**, *18*, 329. [CrossRef] [PubMed]
85. Sytykiewicz, H. Influence of bird cherry-oat aphid (Rhaphalosiphum padi (Linnaeus, 1758))/Hemiptera, Aphidoidea/feeding on the activity of β-glucosidase within tissues of its primary host. *Aphids Hem. Insects* **2008**, *14*, 155–164.
86. Harborne, J.B. Nature, distribution and function of plant flavonoids. In *Plant Flavonoids in Biology and Medicine: Biochemical, Pharmacological, and Structure-Activity Relationships. Progress in Clinical and Biological Research*; Cody, V., Middleton, E., Harborne, J., Eds.; Alan R. Liss: New York, NY, USA, 1978; p. 1, ISBN 0-8451-5063-4.
87. Jeandet, P. Phytoalexins: Current progress and future prospects. *Molecules* **2015**, *20*, 2770–2774. [CrossRef]
88. Jeandet, P.; Clément, C.; Courut, E.; Cordelier, S. Modulation of phytoalexin biosynthesis in engineered plants for disease resistance. *Int. J. Mol. Sci.* **2013**, *14*, 14136–14170. [CrossRef] [PubMed]
89. Adrian, M.; Jeandet, P.; Douillet-Breuil, A.-C.; Tesson, L.; Bessis, R. Stilbene content of mature *Vitis vinifera* berries in response to UV-C elicitation. *J. Agric. Food Chem.* **2000**, *48*, 6103–6105. [CrossRef] [PubMed]
90. Gangwar, S.; Singh, V.P.; Prasad, S.M.; Maurya, J.N. Modulation of manganese toxicity in *Pisum sativum* L. seedlings by kinetin. *Sci Hortic.* **2010**, *126*, 467–474. [CrossRef]
91. Zhang, Y.; Zheng, G.H.; Liu, P.; Song, J.M.; Xu, G.D.; Cai, M.Z. Morphological and physiological responses of root tip cells to Fe$^{2+}$ toxicity in rice. *Acta Physiol Plant.* **2011**, *33*, 683–689. [CrossRef]
92. Gautam, S.; Anjani, K.; Srivastava, N. In vitro evaluation of excess copper affecting seedlings and their biochemical characteristics in *Carthamus tinctorius* L. (variety PBNS−12). *Physiol. Mol. Biol. Plants.* **2016**, *22*, 121–129. [CrossRef] [PubMed]
93. Ivanov, Y.V.; Kartashov, A.V.; Ivanova, A.I.; Savochkin, Y.V.; Kuznetsov, V.V. Effects of zinc on Scots pine (*Pinus sylvestris* L.) seedlings grown in hydroculture. *Plant Physiol. Biochem.* **2016**, *102*, 1–9. [CrossRef] [PubMed]

94. Mathur, S.; Kalaji, H.M.; Jajoo, A. Investigation of deleterious effects of chromium phytotoxicity and photosynthesis in wheat plant. *Photosynthetica* **2016**, *54*, 1–9. [CrossRef]

95. Gwóźdź, E.A.; Przymusiński, R.; Rucińska, R.; Deckert, J. Plant responses to heavy metals: Molecular and physiological aspects. *Acta Physiol. Plant.* **1997**, *19*, 459–465. [CrossRef]

96. Rucińska, R.; Waplak, S.; Gwóźdź, E. Free radical formation and activity of antioxidant enzymes in lupin roots exposed to lead. *Plant. Physiol. Biochem.* **1999**, *37*, 187–194. [CrossRef]

97. Małecka, A.; Piechalak, A.; Tomaszewska, B. Reactive oxygen species production and antioxidative defense system in pea root tissues treated with lead ions: The whole roots level. *Acta Physiol. Plant.* **2009**, *31*, 1053–1063. [CrossRef]

98. Shahid, M.; Pinelli, E.; Dumat, C. Review of Pb availability and toxicity to plants in relations with metal speciation; role of synthetic and natural organic ligands. *J. Hazard. Mater.* **2012**, *15*, 1–12. [CrossRef] [PubMed]

99. Hussain, A.; Abbas, N.; Arshad, F.; Akram, M.; Khan, Z.I.; Ahmad, K.; Mansha, M.; Mirzaei, F. Effects of diverse doses of lead (Pb) on different growth attributes of *Zea mays* L. *Agric. Sci.* **2013**, *4*, 262–265.

100. Malar, S.; Vikram, S.S.; Favas, P.J.C.; Perumal, V. Lead heavy metal toxicity induced changes on growth and antioxidative enzymes level in water hyacinths [*Eichhornia crassipes* (Mart.)]. *Bot. Stud.* **2014**, *55*, 54–65. [CrossRef] [PubMed]

101. Małecka, A.; Piechalak, A.; Morkunas, I.; Tomaszewska, B. Accumulation of lead in root cells of *Pisum sativum*. *Acta Physiol. Plant.* **2008**, *30*, 629–637. [CrossRef]

102. Krzesłowska, M.; Rabęda, I.; Basińska, A.; Lewandowski, M.; Mellerowicz, E.J.; Napieralska, A.; Samardakiewicz, S.; Woźny, A. Pectinous cell wall thickenings fotmation—A common defense of plants to cope with Pb. *Environ. Pollut.* **2016**, *214*, 354–361. [CrossRef] [PubMed]

103. Rabęda, I.; Bilski, H.; Mellerowicz, E.J.; Napieralska, A.; Suski, Sz.; Woźny, A.; Krzesłowska, M. Colocalization of low-methylesterified pectins and Pb deposits in the apoplast of Aspen roots exposed to lead. *Environ. Pollut.* **2015**, *205*, 315–326. [CrossRef] [PubMed]

104. Erofeeva, E.A. Hormesis and paradoxical effects of wheat seedling (*Triticum Aestivum* L.) Parameters upon exposure to different pollutants in a wide range of doses. *Dose-Response* **2014**, *12*, 121–135. [CrossRef] [PubMed]

105. Jia, L.; Liu, Z.; Chen, W.; Ye, Y.; Yu, S.; He, H. Hormesis effects induced by cadmium on growth and photosyntheticperformance in hypperaccumulator, *Jonicera japonica* Thunb. *J. Plant Growth Regul.* **2015**, *34*, 13–21. [CrossRef]

106. Görür, G. Effects of heavy metals accumulation in host plants to cabbage aphid (*Brevicoryne brassicae*)—morphology. *Ekologia (Bratislava)* **2006**, *25*, 314–321.

107. Görür, G. Developmental instability in cabbage aphid (*Brevicoryne brassicae*) populations exposed to heavy metal accumulated host plants. *Ecol. Indic.* **2006**, *6*, 743–748. [CrossRef]

108. Mai, V.C.; Bednarski, W.; Borowiak-Sobkowiak, B.; Wilkaniec, B.; Samardakiewicz, S.; Morkunas, I. Oxidative stress in pea seedling leaves in response to *Acyrthosiphon pisum* infestation. *Phytochemistry* **2013**, *93*, 49–62. [CrossRef] [PubMed]

109. Yalpani, N.; León, J.; Lawton, M.A.; Raskin, I. Pathway of salicylic acid biosynthesis in healthy and virus-inoculated tobacco. *Plant Physiol.* **1993**, *103*, 315–321. [CrossRef] [PubMed]

110. Vine, J.H.; Niton, D.; Plummer, J.A.; Baleriola-Lucas, C.; Mullins, M.G. Simultaneous quantitation of indole-3-acetic acid and abscisic acid in small samples of plant tissue by gas chromatography/mass spectrometry/selected ion monitoring. *Plant Physiol.* **1987**, *85*, 419–422. [CrossRef] [PubMed]

111. Morkunas, I.; Formela, M.; Floryszak-Wieczorek, J.; Marczak, Ł.; Narożna, D.; Nowak, W.; Bednarski, W. Cross-talk interactions of exogenous nitric oxide and sucrose modulates phenylpropanoid metabolism in yellow lupine embryo axes infected with *Fusarium oxysporum*. *Plant Sci.* **2013**, *211*, 102–121. [CrossRef] [PubMed]

112. Stobiecki, M.; Wojtaszek, P.; Gulewicz, K. Application of solid phase extraction for profiling of quinolisidine alkaloids and phenolic compounds in *Lupinus albus*. *Phytochem. Anal.* **1997**, *8*, 153–158. [CrossRef]

113. Morkunas, I.; Stobiecki, M.; Marczak, Ł.; Stachowiak, J.; Narożna, D.; Remlein-Starosta, D. Changes in carbohydrate and isoflavonoid metabolism in yellow lupine in response to infection by *Fusarium oxysporum* during the stages of seed germination and early seedling growth. *Physiol. Mol. Plant Pathol.* **2010**, *75*, 46–55. [CrossRef]

114. Nichols, E.J.; Beckman, J.M.; Hadwiger, L.A. Glycosidic enzyme activity in pea tissue and pea-*Fusarium solani* interactions. *Plant Physiol.* **1980**, *66*, 199–204. [CrossRef] [PubMed]

115. Morkunas, I.; Kozłowska, M.; Ratajczak, L.; Marczak, Ł. Role of sucrose in the development of Fusarium wilt in lupine embryo axes. *Physiol. Mol. Plant Pathol.* **2007**, *70*, 25–37. [CrossRef]

116. Cahill, D.M.; McComb, J.A. A comparison of changes in phenylalanine ammonia-lyase activity, lignin and phenolic synthesis in the roots of *Eucalyptus calophylla* (field resistant) and *E. marginata* (susceptible) when infected with *Phytophtora cinnamon*. *Physiol. Mol. Plant Pathol.* **1992**, *40*, 315–332. [CrossRef]

**Sample Availability:** Samples of the compounds are not available

*molecules*

MDPI

*Article*

# Distinct Mechanisms of Biotic and Chemical Elicitors Enable Additive Elicitation of the Anticancer Phytoalexin Glyceollin I

**Kelli Farrell [1,†], Md Asraful Jahan [2,†] and Nik Kovinich [2,\*]**

1   Department of Biology, West Virginia University, Morgantown, WV 26506, USA; kcfarrell@mix.wvu.edu
2   Division of Plant and Soil Sciences, West Virginia University, Morgantown, WV 26506, USA; mjahan@mix.wvu.edu
\*   Correspondence: nikovinich@mail.wvu.edu; Tel.: +1-304-293-9240
†   These Authors contributed equally to this article.

Received: 16 June 2017; Accepted: 25 July 2017; Published: 27 July 2017

**Abstract:** Phytoalexins are metabolites biosynthesized in plants in response to pathogen, environmental, and chemical stresses that often have potent bioactivities, rendering them promising for use as therapeutics or scaffolds for pharmaceutical development. Glyceollin I is an isoflavonoid phytoalexin from soybean that exhibits potent anticancer activities and is not economical to synthesize. Here, we tested a range of source tissues from soybean, in addition to chemical and biotic elicitors, to understand how to enhance the bioproduction of glyceollin I. Combining the inorganic chemical silver nitrate ($AgNO_3$) with the wall glucan elicitor (WGE) from the soybean pathogen *Phytophthora sojae* had an additive effect on the elicitation of soybean seeds, resulting in a yield of up to 745.1 $\mu$g gt$^{-1}$ glyceollin I. The additive elicitation suggested that the biotic and chemical elicitors acted largely by separate mechanisms. WGE caused a major accumulation of phytoalexin gene transcripts, whereas $AgNO_3$ inhibited and enhanced the degradation of glyceollin I and 6″-O-malonyldaidzin, respectively.

**Keywords:** bioproduction; phytoalexin; isoflavonoid; glyceollin; soybean [*Glycine max* (L.) Merr.]

## 1. Introduction

Plants, like other organisms, have metabolic pathways that remain silent until activated by stresses. Phytoalexins are defense metabolites biosynthesized in response to pathogens, but which for unknown reasons also accumulate in response to specific environmental stresses and inorganic chemicals, such as heavy metals [1–3]. Much of what is known about phytoalexin elicitation mechanisms comes from studies of the glyceollins in soybean, camalexins in Arabidopsis, diterpenoids and flavonoids in rice, stilbenes in grapevine, alkaloids in California poppy, and the 3-deoxyanthocyanidins, terpenoids, and phytodienoic acids in maize. However, very few studies have attempted to distinguish the elicitation mechanisms of biotic and chemical elicitors.

Biotic elicitation begins when a microbial derived pathogen-associated molecular pattern (PAMP) or effector binds to a pattern recognition receptor at the plasma membrane of the plant cell. Mitogen-activated protein kinase (MAPK) or phospholipase signaling ultimately results in the expression of transcription factors (TFs) that directly activate the transcription of phytoalexin biosynthesis genes. MYB-, bHLH-, or WRKY-type TFs directly activate some or all of the phytoalexin biosynthesis genes in cotton, sorghum, rice, Arabidopsis, and grapevine [4–6]. In soybean, no phytoalexin TF has been identified, but transcription of glyceollin biosynthesis genes was coordinately induced in response to the pathogen *Phytophthora sojae* [7,8].

Heavy metals, such as silver nitrate ($AgNO_3$), have elicited chemically diverse phytoalexins in many plant species. The molecular target(s) of these heavy metals remain(s) unknown. $AgNO_3$ was shown to inhibit developmental processes triggered by exogenous ethylene treatment, and thus has been considered a potent inhibitor of ethylene perception [9]. Some evidence has suggested that AgNO$_3$ and pathogens elicit phytoalexins primarily by different mechanisms. The *P. sojae*-resistant soybean variety Harosoy 63 elicited glyceollins rapidly in response to race 1 *P. sojae*, but the susceptible variety Harosoy did not, whereas both varieties responded similarly to AgNO$_3$ [10]. Feeding AgNO$_3$-elicited soybean cotyledons the radiolabeled intermediate phenylalanine did not result in radiolabeled glyceollins, but AgNO$_3$ treatment reduced the degradation of radiolabeled glyceollins [11].

Glyceollins are the major phytoalexins of the soybean. They belong to the pterocarpan subclass of isoflavonoids, which possesses great potential as scaffolds for pharmaceutical development [12]. Glyceollins are biosynthesized from the isoflavonoid daidzein, which can result from de novo biosynthesis beginning at phenylalanine, or possibly from the hydrolysis of preformed isoflavonoid-glycoside conjugates (Figure 1). Glyceollins have broad-spectrum antiproliferative or antitumor activities against human lung, breast, prostate, ovary, skin, and kidney cancers. Glyceollin I is the most potent, and directly antagonizes human estrogen receptor α (ERα) and ERβ [13]. Glyceollin I also exhibits a rare ER-independent mode-of-action via a mechanism that is not yet fully understood [14]. In light of the therapeutic potential of glyceollin I, studies have attempted to produce large-scale (gram) amounts by chemical synthesis or by the elicitation of soybean [15,16]. However, the yield by chemical synthesis was up to 12%, and required a highly concerted effort of specialists six months to complete, rendering it uneconomical for commercial production [15]. Here, we aimed to identify which soybean tissues and treatments provide optimal glyceollin I bioproduction in vitro. Our study provides novel insight into how biotic and chemical elicitation pathways are distinct, and how this can be exploited to enhance the bioproduction of glyceollin I.

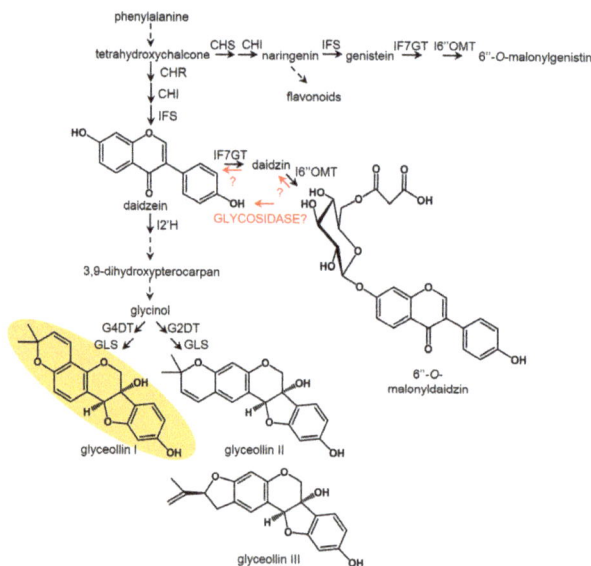

**Figure 1.** Glyceollin I biosynthetic pathway. In addition to de novo biosynthesis, the constitutively accumulating isoflavone conjugate 6″-O-malonyldaidzin may be hydrolyzed to provide daidzein intermediates for glyceollin I biosynthesis. CHS, chalcone synthase; CHR, chalcone reductase; CHI, chalcone isomerase; IFS, isoflavone synthase; I2′H, isoflavone 2′-hydroxylase; G4DT, glycinol 4-dimethylallyl transferase; G2DT, glycinol 2-dimethylallyl transferase; GLS, glyceollin synthase; UF7GT (UGT88E3) UDP-glucose:isoflavone-7-O-glucosyltransferase; I6″OMT (GmMT7) isoflavone-7-O-glucoside-6″-O-methyltransferase.

## 2. Results and Discussion

### 2.1. Imbibing Soybean Seeds Are the Most Abundant Source of Glyceollin I

Prior studies have found that glyceollins were readily elicited up to the first-true-leaf stage of development, when they were essential for defense against *P. sojae* [10,17,18]. They were also elicited in transgenic hairy roots and soybean seeds [19–21]. Yet, no single study has compared different soybean tissues to determine which produces the most glyceollin I.

We compared the organs of seedlings at the first-true-leaf stage, imbibing seeds, and hairy roots. The samples were treated with the well-characterized wall glucan elicitor (WGE) isolated from the mycelium of *P. sojae* [22]. The organs of the seedlings and imbibing seeds were spot-treated on a wound of equivalent size with the same amount of WGE, and tissue of roughly the same mass encompassing the wound site was collected for the measurement of glyceollin I (see Methods).

The imbibing seeds produced the highest amount of glyceollin I: 421.7 µg per gram fresh tissue $(gt^{-1})$ (Figure 2A). This amount was roughly sixfold greater than that of hairy roots and hypocotyls, and 16-fold greater than cotyledons and roots. The glyceollin I from the imbibing seeds represented a major peak at 283 nm absorbance (Figure 2B).

Since glyceollin I is costly to purify from other glyceollins [23], we determined which tissue had the greatest purity of glyceollin I relative to the other glyceollins. The roots produced 85.9% glyceollin I (Supplementary Figure S1), but produced the lowest total amount with our elicitation parameters (Figure 2A). Since imbibing seeds produced the greatest total amounts of glyceollin I and exhibited the second greatest purity (68.5%), we decided to focus on imbibing seeds to compare the biotic and chemical elicitation mechanisms of glyceollin I.

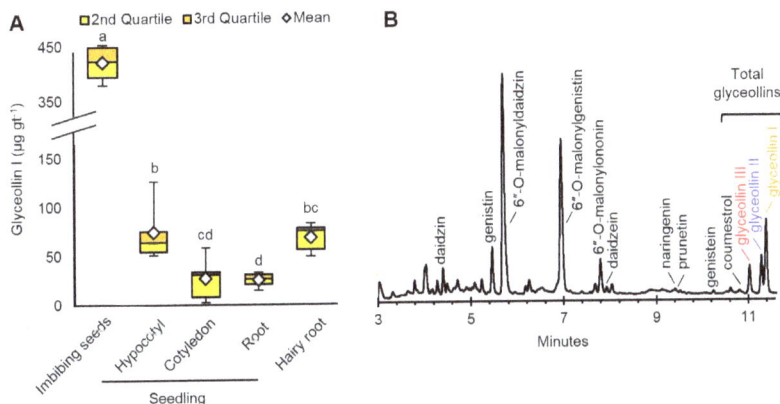

**Figure 2.** (**A**) Amounts of glyceollin I from soybean organs treated with wall glucan elicitor (WGE) from *Phytophthora sojae*. Two-way ANOVA, Tukey post hoc test ($p < 0.001$); (**B**) UPLC-PDA chromatogram at 283 nm of isoflavonoids from imbibing seeds. Isoflavonoid identities were determined by UPLC-PDA-MS$^n$ retention time, fragmentation pattern, and absorbance features by comparison to standards (commercial or purified) and by comparison to the literature (Supplementary Table S1). Different letters show significant differences by ANOVA.

### 2.2. Wall Glucan Elicitors from P. sojae and Pythium Elicit More Glyceollin I Than Rhizopus, Aspergillus, and Fusarium Microspores at Standard Treatment Concentrations

A wide variety of fungal, nematode, and oomycete pathogens elicit glyceollins. The commonly studied microspore elicitors are from the *Rhizopus* and *Aspergillus* species, whereas WGE is most commonly from *P. sojae* [24–26]. To compare the elicitation features of several biotic elicitors at typical treatment concentrations, we treated imbibing soybean seeds with $7 \times 10^7$ CFU mL$^{-1}$ microspores from *Aspergillus fumigatus*, *Rhizopus nigricans*, and *Fusarium tricinctum*, or WGEs from *P. sojae* and *Pythium* at 20 mg mL$^{-1}$.

WGE from *P. sojae* induced the highest amounts of glyceollin I (428.3 µg gt$^{-1}$), followed by *Pythium* (199.0 µg gt$^{-1}$) (Figure 3A). Glyceollin I comprised more than 50% of the total glyceollin amounts elicited by *P. sojae* and *Pythium*, whereas it was less than 50% for the fungal microspore elicitors (Supplementary Figure S1).

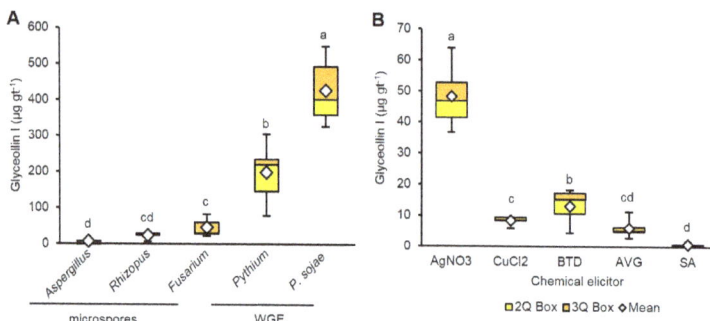

Figure 3. (A) Amounts of glyceollin I from soybean seeds treated with biotic elicitors for 24 h. Two-way ANOVA, Tukey post hoc test ($p < 0.001$); (B) Treatment of seeds with chemical elicitors (AgNO$_3$), copper chloride (CuCl$_2$), benzothiadiazole (BTD), aminoethoxyvinyl glycine (AVG), and salicylic acid (SA) at 1 mM. Different letters show significant differences by ANOVA.

### 2.3. AgNO$_3$ Elicits More Glyceollin I Than CuCl$_2$, BTD, AVG, and SA at Equivalent Treatment Concentrations

Chemical elicitors, such as heavy metals or inorganic compounds, that stimulate or inhibit components of the plant immune system can function as phytoalexin elicitors. We chose to compare under the same conditions as our biotic elicitors the elicitation potential of the heavy metals AgNO$_3$ and CuCl$_2$ and inorganic compounds that affect the immune system. The duration of elicitation for our experiments was initially set to 24 h, since that is the duration typically reported in previous studies. Benzothiadiazole (BTD) stimulated the immune system by functioning as a mimic of the plant hormone salicylic acid (SA) [27]. Aminoethoxyvinyl glycine (AVG) inhibited the biosynthesis of ethylene [28], and thus could be compared to the putative inhibitor of ethylene perception, AgNO$_3$.

At 1 mM treatment concentration, AgNO$_3$ was the most potent and elicited 48.2 µg gt$^{-1}$ glyceollin I (Figure 3B). That was eightfold more than AVG. BTD was the second most potent, eliciting 48.2 µg gt$^{-1}$, which was 17-fold more than SA. Only AgNO$_3$ preferentially elicited glyceollin I (Supplementary Figure S1).

### 2.4. AgNO$_3$ and P. sojae WGE Elicit Glyceollin I with Different Dynamics

Since AgNO$_3$ and WGE from *P. sojae* were the most potent chemical and biotic elicitors at the treatment concentrations tested, we chose to investigate in more detail their mechanisms, beginning with elicitation dynamics. AgNO$_3$ elicited the accumulation of glyceollin I in a biphasic fashion, with peaks at 24 h and 96 h after treatment (Supplementary Figure S2). By contrast, WGE reached a maximum at 24 h that was sustained. To distinguish effect of each elicitor from that of the wounding, we calculated the fold change in metabolite levels of each elicitor relative to the water control. WGE caused the greatest induction over the wounded control at 48 h, whereas AgNO$_3$ was at 72 h (Supplementary Figure S2).

### 2.5. P. sojae WGE and AgNO$_3$ Elicit the Accumulation of Glyceollin I Mainly by Distinct Mechanisms

Since the elicitation dynamics were different for WGE compared to AgNO$_3$, we anticipated that each elicitor functioned by a different mechanism. We tested this hypothesis by first identifying the optimal concentration of each elicitor required to elicit the maximum accumulation of glyceollin I, and then combined the elicitors at these maximum concentrations. The rationale was that the combined treatment would result in a greater accumulation of glyceollin I only if the elicitation mechanisms were different. To determine the maximum level of glyceollin I that WGE and $AgNO_3$ can respectively elicit, we conducted dose response analyses. Germinating soybean seeds were treated with up to 10 mM $AgNO_3$ and 60 mg mL$^{-1}$ WGE.

At 48 h after treatment, 5 mM $AgNO_3$ elicited a maximum mean concentration of glyceollin I of 202.5 µg gt$^{-1}$ (Supplementary Figure S3). By contrast, 20 mg mL$^{-1}$ WGE elicited more than double this amount, with a mean concentration of 449.8 µg gt$^{-1}$. The combined treatment elicited a mean glyceollin I concentration of 635.8 µg gt$^{-1}$ and a maximum of 745.1 µg gt$^{-1}$ (Figure 4). This mean concentration was significantly greater than that observed for the single treatments, and was very close empirically to the sum of the individual elicitor treatments (640.0 µg gt$^{-1}$). This strongly suggests that WGE and $AgNO_3$ function by distinct elicitation mechanisms.

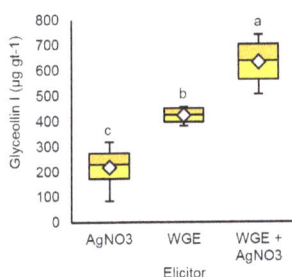

**Figure 4.** Elicitation with $AgNO_3$ (5 mM) and WGE (20 mg mL$^{-1}$) separately and in combination. Two-way ANOVA, Tukey post hoc test, $p < 0.001$. Different letters show significant differences by ANOVA.

The combined treatment also enhanced the purity of glyceollin I to 89.4% of the total glyceollins and to 34.6% of the total seed isoflavonoids (Supplementary Figures S1 and S4, respectively).

### 2.6. P. sojae WGE and not AgNO$_3$ Induces Major Accumulation of Glyceollin Gene Transcripts

Since glyceollin biosynthesis genes were regulated at the level of transcription in response to *P. sojae* [7,8,29], we determined whether the elicitors WGE and $AgNO_3$ induced glyceollin biosynthesis at the mRNA level. We measured mRNA accumulation for genes spanning from general isoflavonoid to late-stage glyceollin I biosynthesis. The homologous genes *IFS1* and *IFS2* encode isoflavone synthase (IFS) enzymes that catalyze the first committed step in isoflavonoid biosynthesis (Figure 1). The enzyme isoflavone 2'-hydroxylase encoded by *I2'H* and the glycinol 4-dimethylallyltransferase encoded by *G4DT* catalyze the first committed steps for the biosynthesis of all glyceollins and glyceollin I, respectively [7,30].

WGE induced the accumulation of the mRNAs of all four genes compared to the solvent control at 48 h after elicitation (Figure 5A). By contrast, $AgNO_3$ did not significantly induce any genes at this time point. Further, the combined treatment did not have elevated levels of mRNAs compared to the WGE treatment, and even exhibited a reduced expression of IFS1. These results suggest that $AgNO_3$ did not elicit glyceollins by stimulating the accumulation of glyceollin biosynthesis gene mRNAs.

Since the elicitation of phytoalexins is a transient process subject to feedback mechanisms, we also tested whether $AgNO_3$ elicited the accumulation of biosynthesis gene mRNAs more rapidly than WGE. To test this, we measured mRNA levels at 8 h after treatment, a time point that preceded the measurable accumulation of glyceollin I. Again, $AgNO_3$ did not elicit the accumulation of any of the mRNAs (Figure 5B). Further, the combined treatment exhibited less expressions of *IFS2* and *G4DT* compared to the WGE treatment, suggesting that the elevated levels of glyceollin I observed at 48 h were not due to regulation of biosynthesis at the mRNA level.

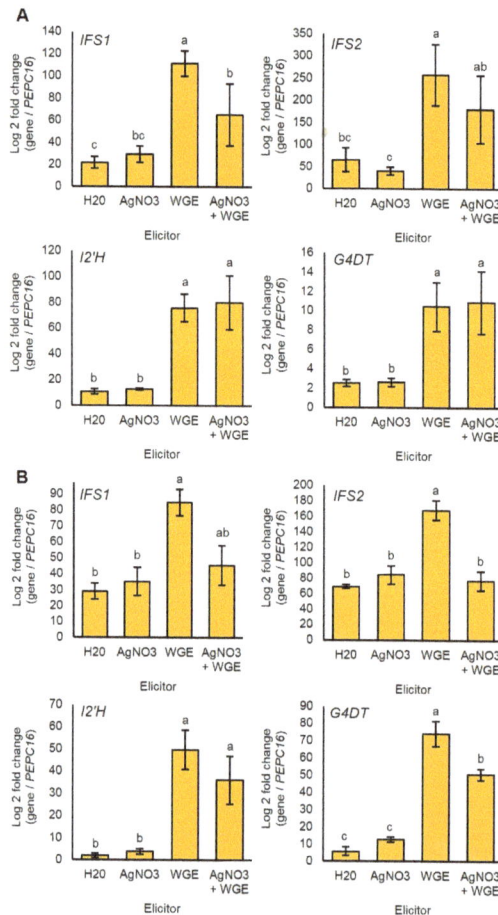

**Figure 5.** (**A**) Relative gene expression levels 48 h after elicitor treatment with elicitor measured by qRT-PCR; (**B**) Relative gene expression levels at 8 h after elicitor treatment. Expressions in each sample were measured relative to the endogenous reference gene *PEPC16*. Two-way ANOVA, Tukey post hoc test, $p < 0.01$. Two independent experiments with four biological replicates were conducted with similar results. Different letters show significant differences by ANOVA.

## 2.7. AgNO₃ Inhibits the Degradation of Glyceollin I and Enhances the Specific Hydrolysis of 6''-O-Malonyldaidzin

Our gene expression measurements demonstrated that $AgNO_3$ did not induce the expression of glyceollin biosynthesis genes. This suggested that $AgNO_3$ acted via another mechanism to elicit glyceollin I in the soybean seed. Heavy metal elicitors have been suggested to elicit isoflavonoid phytoalexins by inhibiting their degradation or by increasing the rates of the hydrolysis of isoflavone conjugates to provide metabolic intermediates for phytoalexin biosynthesis [11,31]. To investigate whether $AgNO_3$ inhibited the degradation of glyceollin I, we incubated dissected seeds in water or 5 mM $AgNO_3$ for 2 h, then transferred them to an extract containing partially purified glyceollin I.

The incubation of the dissected seed with 115 umol mL$^{-1}$ of glyceollin I resulted in the degradation of 54.9 μmol mL$^{-1}$ (Figure 6A). The pretreatment of the seed with $AgNO_3$ for 2 h did not cause a detectable induction of glyceollin I (not shown) and reduced this degradation by 25.7%, rendering the final amount statistically equivalent to the initial extract. A similar inhibition of degradation was observed for the intermediate daidzein (Figure 6B). By contrast, AgNO$_3$ accelerated the hydrolysis of the isoflavone conjugate 6''-O-malonyldaidzin (Figure 6C). The levels of daidzin were unchanged compared to the control (Figure 6D). AgNO$_3$ did not affect the levels of any of the metabolites in the absence of the seed (not shown), indicating that the changes in metabolite levels were catalyzed by seed enzymes. During the course of the experiment, no de novo biosynthesis of glyceollin I was detected from seeds incubated in the absence of the glyceollin I extract (Figure 6A). In summary, these results show that AgNO$_3$ acts by inhibiting the degradation of glyceollin I and by accelerating the hydrolysis of 6''-O-malonyldaidzin, possibly to provide daidzein intermediates for glyceollin I biosynthesis.

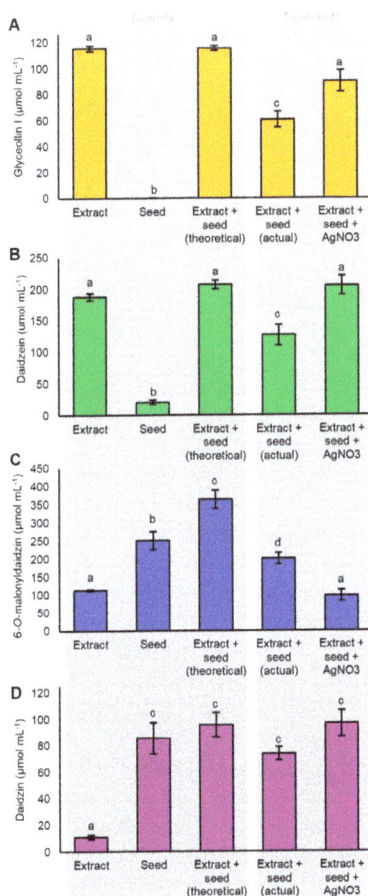

**Figure 6.** Degradation of externally supplied glyceollin I (**A**); daidzein (**B**); 6''-O-malonyldaidzin (**C**); and daidzin (**D**) by soybean seeds. Isoflavonoids were partially purified from an ethanolic extract of WGE-elicited soybean seeds, and were incubated with imbibed soybean seeds pretreated with AgNO$_3$ or water to test for effects on metabolite degradation. Shown is the amount of metabolite from the initial partially purified ethanolic extract (Extract), the amount from imbibed seeds in incubated in water only (Seed), the theoretical amount from empirically adding the values from Extract and Seed (theoretical), the amount observed from incubating the seed in water for 2 h followed by metabolite extract for 4 h (actual), or from incubating the seed in 5 mM AgNO$_3$ for 2 h followed by metabolite extract for 4 h (Extract + seed + AgNO$_3$). Two-way ANOVA, Tukey post hoc test ($p < 0.001$). The results represent four biological replicates. Different letters show significant differences by ANOVA.

## 2.8. Discussion

Previous attempts were made to biosynthesize and synthesize glyceollin I on a preparative scale. The most effective approach yielded 12% glyceollin I, but its scale up to produce gram amounts required ~6 months and a highly concerted team effort of highly experienced chemists [15]. A creative approach to biosynthesize prenylated isoflavonoids on a preparative scale involved the malting (as in the brewing industry) and simultaneous elicitation of 4.0 kg of germinating soybeans with the fungus *Rhizopus microsporus* [16]. This yielded up to 335 µg gt$^{-1}$ glyceollin I in nine days. Thus, the overall yield from 4.0 kg of seeds could be 1.34 g in a fraction of the time needed for synthesis. Here, we demonstrated on an analytical scale that combining biotic and chemical elicitors resulted in an additive effect on the accumulation of glyceollin I, with WGE and AgNO$_3$ eliciting almost double that amount (635.8 µg gt$^{-1}$) in 48 h (Figure 4). The approach was labor-intensive and could benefit from mechanization, but if amenable to direct scale up would yield almost double the amount (2.54 g) of glyceollin I from 4.0 kg of seeds in 2–3 days.

Several lines of evidence suggested that this occurred because the chemical and biotic elicitors functioned by distinct mechanisms. Firstly, AgNO$_3$ and WGE elicited different patterns of glyceollin I accumulation over time (Supplementary Figure S2). Second, our gene expression experiments demonstrated that AgNO$_3$ did not increase the accumulation of essential glyceollin genes, whereas WGE from *P. sojae* highly upregulated both the isoflavonoid and glyceollin genes (Figure 5). Finally, AgNO$_3$ inhibited the degradation of glyceollin I (Figure 6), whereas WGE did not [11]. These results suggested distinct genetic targets for enhancing biotic and chemical elicitation pathways by genetic engineering. Transcription factors (TFs) of the MYB, WRKY, and bHLH families directly activated the transcription of phytoalexin biosynthesis gene promoters, and thus their overexpression could be used to enhance the biotic elicitation pathway. Yet to our knowledge, TFs that activate specifically the heavy metal elicitation pathway have not been identified.

AgNO$_3$ was shown to be a potent inhibitor of ethylene perception [9]. Yet, even though we applied the ethylene biosynthesis inhibitor AVG at four times the concentration found to fully inhibit ethylene biosynthesis in soybean [28], we observed 8.1-fold less elicitation compared to AgNO$_3$. This suggested that the mechanism-of-action of AgNO$_3$ on glyceollin I was mostly distinct from the inhibition of ethylene biosynthesis. AgNO$_3$ and several other heavy metals inhibited the degradation of glyceollins in soybean cotyledons, suggesting that heavy metal elicitors may target a glyceollin degrading enzyme or pathway [11]. Our results confirmed this, but we also found that AgNO$_3$ enhanced the degradation of the major isoflavone conjugate, 6''-O-malonyldaidzin, and reduced the consumption of daidzein (Figure 6). This suggested that AgNO$_3$ acted by stimulating the expression or activity of an enzyme that catalyzed the hydrolysis of 6''-O-malonyldaidzin to provide daidzein intermediates for the biosynthesis of glyceollin I, rather than by solely inhibiting glyceollin degradation. Future studies should address whether AgNO$_3$ directly enhances the activity of a glycosidase that hydrolyzes 6''-O-malonyldaidzin, or whether AgNO$_3$ stimulates the upregulation of the corresponding glycosidase gene(s). Our results suggest that overexpressing a glycosidase with specificity for 6''-O-malonyldaidzin would enhance the heavy metal elicitation pathway. Glycosidase activities towards malonyl conjugates of daidzin and genistin were detected in the *f* seed and root exudates of soybean [32]. The jasmonic acid (JA)-inducible glycosidase G2 from *Medicago truncatula* has high specificity for daidzin (i.e., daidzein-7-O-glucoside) when expressed in yeast [31], but 6''-O-malonyldaidzin was not tested as a substrate. The significance of AgNO$_3$ preferentially stimulating the degradation of 6''-O-malonyldaidzin and not daidzin remains unclear. Studies may have to address the subcellular localization of these metabolites relative to the glycosidase(s) capable of hydrolyzing them to understand this mechanism.

## 3. Experimental Section

### 3.1. Chemicals

Stocks of the chemical elicitors AgNO$_3$ (Sargent Welch, Buffalo, NY, USA), CuCl$_2$, and 2,1,3-benzothiadiazole were 5 M in water. Salicylic acid was 5 M in DMSO. Elicitors were purchased from Sigma. The glyceollin I synthetic standard was from Paul Erhardt (University of Toledo, Bancroft, MI, USA), daidzein from Cayman Chemical and daidzin, genistin (Ann Arbor, MI, USA), and genistein were from Indofine (Hillsborough, CA, USA). All UPLC solvents were LC-MS grade from Fisher (Hampton, FL, USA).

### 3.2. Plant Growth and Elicitation

Harosoy 63 seeds were sterilized in 70% ethanol 0.2% triton X (*v*/*v*) for 5 min, rinsed thrice with 95% ethanol, and imbibed in sterile water overnight. The imbibate was then discarded to remove growth inhibitors. For the elicitation of seeds, the seed coat was carefully removed to avoid damaging the embryo. The distal 2–3 mm of the cotyledons (the tips) were excised, and an incision was made through the axis of the central vein two thirds of the way towards the radical. Elicitor was applied to the wound, and the embryo was placed upright on its distal end on sterile wetted filter paper in a petri dish. For the elicitation of seedlings, sterilized seeds were transferred to water-soaked sterile vermiculite in beakers. The beakers were covered with a sterile cheese that had a circular hole cut into it to permit the passage of light. The cloth was covered with plastic wrap to ensure aseptic growth. The seed coats were carefully removed within 1–2 days. Seedlings were grown until the first-true-leaf stage (~1 week). For elicitation, a knotch ~3 mm in diameter and 1–2 mm deep was cut into the central vein on the abaxial side of the cotyledon or into the hypocotyl, and roughly 250 mg tissues (the weight of a seed) were harvested for the measurement of glyceollin I. Hairy root transformation with *Agrobacterium rhizogenes* strain K599 was done according to [33]. The roots and hairy roots were elicited by cutting ~250 mg into 1 cm pieces and applying elicitor over the wounds. All seedling growth and elicitation treatments were done under 24 h cool white T5 fluorescent lights (500 $\mu$E m$^{-2}$ s$^{-1}$) at 24 °C. The samples were treated with 50 $\mu$L of elicitor for 48 h unless otherwise specified. The hairy roots and roots were treated for 24 h. WGE was purified from race 1 *P. sojae* according to [22], with the exception that suction filtration was replaced by centrifugation at all steps.

### 3.3. Isoflavonoid Analyses

For the extraction of isoflavonoids, elicited seeds were pulverized for 3 min using 5 mm stainless steel grinding balls in a MM400 mixer mill (Retsch, Newtown, CT, USA) at frequency of 30/s, then again following the addition of 80% ethanol (1 $\mu$L mg$^{-1}$ fresh tissue). Seedling organs were ground using a mortar and pestle following freeze-drying, and were extracted similarly to seeds. The samples were centrifuged at 17,000× *g* for 3 min, clarified at −20 °C overnight, then centrifuged again. The supernatant was filtered through a 0.2 $\mu$m, and 1 $\mu$L was analyzed by UPLC-PDA-MS$^n$ adapted from [24]. The quantity of each isoflavonoid was determined by comparison of the area under their peak at 283 nm absorbance in comparison to a standard curve of authentic standard. To estimate the amount of glyceollin I relative to all other seed isoflavonoids, unknown compounds and compounds for which we did not have standards were quantified in comparison a standard curve of daidzin.

### 3.4. UPLC-PDA-MS$^n$

UPLC-PDA-MS$^n$ was conducted using an Accela system (Thermo Scientific, San Jose, CA, USA) consisting of a 1250 pump, Open AS autosampler, and photodiode array (PDA) detector connected to a Q-Exactive–Oribitrap MS containing a HESI. Separations were achieved with an Acquity UPLC BEH shield RP18 column (2.1 mm i.d. × 150 mm, 1.7 $\mu$m particle size; Waters, Milford, MA, USA) with an Acquity UPLC BEH shield RP18 VanGuard precolumn (2.1 mm i.d. × 5 mm, 1.7 $\mu$m particle size; Waters). The solvents, at a flow rate of 300 $\mu$L/min, were water acidified with 0.1% (*v*/*v*) acetic acid, eluent A, and ACN acidified with 0.1% (*v/v*) acetic acid, eluent B. The temperature of the column oven was 35 °C. The elution profile was as follows: 0−2 min, from 10% to 25% (*v/v*) B; 2−9 min, from 25% to 50% B; 9−12 min, 50% B; 12−17 min, from 50% to 100% B; 22−24 min, 100% to 10% B; and 24−30 min, 10% B. MS analysis was performed in Full MS/AIF mode in both positive and negative polarities. The Full MS properties were: Resolution 70,000, AGC target 3e6, Maximum IT 200 ms, and Scan range 120−1200 *m/z*. AIF properties were: Resolution 70,000, AGC target 3e6, Maximum IT 200 ms, NCE 35, and Scan range 80−1200 *m/z*. Nitrogen was used as sheath and auxiliary gas.

*3.5. qRT-PCR*

For the gene expression measurements, seeds were harvested into liquid nitrogen and lyophilized for at least 3 days. Tissues were ground to a fine powder using a mixer mill as indicated above. Total RNA was extracted using the Spectrum Plant Total RNA Kit (Sigma-Aldrich, St. Louis, MO, USA) following the manufacturer's protocol with the exception that 700 μL lysis buffer was used per seed. RNA (500 ng) was treated with DNase I (Amplification grade, Invitrogen, Carlsbad, CA, USA) to remove contaminating DNA. First-strand cDNA was synthesized using SuperScript II Reverse Transcriptase (Invitrogen). Parallel reactions were performed in the absence of Superscript II to test for genomic DNA contamination. Gene expressions from each cDNA sample were normalized to the endogenous reference PEPC16 [34]. The reactions (5 μL) consisted of 1 μL of first-strand cDNA (or untreated RNA) diluted 1/4 to 1/10, 250 nM of forward and reverse primers, and 2.5 μL of the iQ SYBR Green Supermix (BioRad, Hercules, CA, USA). qRT-PCR was performed on cDNA from four biological replicates or untreated RNA using a 7500 Realtime-PCR System (Applied Biosystems, Foster City, CA, USA). To verify the specificity of the qRT-PCR reactions, melting curves were determined subsequent to each reaction, and RT-PCR products for each primer set were fractionated on agarose gels. For the primer sequences, see Supplementary Table S2.

*3.6. Degradation of Isoflavonoids*

To obtain an extract that was concentrated in glyceollin I, or 6″-*O*-malonyldaidzin and daidzin, Harosoy 63 seeds (~500 g) were elicited for 48 h with 20 mg mL$^{-1}$ *P. sojae* WGE, and the total isoflavonoids were extracted with 80% ethanol (2 μL mg$^{-1}$ fresh tissue) using a Waring blender. Ethanol was removed by rotary evaporation and water by freeze-drying. The powder was reconstituted in 100 mL ethyl acetate and fractionated on ~20 preparative TLC plates according to [35]. The top band ($R_f$ = 0.32) contained ~60% of glyceollin I and minor amounts of glyceollin II, III, coumestrol, and genistein, and was extracted separately from the origin containing 39% and 34% 6″-*O*-malonyldaidzin and daidzin, respectively. The extracts were resuspended in deionized water to an absorbance of 0.5–0.6 to be of physiologically relevant concentrations similar to that typically extracted from soybean seeds. The extract was sterilized by passage through nylon (0.22 μm). To test whether the seeds catalyzed the degradation of metabolites, seed coats were removed from surface-sterilized seeds that were imbibed overnight in water. The seeds were cut in half through the radical to separate the two cotyledons, and each cotyledon was incised once from the distal tip along the adaxial plane to ~3 mm from the radical. In 24-well plates in the dark, dissected half-seeds were incubated in 600 μL of sterile $H_2O$ or 5 mM $AgNO_3$ for 2 h. These were then replaced with either extract enriched in glyceollin I, or 6″-*O*-malonyldaidzin and daidzin, or water, and incubated for 4 h. The seeds were extracted with 80% ethanol (1 μL mg$^{-1}$ fresh tissue) as indicated above. The liquid surrounding the seeds was extracted three times with an equal volume of ethyl acetate. The pooled extracts were evaporated to dryness with nitrogen gas and reconstituted with the extract from the corresponding seed. Isoflaonoids were quantified from the extract by UPLC-PDA-MS$^n$ as indicated above.

## 4. Conclusions

We have identified that combining the biotic elicitor WGE from *P. sojae* and the chemical elicitor $AgNO_3$ stimulated the accumulation of the anticancer phytoalexin glyceollin I in an additive fashion, because they

*Molecules* **2017**, *22*, 1261

functioned by largely distinct elicitation mechanisms. WGE elicited a massive accumulation of biosynthesis gene mRNAs, and $AgNO_3$ stimulated the hydrolysis of the isoflavone conjugate 6''-*O*-malonyldaidzin. Thus, our work suggests that overexpressing TFs that activate the transcription of phytoalexin biosynthesis genes and glycosidases that generate biosynthetic intermediates should be targets to genetically enhance biotic and chemical elicitation pathways, respectively. The combined elicitation approach represents an important discovery towards the economical bioproduction of glyceollin I and potentially other phytoalexins of medicinal or agricultural value.

**Supplementary Materials:** Supplementary materials are available online.

**Acknowledgments:** We thank Elroy Cober (AAFC, Ottawa) for kindly providing the Harosoy 63 seeds, Gustavo MacIntosh (Iowa State University) for the *A. Rhizogenes*, and Brett Tyler and Mannon Gallegly for the race 1 *P. sojae*. We thank Huiyuan Li for establishing the UPLC-PDA-MS$^n$ method, Maisha Huq for partially purifying the isoflavonoids, and Matt Kasson for identifying the fungal species isolated from soybean. This work was supported by WVU start-up funds to NK and is based upon work that is supported by the NIFA, USDA, and Hatch project under 1010200.

**Author Contributions:** K.F., M.A.J. and N.K. conducted the experiments and analyzed the data. N.K. conceived and designed the research, and wrote the manuscript.

**Conflicts of Interest:** The authors declare no conflicts of interest.

## References

1. Ahuja, I.; Kissen, R.; Bones, A.M. Phytoalexins in defense against pathogens. *Trends Plant Sci.* **2012**, *17*, 73–90. [CrossRef] [PubMed]
2. Großkinsky, D.K.; van der Graaff, E.; Roitsch, T. Phytoalexin transgenics in crop protection—Fairy tale with a happy end? *Plant Sci.* **2012**, *195*, 54–70. [CrossRef] [PubMed]
3. Jeandet, P.; Hébrard, C.; Deville, M.-A.; Cordelier, S.; Dorey, S.; Aziz, A.; Crouzet, J. Deciphering the role of phytoalexins in plant-microorganism interactions and human health. *Molecules* **2014**, *19*, 18033–18056. [CrossRef] [PubMed]
4. Ibraheem, F.; Gaffoor, I.; Tan, Q.; Shyu, C.-R.; Chopra, S. A sorghum MYB transcription factor induces 3-deoxyanthocyanidins and enhances resistance against leaf blights in maize. *Molecules* **2015**, *20*, 2388–2404. [CrossRef] [PubMed]
5. Xu, Y.-H.; Wang, J.-W.; Wang, S.; Wang, J.-Y.; Chen, X.-Y. Characterization of gawrky1, a cotton transcription factor that regulates the sesquiterpene synthase gene (+)-δ-cadinene synthase-a. *Plant Physiol.* **2004**, *135*, 507–515. [CrossRef] [PubMed]
6. Yamamura, C.; Mizutani, E.; Okada, K.; Nakagawa, H.; Fukushima, S.; Tanaka, A.; Maeda, S.; Kamakura, T.; Yamane, H.; Takatsuji, H.; et al. Diterpenoid phytoalexin factor, a bHLH transcription factor, plays a central role in the biosynthesis of diterpenoid phytoalexins in rice. *Plant J.* **2015**, *84*, 1100–1113. [CrossRef] [PubMed]
7. Akashi, T.; Sasaki, K.; Aoki, T.; Ayabe, S.-I.; Yazaki, K. Molecular cloning and characterization of a cdna for pterocarpan 4-dimethylallyltransferase catalyzing the key prenylation step in the biosynthesis of glyceollin, a soybean phytoalexin. *Plant Physiol.* **2009**, *149*, 683–693. [CrossRef] [PubMed]
8. Moy, P.; Qutob, D.; Chapman, B.P.; Atkinson, I.; Gijzen, M. Patterns of gene expression upon infection of soybean plants by *Phytophthora sojae*. *Mol. Plant Microbe Interact.* **2004**, *17*, 1051–1062. [CrossRef] [PubMed]
9. Beyer, E.M. A potent inhibitor of ethylene action in plants. *Plant Physiol.* **1976**, *58*, 268–271. [CrossRef] [PubMed]
10. Bhattacharyya, M.; Ward, E. Resistance, susceptibility and accumulation of glyceollins I–III in soybean organs inoculated with *Phytophthora megasperma* f. sp. *glycinea*. *Physiol. Mol. Plant Path.* **1986**, *29*, 227–237. [CrossRef]
11. Yoshikawa, M. Diverse modes of action of biotic and abiotic phytoalexin elicitors. *Nature* **1978**, *275*, 546–547. [CrossRef]
12. Selvam, C.; Jordan, B.C.; Prakash, S.; Mutisya, D.; Thilagavathi, R. Pterocarpan scaffold: A natural lead molecule with diverse pharmacological properties. *Eur. J. Med. Chem.* **2017**, *128*, 219–236. [CrossRef] [PubMed]

13. Zimmermann, M.C.; Tilghman, S.L.; Boue, S.M.; Salvo, V.A.; Elliott, S.; Williams, K.Y.; Skripnikova, E.V.; Ashe, H.; Payton-Stewart, F.; Vanhoy-Rhodes, L.; et al. Glyceollin I, a novel antiestrogenic phytoalexin isolated from activated soy. *J. Pharm. Exp. Ther.* **2010**, *332*, 35–45. [CrossRef] [PubMed]

14. Rhodes, L.V.; Tilghman, S.L.; Boue, S.M.; Wang, S.; Khalili, H.; Muir, S.E.; Bratton, M.R.; Zhang, Q.; Wang, G.; Burow, M.E. Glyceollins as novel targeted therapeutic for the treatment of triple-negative breast cancer. *Oncol. Lett.* **2012**, *3*, 163–171. [PubMed]

15. Luniwal, A.; Khupse, R.; Reese, M.; Liu, J.; El-Dakdouki, M.; Malik, N.; Fang, L.; Erhardt, P. Multigram synthesis of glyceollin I. *Org. Process Res. Dev.* **2011**, *15*, 1149–1162. [CrossRef]

16. Simons, R.; Vincken, J.P.; Roidos, N.; Bovee, T.F.H.; van Iersel, M.; Verbruggen, M.A.; Gruppen, H. Increasing soy isoflavonoid content and diversity by simultaneous malting and challenging by a fungus to modulate estrogenicity. *J. Agric. Food Chem.* **2011**, *59*, 6748–6758. [CrossRef] [PubMed]

17. Bhattacharyya, M.; Ward, E. Expression of gene-specific and age-related resistance and the accumulation of glyceollin in soybean leaves infected with *Phytophthora megasperma* f. sp. *glycinea. Physiol. Mol. Plant Path.* **1986**, *29*, 105–113. [CrossRef]

18. Graham, T.L.; Graham, M.Y. Signaling in soybean phenylpropanoid responses (dissection of primary, secondary, and conditioning effects of light, wounding, and elicitor treatments). *Plant Physiol.* **1996**, *110*, 1123–1133. [CrossRef] [PubMed]

19. Burow, M.E.; Boue, S.M.; Collins-Burow, B.M.; Melnik, L.I.; Duong, B.N.; Carter-Wientjes, C.H.; Li, S.F.; Wiese, T.E.; Cleveland, T.E.; McLachlan, J.A. Phytochemical glyceollins, isolated from soy, mediate antihormonal effects through estrogen receptor alpha and beta. *J. Clin. Endocr. Metab.* **2001**, *86*, 1750–1758. [PubMed]

20. Cheng, Q.; Li, N.; Dong, L.; Zhang, D.; Fan, S.; Jiang, L.; Wang, X.; Xu, P.; Zhang, S. Overexpression of soybean isoflavone reductase (GmIFR) enhances resistance to *Phytophthora sojae* in soybean. *Front. Plant Sci.* **2015**, *6*, 1024–1035. [CrossRef] [PubMed]

21. Lygin, A.V.; Zernova, O.V.; Hill, C.B.; Kholina, N.A.; Widholm, J.M.; Hartman, G.L.; Lozovaya, V.V. Glyceollin is an important component of soybean plant defense against *Phytophthora sojae* and macrophomina phaseolina. *Phytopathology* **2013**, *103*, 984–994. [CrossRef] [PubMed]

22. Ayers, A.R.; Ebel, J.; Valent, B.; Albersheim, P. Host-pathogen interactions 10. Fractionation and biological-activity of an elicitor isolated from mycelial walls of *Phytophthora-megasperma* var. *sojae. Plant Physiol.* **1976**, *57*, 760–765. [CrossRef] [PubMed]

23. Van De Schans, M.G.; Vincken, J.-P.; De Waard, P.; Hamers, A.R.; Bovee, T.F.; Gruppen, H. Glyceollins and dehydroglyceollins isolated from soybean act as serms and er subtype-selective phytoestrogens. *J. Steroid Biochem. Mol. Biol.* **2016**, *156*, 53–63. [CrossRef] [PubMed]

24. Aisyah, S.; Gruppen, H.; Madzora, B.; Vincken, J.-P. Modulation of isoflavonoid composition of rhizopus oryzae elicited soybean (*Glycine max*) seedlings by light and wounding. *J. Agric. Food Chem.* **2013**, *61*, 8657–8667. [CrossRef] [PubMed]

25. Boue, S.M.; Carter, C.H.; Ehrlich, K.C.; Cleveland, T.E. Induction of the soybean phytoalexins coumestrol and glyceollin by aspergillus. *J. Agric. Food Chem.* **2000**, *48*, 2167–2172. [CrossRef] [PubMed]

26. Simons, R.; Vincken, J.P.; Bohin, M.C.; Kuijpers, T.F.M.; Verbruggen, M.A.; Gruppen, H. Identification of prenylated pterocarpans and other isoflavonoids in *Rhizopus* spp. Elicited soya bean seedlings by electrospray ionisation mass spectrometry. *Rapid Commun. Mass Spectrom.* **2011**, *25*, 55–65. [CrossRef] [PubMed]

27. Bektas, Y.; Eulgem, T. Synthetic plant defense elicitors. *Front. Plant Sci.* **2014**, *5*, 1–17. [CrossRef] [PubMed]

28. Paradies, I.; Konze, J.R.; Elstner, E.F.; Paxton, J. Ethylene: Indicator but not inducer of phytoalexin synthesis in soybean. *Plant Physiol.* **1980**, *66*, 1106–1109. [CrossRef] [PubMed]

29. Ward, E.W.; Cahill, D.M.; Bhattacharyya, M.K. Abscisic acid suppression of phenylalanine ammonia-lyase activity and mrna, and resistance of soybeans to *Phytophthora megasperma* f. sp. *glycinea. Plant Physiol.* **1989**, *91*, 23–27. [CrossRef] [PubMed]

30. Liu, C.-J.; Huhman, D.; Sumner, L.W.; Dixon, R.A. Regiospecific hydroxylation of isoflavones by cytochrome p450 81e enzymes from medicago truncatula. *Plant J.* **2003**, *36*, 471–484. [CrossRef] [PubMed]

31. Naoumkina, M.; Farag, M.A.; Sumner, L.W.; Tang, Y.; Liu, C.-J.; Dixon, R.A. Different mechanisms for phytoalexin induction by pathogen and wound signals in medicago truncatula. *Proc. Natl. Acad. Sci. USA* **2007**, *104*, 17909–17915. [CrossRef] [PubMed]

32. Graham, T.L. Flavonoid and isoflavonoid distribution in developing soybean seedling tissues and in seed and root exudates. *Plant Physiol.* **1991**, *95*, 594–603. [CrossRef] [PubMed]

33. Jacobs, T.B.; LaFayette, P.R.; Schmitz, R.J.; Parrott, W.A. Targeted genome modifications in soybean with crispr/cas9. *BMC Biotechnol.* **2015**, *15*, 16–26. [CrossRef] [PubMed]

34. Kovinich, N.; Saleem, A.; Arnason, J.T.; Miki, B. Combined analysis of transcriptome and metabolite data reveals extensive differences between black and brown nearly-isogenic soybean (*Glycine max*) seed coats enabling the identification of pigment isogenes. *BMC Genom.* **2011**, *12*, 381–400. [CrossRef] [PubMed]

35. Ayers, A.R.; Ebel, J.; Finelli, F.; Berger, N.; Albersheim, P. Host-pathogen interactions 9. Quantitative assays of elicitor activity and characterization of elicitor present in extracellular medium of cultures of *Phytophthora-mega*-sperma var. *sojae. Plant Physiol.* **1976**, *57*, 751–759. [CrossRef] [PubMed]

**Sample Availability:** Samples of the compounds used in this study are not available from the authors.

*molecules*

MDPI

Article

# Systemic Induction of the Defensin and Phytoalexin Pisatin Pathways in Pea (*Pisum sativum*) against *Aphanomyces euteiches* by Acetylated and Nonacetylated Oligogalacturonides

Sameh Selim [1],*, Jean Sanssené [2], Stéphanie Rossard [3] and Josiane Courtois [4]

[1]  HydrISE, UniLaSalle, Beauvais, SFR Condorcet 3417, 19 Rue Pierre Waguet, BP 30313,
    F-60026 Beauvais CEDEX, France
[2]  Current address: JS Consulting, 17c Avenue Jean Jaurès, 31290 Villefranche de Lauragais, France;
    j.sanssene@js-consult.fr
[3]  Current address: University of Technology of Compiègne (UTC), Centre Pierre Guillaumat,
    Rue du Docteur Schweitzer, F-60203 Compiègne CEDEX, France; stephanie.rossard@utc.fr
[4]  Laboratoire des Polysaccharides Microbiens et Végétaux, Université de Picardie Jules Verne,
    Avenue des Facultés, Le Bailly, F-80025 Amiens CEDEX, France; josiane.courtois@u-picardie.fr
*   Correspondence: sameh.selim@unilasalle.fr; Tel.: +33-034-406-3825

Received: 24 May 2017; Accepted: 17 June 2017; Published: 19 June 2017

**Abstract:** Oligogalacturonides (OGs) are known for their powerful ability to stimulate the plant immune system but little is known about their mode of action in pea (*Pisum sativum*). In the present study, we investigated the elicitor activity of two fractions of OGs, with polymerization degrees (DPs) of 2–25, in pea against *Aphanomyces euteiches*. One fraction was nonacetylated (OGs − Ac) whereas the second one was 30% acetylated (OGs + Ac). OGs were applied by injecting the upper two rachises of the plants at three- and/or four-weeks-old. Five-week-old roots were inoculated with $10^5$ zoospores of *A. euteiches*. The root infection level was determined at 7, 10 and 14 days after inoculation using the quantitative real-time polymerase chain reaction (qPCR). Results showed significant root infection reductions namely 58, 45 and 48% in the plants treated with 80 µg OGs + Ac and 59, 56 and 65% with 200 µg of OGs − Ac. Gene expression results showed the upregulation of genes involved in the antifungal defensins, lignans and the phytoalexin pisatin pathways and a priming effect in the basal defense, SA and ROS gene markers as a response to OGs. The reduction of the efficient dose in OGs + Ac is suggesting that acetylation is necessary for some specific responses. Our work provides the first evidence for the potential of OGs in the defense induction in pea against *Aphanomyces* root rot.

**Keywords:** pea root rot; *Aphanomyces euteiches*; oligogalacturonides; real-time qPCR; gene expression; pea defense pathways; defensins; pisatins; phytoalexins

---

## 1. Introduction

Plants need to be able to recognize pathogen attacks in a timely manner in order to activate their defenses that provide protection against the infection process. The plant cell wall is the site of initial interaction with microbial pathogens. Pectin is one of the most accessible components of the cell wall and, therefore, is among the first structures to be altered upon pathogen invasion. The oligogalacturonides (OGs) are produced upon partial degradation of the pectin homogalacturon by pathogen pectinases and polygalacturonases [1]. However, OGs have been indicated as damage-associated molecular patterns (DAMPs) which may trigger plant defenses against pathogens. OGs endogenous elicitors consist of linear chains of α-(1-4)-linked D-galacturonic acid [2,3] and those with high biological activity have often a degree of polymerization (DP) between 10 and 15 [4].

This size is optimal for the formation of $Ca^{2+}$ mediated inter molecular cross links resulting in structures called "egg boxes" that are thought to be necessary for OGs activity [5,6]. The short OGs with a DP of 2–6 have been reported in few cases to exhibit elicitor activity in tomato [7,8]; however, they appear to suppress defense responses in wheat [9]. It has been shown that the degree of OGs methylation clearly influences plant defense responses [10–12]. In wild strawberry (*Fragaria vesca*), partial demethylation of OGs in transgenic fruit enhanced resistance to *Botrytis cinerea* [13]. Furthermore, in wheat, Wietholter et al. [14] found a significant difference in the methyl ester distribution in OGs from cultivars susceptible or resistant to stem rust. Recently, we reported more than 57% protection in wheat against *Blumeria graminis* f. sp. *tritici* using 30% chemically acetylated or nonacetylated citrus OGs with DPs of 2–25 [15]. We found that only the acetylated OGs led to an increase in papilla-associated fluorescence and a reduction in the fungal haustoria formation, suggesting that acetylation is necessary for some specific responses. However, OGs elicit in several plant species [16–18] a wide range of defense responses, including induction of polygalacturonase-inhibiting protein (PGIP) [19,20], accumulation of phytoalexins [21], glucanase, and chitinase [22,23], deposition of callose, production of reactive oxygen species [15], and nitric oxide [24]. Root rot caused by the oomycete *Aphanomyces euteiches* is the major destructive soil-borne fungal disease of pea (*Pisum sativum*) with up to 80% yield loss per year. It is widespread in North America, Europe, Japan, Australia and New Zealand [25]. Oospores released from infected roots into the rhizosphere constitute the primary source of inoculum. *A. euteiches* invades the root system leading to a complete arrest of root growth and ultimately plant death. To date, disease control measures are limited to crop rotation and no resistant pea lines are available. The fact that oospores are able to remain dormant in the soil for up to 10 years reduces the effectiveness of crop rotation in decreasing the propagation of this pathogen [26]. Moreover, oomycetes are distantly related to true fungi and their particular physiology makes them insensitive to most fungicides [27]. Therefore, the development of alternative control methods against oomycetes is becoming urgent. In the present work, the efficiency of two distinct biochemical fractions of OGs to protect pea against *A. euteiches* was studied. These fractions consisted of OGs with DPs ranging from 2 to 25 (OGs − Ac) and one fraction that was chemically 30% acetylated (OGs + Ac). The systemic defense mechanisms elicited in pea roots as a response to rachis injections with OGs are discussed.

## 2. Results

### 2.1. Elicitor Effect of OG Compounds

*A. euteiches* oospores were observed within necrotic tissues between 7 and 14 days after inoculation (dai). However, at ten dai, high *P. sativum-A. euteiches* compatibility was observed with the disease severity index (DSI) 3.5 and 76% of root fragments with more than 50 oospores (Figure 1).

No significant difference was observed between controls injected with water at one and/or two weeks before inoculation (wbi). At 10 dai, OGs at the dose of 20 µg/plant and all other tested elicitors did not lead to any significant protection compared to the control injected with water (Figure 2). The protection level conferred on pea against *A. euteiches* was significant and increased to 43.5% and 47.8% as a response to the increase in the injected dose of OGs − Ac and OGs + Ac, respectively, to 40 µg/plant at two wbi (Figure 3a). This protection level was associated with a significant reduction (50.7% and 60%, respectively) in the percentages of root fragments containing more than 50 oospores (Figure 3b).

As the classic methods using the DSI and the counting of oospores are time-consuming and tedious, especially with the large number of root samples, we developed primers and probes to evaluate the disease severity using qPCR. Figure 4 shows the relation between the *A. euteiches* specific gene and the qPCR threshold cycles. The efficiency of the qPCR was 99.58% with high sensitivity to detecting one copy of the *A. euteiches* specific gene (Figure 4).

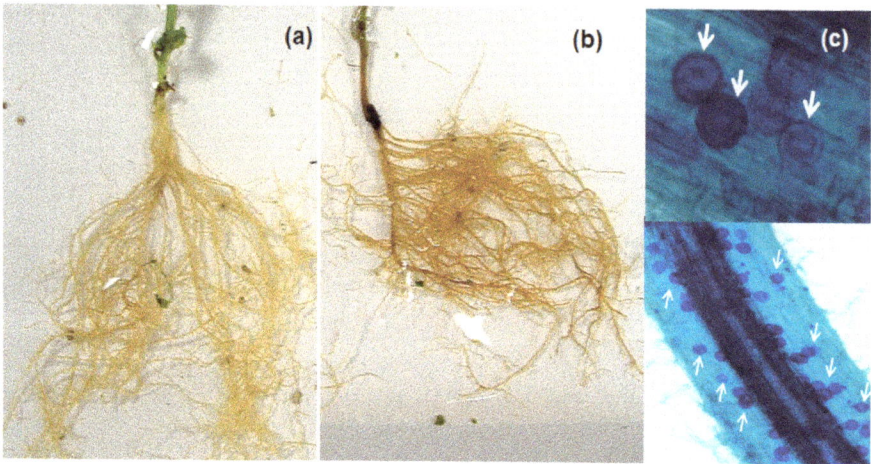

**Figure 1.** Pea stem basal part and root necrosis at 10 days after inoculation with $10^5$ zoospores of *Aphanomyces euteiches* at 5-weeks-old; (**a**) Roots of plant injected with a solution of nonacetylated oligogalacturonides (OGs − Ac) elicitor in the upper two rachises (20 µg/rachis = 40 µg/plant) at two weeks before inoculation; (**b**) Controls injected with water; (**c**) Plant roots colored with lactophenol cotton blue. *A. euteiches* oospores indicated with white arrows.

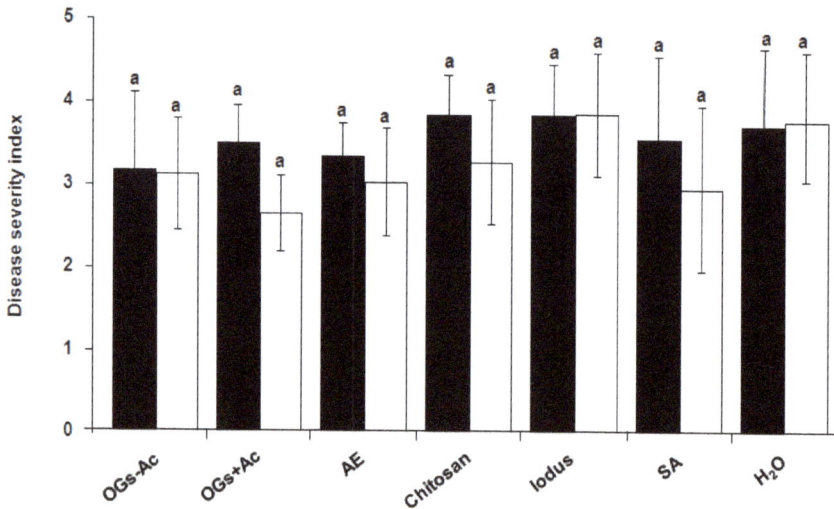

**Figure 2.** Pea root rot disease severity index at 10 days after inoculation with *Aphanomyces euteiches* at $10^5$ zoospores/plant at 5-week-old. The upper plant rachis was injected with elicitors (20 µg/rachis) at two (white) or one (black) weeks before inoculation. OGs − Ac; nonacetylated oligogalacturonides, OGs + Ac; acetylated oligogalacturonides, AE; an inoculum of *A. euteiches* zoospores ($10^5$ zoospores. $mL^{-1}$) heated at 100 °C for 10 min, Chitosan, Iodus, SA; salicylic acid. Controls were injected with water. The values shown are means with SD ($n = 5$). Different lower-case letters indicate significant differences between treatments according to the Tukey test ($p \leq 0.05$).

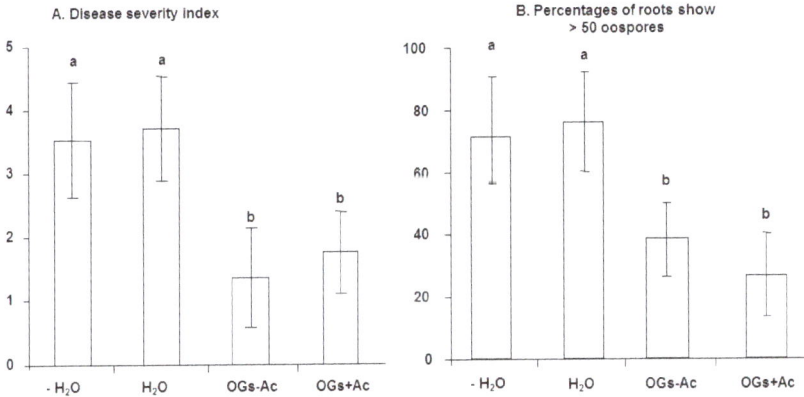

**Figure 3.** (**A**) Pea root rot disease severity index; (**B**) Percentage of roots showing >50% of oospores, at 10 days after inoculation with *Aphanomyces euteiches* at $10^5$ zoospores/plant at 5-weeks-old. The upper two rachises of plants were injected with 20 μg/rachis (=40 μg/plant) of acetylated oligogalacturonides (OGs + Ac) or nonacetylated OGs (OGs − Ac) two weeks before inoculation. $H_2O$; controls injected with water, $-H_2O$; controls without water injection. The values shown are means with SD ($n = 5$). Different lower-case letters indicate significant differences between treatments according to the Tukey test ($p \leq 0.05$).

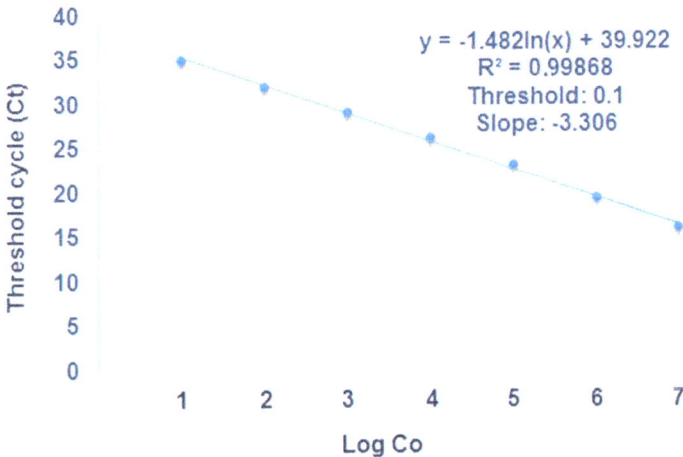

**Figure 4.** Standard curve using known copies ($10^1$ to $10^7$) of the appropriate cloned target sequence of the *Aphanomyces euteiches* specific gene (GenBank accession No. AF228037.1). For each reaction, the cycle threshold (Ct), the initial cycle number at which an increase in fluorescence above a baseline can be detected, is plotted against the log10 (Log Co) of the *A. euteiches* specific gene copies. Three technical PCR replicates were performed for each concentration.

The results in Figure 5 show the *A. euteiches* specific gene copy numbers in 100 ng of total DNA (AESG$_{100ng}$) extracted from root samples collected at 7, 10 and 14 dai. Elicitor efficiency was calculated by comparing the levels of AESG$_{100ng}$ in the inoculated plants after the elicitor pretreatment with those in the inoculated control plants without elicitor pretreatment. The well-known endogenous elicitor salicylic acid (SA) was used as a reference. No significant differences in AESG$_{100ng}$ were observed between all the inoculated control modalities, injected or non-injected with water, at the three observation dates. These controls were grouped and used as repetitions for non-treated inoculated controls. Plants treated with the high doses (200 and 400 µg/plant) of SA showed phytotoxicity symptoms of brown necrotic lesions on the leaves and the injected rachises were dead. On the other hand, unstable protection efficiencies (50, 38 and 4%) and (48, 32 and 7%) were recorded in root samples collected at 7, 10 and 14 dai, respectively, as a response to the injection with SA (40 µg/plant) once at two wbi or twice (= 80 µg/plant) at two and one wbi (Figure 5). The same phytotoxicity symptoms, but less severe than those seen with SA, were observed in plants injected with the high dose (400 µg/plant) of OGs − Ac and OGs + Ac. Unstable protection was observed with the one-date injection modalities (two wbi) with 40 µg/plant of OGs − Ac (74, 33 and 6%) and OGs + Ac (38, 35 and 7%) at 7, 10 and 14 dai, respectively (Figure 5). However, significant and stable protection efficiencies of 58, 45 and 55% were recorded at 7, 10 and 14 dai, respectively, in the plants treated twice, at two and one wbi, with 40 µg/plant OGs + Ac at each date (=80 µg/plant) and of 59, 56 and 65% in the plants treated at only one date (two wbi) with 200 µg/plant OGs − Ac (=2.5 folds more than OGs + Ac) (Figure 5). The differences between these two treatments were not significant. The root samples of these two efficient treatments were used for gene expression studies.

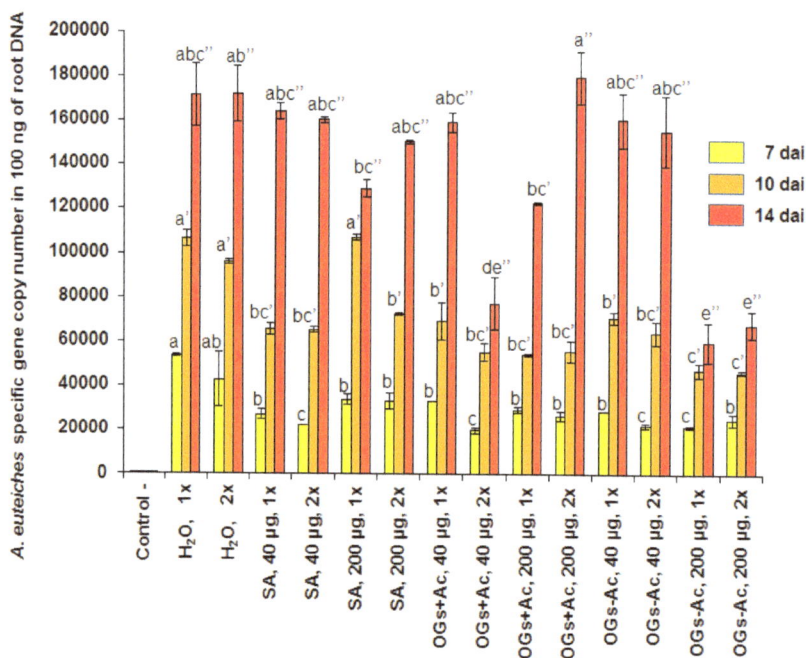

**Figure 5.** *Aphanomyces euteiches* specific gene copy numbers in 100 ng of the total DNA extracted from root samples collected at 7, 10 and 14 dai (days after inoculation). Pea plants injected in the upper two rachises with 20 µL/rachis of salicylic acid (SA), acetylated oligogalacturonides (OGs + Ac) or nonacetylated OGs (OGs − Ac). The elicitor injections were done once (1×) two weeks before inoculation (wbi) or twice (2×) two and one wbi. The final elicitor concentrations were 40, 80, 200 or 400 µg/plant. Controls were injected with water. The values shown are means with SD ($n$ = 5). Different lower-case letters indicate significant differences between treatments according to the Tukey test ($p \leq 0.05$).

## 2.2. Gene Expression

The expression of Pathogenesis Related protein 1 (*PR1*), *1,3 β glucanase* and phenylalanine ammonia-lyase (*PAL*) genes was followed at 3, 6, 12, 24, 48, 96, 168 and 336 hat and using the same time course after inoculation (hai) in root samples harvested from plants injected twice at two and one wbi with 80 µg/plant of OGs + Ac (40 µg/plant/date) or once at two wbi with OGs − Ac (200 µg/plant). No upregulation of the three tested genes was detected in the controls injected with water in comparison with the control non-injected and non-inoculated. Neither elicitor treatments nor inoculation with *Aphanomyces* showed any expression changes for the *PR1* and *1,3 β glucanase* genes over the tested time course. However, significant upregulation (≥2 folds) of the *PAL* gene was recorded at 6, 12, 48, 168 and 336 hai with *A. euteiches* (Figure 6). On the other hand, high and early induction (3 hat) of *PAL* was observed in roots pretreated with both OG compounds. This active response of *PAL* over the tested time course in non-inoculated plants was higher and more stable with OGs − Ac (until 360 hat) than with OGs + Ac (until 168 hat) (Figure 6). The highest recorded *PAL* expression value was found at 168 hat in the plant roots pretreated with OGs + Ac (12.9-fold) and 31-fold at 24 hat in the plant roots pretreated with OGs − Ac. At the inoculation time (336 hat), *PAL* expression in OGs − Ac samples was upregulated 2.8-fold and this value increased strongly after inoculation to reach 13.8, 56.7, 8.0, 10.8 and 8.4-fold at 3, 6, 12, 24, and 96 hai, respectively (Figure 6). These values were significantly higher than those in the inoculated non-treated control which were up-regulated 1.4, 3.6, 2.7, 1.5 and 1.3-fold, respectively. In the case of OGs + Ac, *PAL* expression was 0.3-fold at the time of inoculation and upregulated after inoculation to reach 5.7-, 18.2-, 2.9- and 5.9-fold at 3, 6, 24 and 48 hai, respectively (Figure 6). These values were significantly higher than those in the inoculated non-treated plants which were up-regulated 1.4, 3.6, 1.5 and 2.5-fold, respectively.

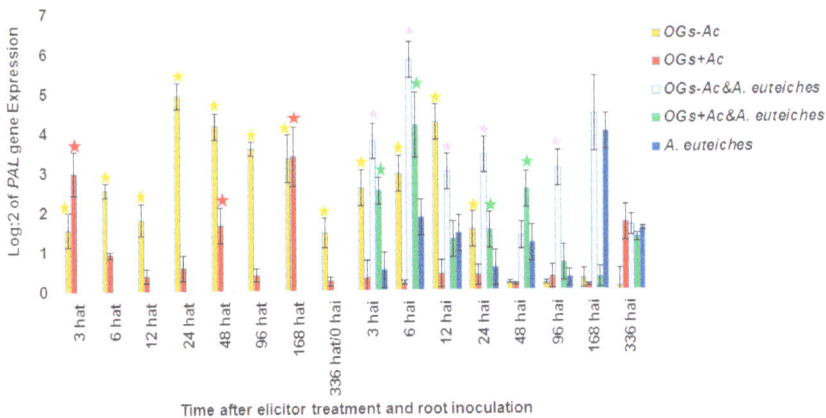

**Figure 6.** *PAL* gene expression ratio at 3, 6, 12, 24, 48, 96, 168 and 336 hours after treatment (hat) with water, 80 µg/plant of acetylated oligogalacturonides (OGs + Ac) or 200 µg/plant of nonacetylated OGs (OGs − Ac). The elicitor injections were done two weeks before inoculation (wbi) and two and one wbi for OGs − Ac and OGs + Ac, respectively. After inoculation with $10^5$ *Aphanomyces euteiches* zoospores/plant, *PAL* gene expression was followed over the same time course (hours after inoculation (hai)). The values shown are means of five repetitions. ☆ Stars indicate gene induction ≥2-folds and significant differences between elicitor treatments and inoculated non-treated control according to the Tukey test ($p ≤ 0.05$).

The expression changes in twenty-two pea defense genes as a response to OG pretreatments were followed using RT-PCR in root samples at 0 (just at the time of root inoculation), 7, 10 and 14 dai with *A. euteiches* in three modalities: inoculated without OG elicitor pretreatment; inoculated and pretreated with 80 µg/plant OGs + Ac; and inoculated and pretreated with 200 µg/plant OGs − Ac. These modalities were compared with the inoculated untreated controls. The results in Figure 7 show that at the inoculation time (0 dai = 14 dat), except for *GST* with OGs − Ac pretreatment, no significant induction of the genes involved in the ROS pathway (*SOD*, *POX*, *Catalase*, *NOS*, *Metallothionein*) was observed. After inoculation, only the *Catalase* and *NOS* genes were significantly upregulated at 10 dai in plants pretreated with OGs − Ac compared to their levels in the inoculated non-treated control. In addition, no significant induction was observed for the genes coding for *MAPK* and *PRP* (Figure 8). For the genes coding for pathogenesis-related proteins, *PR1*, *β1,3 glucanase*, *chitinase*, *DRR230*, *DRR276*, *DRR206* and *DRR49*, significant inductions of *chitinase* expression, at 0 and 10 dai, were recorded in the plant roots pretreated with OGs − Ac and OGs + Ac, respectively. At 7 dai, *DRR276* and *DRR49* were recorded in the plant roots pretreated with OGs − Ac (Figures 7 and 8). The *DRR206* and *DRR230* genes showed significant upregulations at 0 dai and at 0 and 7 dai respectively, in the plants pretreated with the two OG compounds (Figure 8). However, no significant activation was observed for *PR1* and *β1,3 glucanase* genes with either of the two OG elicitors. The *PGIP* gene, coding for polygalacturinase inhibitor protein enzyme (marker gene for the basal defense), and *LOX*, coding for lipoxygenase enzyme (marker gene of the jasmonic acid pathway), showed no significant upregulation except for *PGIP* at 10 dai and only with OGs − Ac.

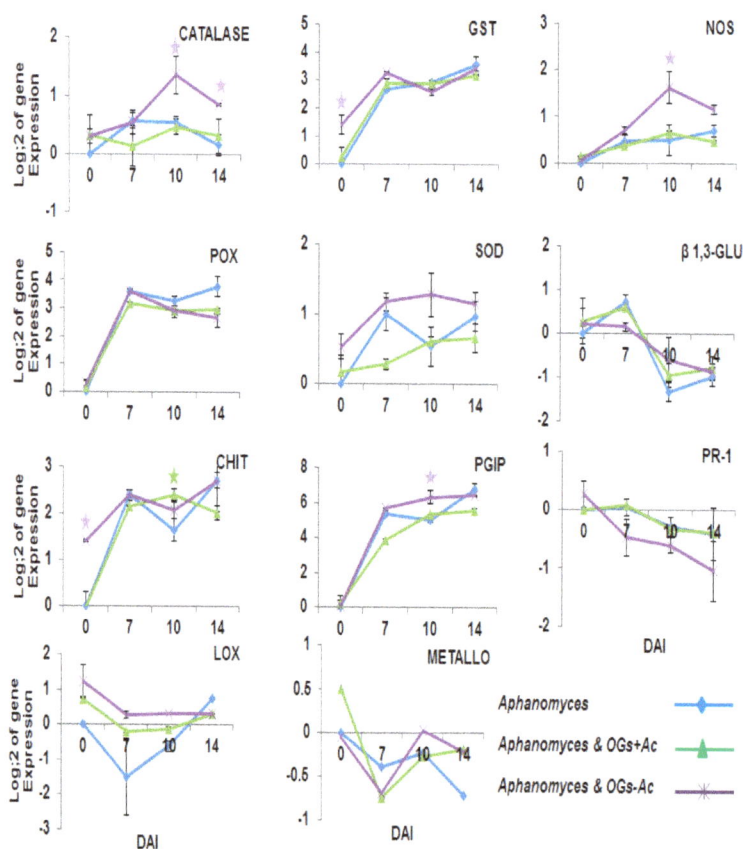

**Figure 7.** Pea gene expression ratio at 0 (just at the time of root inoculation), 7, 10 and 14 days after inoculation with $10^5$ *Aphanomyces euteiches* zoospores/plant. Pea plants were injected with acetylated oligogalacturonides (OGs + Ac) one and two weeks before inoculation (final concentration = 80 μg/plant) or with nonacetylated OGs (OGs − Ac) two (wbi) (final concentration = 200 μg/plant). Controls were injected with water. The values shown are means of 5 repetitions. ☆ Stars indicate gene induction ≥2-folds and significant differences between elicitor treatments and inoculated non-treated control according to the Tukey test ($p \leq 0.05$).

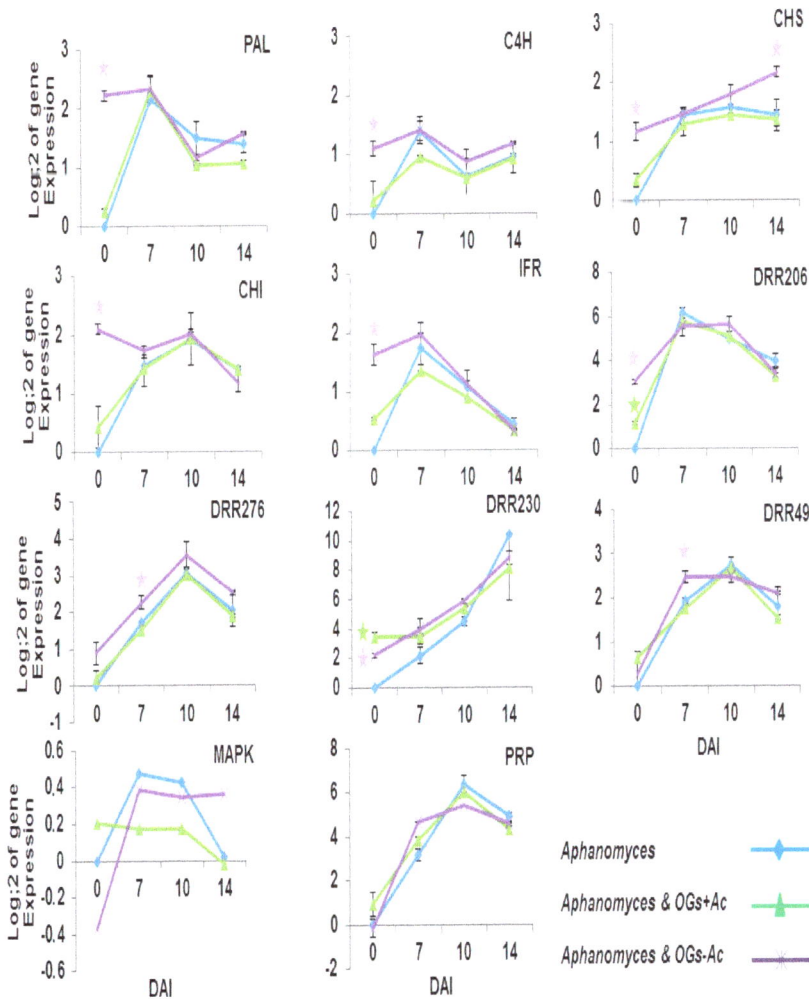

**Figure 8.** Pea gene expression ratio at 0 (at the time of root inoculation), 7, 10 and 14 days after inoculation with $10^5$ *Aphanomyces euteiches* zoospores/plant. Pea plants were injected with acetylated oligogalacturonides (OGs + Ac) one and two weeks before inoculation (final concentration = 80 µg/plant) or with nonacetylated OGs (OGs − Ac) two (wbi) (final concentration = 200 µg/plant). Controls were injected with water. The values shown are means of 5 repetitions. ✰ Stars indicate gene induction ≥2-folds and significant differences between elicitor treatments and inoculated non-treated control according to the Tukey test ($p \leq 0.05$).

For the genes involved in the phenylpropanoid and phytoalexin pathways (*PAL*, cinnamate-4-hydroxylase (*C4H*), chalcone synthase (*CHS*), chalcone isomerase (*CHI*), and isoflavone reductase (*IFR*)), and known as SA gene markers, these were all significantly upregulated at the time just before inoculation (0 dai) in the plants pretreated with OGs − Ac. Only the *CHS* gene was significantly upregulated at 14 dai in roots from plants pretreated with OGs − Ac compared to those untreated and inoculated with *A. euteiches*. None of these genes showed upregulation at the time of inoculation in the plants pretreated with OGs + Ac.

## 3. Discussion

The oomycete *Aphanomyces euteiches* causes up to 80% crop loss in pea (*P. sativum*). To date, disease control measures are limited to crop rotation and no resistant pea lines or efficient fungicides are available. The present study aimed to investigate the potential of citrus-derived OGs to stimulate defense mechanisms in pea roots against *A. euteiches*. We tested two OGs fractions with DPs ranging from 2 to 25, nonacetylated (OGs − Ac) and acetylated at 30% (OGs + Ac).

Our results revealed no protection efficiency against *A. euteiches* root rot as a response to the commercial elicitors tested, Chitosan® and Iodus®, as well as the heated inoculum of *A. euteiches* zoospores. In parallel, SA showed unstable protection efficiencies and phytotoxicity symptoms with the higher doses tested.

In contrast, plants injected with the OGs showed a significant disease reduction. High and stable protection efficiencies (>45%) were recorded in plants pretreated with OGs. On the other hand, the acetylation of OGs reduced the efficient dose more than two times compared to the nonacetylated OGs. It has been shown that the degree of OGs methylation clearly influences plant defense responses in wild strawberry against *B. cinerea* [13], in tomato against *Ralstonia solanacearum* [11], in wheat against *Puccinia graminis* f. sp. *tritici* [14] and against *B. graminis* f. sp. *tritici* [15]. Indeed, the ability of bacterial or fungal necrotrophs to produce pectin methylesterases (PME) is often related to a successful initiation of the infective process. Pectin is synthesized in a highly methylesterified form and is subsequently de-esterified in muro by PME. De-esterification makes pectin more susceptible to the degradation by pectic enzymes such as endopolygalacturonases and pectate lyases [28]. Wayra and Bari [11] observed in their immunohistochemical studies constitutive differences between tomato genotypes susceptibility to *R. solanacearum* which manifested in methyl-ester distribution of homogalacturonan (HG), arabinan and galactan side chain composition of rhamnogalacturonan I (RG I) and arabinogalactan-protein (AGP) in the xylem parenchyma and in vessel cell walls. They suggested that *R. solanacearum* PME may act on HGs of the susceptible plant in a non-blockwise deesterification pattern, while in the resistant genotype the constitutive, more blockwise methyl-ester distribution and the increased AGP content and higher side chain branching of RG I in vessel cell walls may inhibit easy degradation. However, it is not yet clear how esterification affects OGs biological activity [29].

Gene expression analysis showed no upregulation of the genes involved in the ROS pathway as a response to OGs except for the *GST* gene with OGs − Ac. However, a priming effect in the expression of the *catalase* and *NOS* genes were observed after challenging the OGs-treated plants with *A. euitches*. Most interestingly, it is known that elicitors such as jasmonic acid, salicylic acid, nitric oxide and superoxides or their precursors do not significantly enhance the resistance induction of pea, and at high concentrations can negatively affect resistance [30]. The induction of the *GST* gene could be explained by its broad spectrum of functions in plants, such as transport and storage of reduced sulfur, detoxification, and antioxidation, as well as a role as cofactor in enzymatic processes, protein reduction, and in phytochelatin by complex-binding heavy metals that have thiol affinity [31,32]. The observed priming induction of the *PGIP* as a response to OGs − Ac indicates the importance of this gene against *A. euteiches*. PGIP are plant extracellular leucine-rich repeat proteins that effectively and specifically bind and inhibit fungal [33] and bacterial [19,20] polygalacturonases and inhibit further invasion of these pathogens. The down-regulation of *PGIP* at early time points after inoculation could be attributed to virulence factors released by the pathogen to suppress the host resistance and facilitate host colonization [34].

Looking at the genes coding for pathogenesis-related proteins, only *DRR276* and *DDR49* genes were upregulated as a response to OGs − Ac. Both the *PR10 DRR49*, *DRR276* pea genes code for a protein homologous with RNase [35,36]. Transgenic 'Shepody' potatoes possessing the pea gene DRR49 displayed resistance to potato early dying disease (*Verticillium dahliae*) [37]. However, the utility of this gene may be related to the close homology of its product to other plant allergens [38].

In parallel, the *chitinase*, *DDR206* and *DDR230* genes were significantly overexpressed as a response to both OGs. The stimulation of *chitinase*, which codes for an enzyme that digests chitin in the fungus cell wall, could be an important mechanism against *A. euteiches*, which contains 10% of chitin in its wall structure [39]. It has been reported that the pea defense gene *DRR206* confers resistance to black leg (*Leptosphaeria maculans*) disease in transgenic canola (*Brassica napus*) by inhibiting fungal germination and decreasing hyphal growth at inoculation sites [40]. Recently, Seneviratne et al. [41] investigated *in planta* the biochemical function of the DRR206 and reported that the metabolite associated with its gene induction is the pinoresinol monoglucoside. The pinoresinol is a member of a large, structurally diverse, class of lignans, which have a wide range of physiological and pharmacologically important properties [42,43]. Because of their pronounced biological (antimicrobial, antifungal, antiviral, antioxidant and anti-feedant) properties, a major role of lignans in vascular plants is to apparently help confer resistance against various opportunistic pathogens and predators [41]. The *DDR230* gene that codes for proteins with a high cysteine content, called defensins [44], has been found previously actively expressed in the pea endocarp during the resistance response to *Fusarium solani* f. sp. *phaseoli* [45]. Defensins are antifungal products of some PR genes and one of the first defensin genes cloned, *DRR230*, was isolated from pea [44]. When it was overexpressed in canola, extracts of these plants inhibited the in vitro germination of *L. maculans* [40]. Pea defensins have also been used in the biological control of blue mold in apple [46]. The mode of action of pea defensins corresponds to that of similar highly-conserved antimicrobial peptides present in a broad range of biological organisms [47] and could consist of inducing the membrane destabilization/permeabilization required for fungal growth, inhibiting protein synthesis, enzyme activity, and ion channels [48].

The earlier induction of the *PAL* gene, marker of the phenylpropanoid and phytoalexin pathway, in pea roots infected with *A. euteiches* indicates the importance of this pathway in pea defense reactions. The preventive induction of this pathway has previously been reported as one of the important strategies to control root legume diseases [49]. Before inoculation with *A. euteiches*, a high and early induction (3 hat) of *PAL* was observed in pea roots as a response to both OGs elicitors, with higher levels with OGs − Ac. After root inoculation with *A. euteiches*, the induction of *PAL* as a response to both OGs was primed to be significantly higher than its levels in inoculated non-treated plants. This priming effect was observed during the first 4 dai. In fact, PAL enzyme catalyzes the first step in the phenylpropanoid pathway toward the biosynthesis of a large variety of products, including antimicrobial phytoalexin compounds such as pisatins, antioxidant protectants such as flavonoid compounds, and precursors of lignin [50–52]. Pisatin production is dependent on PAL and a series of other secondary enzymes, such as TCAH, CHS, CHI, and IFR [49,53]. At the inoculation time, all of these genes were significantly upregulated as a response to OGs − Ac pretreatment. Previous investigations reported that multiple genes control the pathogenicity of fungal isolates on pea and of these, the gene for pisatin demethylase enzyme (PDA) [54], which detoxifies the phytoalexin pisatin [55], is considered the most important. It has also been reported that all isolates without the *PDA* gene were essentially non-pathogenic on peas [30,56] showing the importance of the pisatin pathway in the pea defense against pathogen attack. Interestingly, both pinoresinol monoglucoside and pisatin were found co-localized in pea pod endocarp epidermal cells and associated with *CHS* and *DRR206* gene expression, indicating that both pisatin and pinoresinol monoglucoside function in the overall phytoalexin responses [41].

In conclusion, acetylated and nonacetylated oligogalacturonides confer protection in pea against *A. euteiches* root rot. Acetylation allows a significant reduction of the efficient elicitor dose of OGs, suggesting that acetylation is necessary for some specific responses. The induction of the antifungal defensins, lignans and the phytoalexin pisatin pathways and their priming effect in the expression of the basal defense, SA and ROS gene markers could explain their stable and synergetic protection efficiency. Taken together, acetylated OGs are interesting elicitors to stimulate defense mechanisms in pea.

## 4. Materials and Methods

### 4.1. Elicitor Compounds

OGs were produced at the Laboratoire des Polysaccharides Microbiens et Végétaux (Université Jules Verne, Amiens, France). A mixture of OGs was obtained by thermal degradation of polygalacturonic acid from citrus fruit following the same methods described previously in Randoux et al. (2010). OGs with polymerization degrees (DPs) of 2 to 25 were selected by sequences of purification using acetic acid and isopropanol and then mixed together. This mixture of OGs is hereafter referred to as the nonacetylated OGs (OGs − Ac). Dried OGs − Ac were acetylated using acetic anhydride. After addition of $H_2O$, the preparation was dialyzed and the acetylated galacturonides were freeze-dried. Samples were then dissolved in $D_2O$, 99.96% D; the final concentration was 15 g $L^{-1}$. The degree of acetylation was calculated by integration of the signals in the downfield, upfield, and acetyl regions as described in the literature [57] and OGs with a degree of substitution (DS) of 30% (OGs + Ac) were used in further experiments. OGs + Ac were characterized by the presence of acetyl groups linked on either the C2 or C3 of galacturonan residues, as described in [15]. For all the esterification procedures applied, the DS and the distribution of acetyl groups on the galacturonan residue were always the same.

### 4.2. Plant Material and Growth Conditions

*Pisum sativum* L. commercial cv. Alezan, highly susceptible to pea root rot caused by *Aphanomyces euteiches*, was used in the experiments. The seeds were sterilized by immersing them in 70% ethanol alcohol for 5 min and then in a solution of 1% NaOCl for 15 min with three intervals of washing in sterilized distilled water. Plants were grown from seeds in 0.52 L pots (9 × 9 × 8 cm) filled with autoclaved vermiculite in a growth chamber at 20 ± 2 °C and under a photoperiod of 16 h daylight with a light intensity of 150 μmol m$^{-2}$ s$^{-1}$ photon flux density supplied by high-output white fluorescent tubes (Philips Master Cool White 80 W//865, Lamotte Beuvron, France). Plants were irrigated daily and once a week with 25% Murashige and Skoog nutritive solution (Murashige and Skoog medium, Sigma-Aldrich, Saint Louis, Mo, USA).

### 4.3. Inoculum Preparation and Inoculation

*A. euteiches* was cultured on corn meal agar medium (Sigma-Aldrich, St. Quentin Fallavier, France) at 18 °C in the dark for three days. Then, zoospores of *A. euteiches* were produced in a mineral salt solution as described by Carman and Lockwood [58]. At 5 weeks post-germination, 25 mL of water containing $10^5$ zoospores was added to each pot on top of the vermiculite. Control plants were irrigated with the same amount of water without zoospores. Three, seven, ten and fourteen days after inoculation, plant roots were harvested and colored with lactophenol cotton blue stain for microscopic observations, or conserved at −80 °C until DNA extraction to follow the disease progression using qPCR. *P. sativum-A. euteiches* compatibility was evaluated at 7, 10 and 14 dai using the DSI ratings from 1 to 5 as follows: 1 = no necrosis of roots and hypocotyls; 2 = slight necrosis of roots and hypocotyls; 3 = necrosis of roots and lower hypocotyls, slight chlorosis of cotyledons, and moderate stunting of stem; 4 = extensive necrosis of roots, hypocotyls, cotyledons, and severe stunting of stem; 5 = plant death. In the same samples, the percentage of roots containing more than 50 oospores was determined in samples of one hundred 1-cm root fragments per condition.

### 4.4. Protection Assay

All tested elicitors were dissolved in water. OGs − Ac and OGs + Ac elicitor solutions were prepared at 1 and 5 g $L^{-1}$. Solutions of SA (Sigma-Aldrich, St. Quentin Fallavier, France), Chitosan® (Sigma-Aldrich, St. Quentin Fallavier, France), Iodus® (Goëmare, Saint Malo, France) at 1 g $L^{-1}$ were tested. An inoculum of *A. euteiches* (AE) zoospores ($10^5$ zoospores mL$^{-1}$), heated at 100 °C for 10 min to kill the zoospores, was also tested for its elicitor activity. Plants were injected with each elicitor compound or with water for the control plants on the upper one or two proximal rachises (20 µL/rachis). Different elicitor doses were used: 20, 40, 80, 200 and 400 µg/plant. Depending on the date of elicitor injection, three timing modalities were carried out: at 3-weeks, at 4-weeks, or twice at 3- and 4-weeks post-germination. The plants were then inoculated at 5-weeks-old as mentioned above. Three plant control modalities were used: injected with water and inoculated; non-injected and inoculated; non injected and non-inoculated. The roots from controls and treated plants were harvested at 7, 10 and 14 dai for disease observation and at 3, 6, 12, 24, 48, 96, 168 and 336 hours after treatment (hat) and at the same hours after inoculation (hai) for gene expression studies.

### 4.5. DNA and RNA Extraction

For *A. euteiches* DNA quantification and gene expression experiments, roots were harvested and stored immediately in liquid nitrogen and subsequently used for DNA and RNA extraction. Total DNA and RNA were isolated from *P. sativum* roots with the DNeasy and RNeasy plant mini kits (Qiagen, Les Ulis, France), respectively, in accordance with the manufacturer's recommendations. DNA and RNA concentrations and qualities were evaluated using absorption values at 260 and 280 nm, and RNA quality was also checked by gel electrophoresis.

### 4.6. Real-Time PCR

#### 4.6.1. Real-Time Quantitative PCR (qPCR)

To quantify infection levels of *A. euteiches*, primers and TaqMan minor groove binder probes (Table 1) were designed, using the Primer Express 3 software (Applied Biosystems, Foster, CA, USA), to target a 61-bp fragment of the *A. euteiches* specific gene (GenBank accession No. AF228037.1 [59]). A TaqMan assay was carried out in 25 µL of a reaction mixture containing the following: 12.5 µL of universal TaqMan PCR Master Mix (Applied Biosystems, Foster, CA, USA), 0.3 µM of each primer, 0.2 µM of probe, 200 ng of DNA and water up to a volume of 25 µL. The conditions of qPCR amplification were the following: 10 min at 95 °C, followed by 40 cycles of 15 seconds at 95 °C and 1 min at 60 °C. qPCR analysis of the *A. euteiches* specific gene was calibrated from $10^2$ to $10^7$ copies by serial dilution of the appropriate cloned target sequence.

**Table 1.** Oligonucleotide primer sequences of pea defense genes.

| Gene Name | GenBank Accession N° | Forward, Reverse Primers and Probes (5′–3′) | $T_m$ (°C) | Amplicon Length |
|---|---|---|---|---|
| *Aphanomyces euteiches* | AF228037 | TTTTGGAACACCCAAACGTACTG | 58 | 61 |
| | | AGTCCAAGAGGCATTCGACAA | 58 | |
| | | ACGCTGAGCTTGAC | 68 | |
| **Housekeeping genes** | | | | |
| GAPDH (Mtgap1) [1] | X73150 | GTCTTTGCACACAGGAACCCA | 59 | 123 |
| | | GGCACCACCCTTCAAATGAG | 59 | |
| | | CCCATGGGCCAGCAC | 70 | |
| **Defense and cell rescue** | | | | |
| Pathogenesis protein 1 (PR1) | AJ586324 | CCTTCCCCTCATGGCTATCC | 59 | 69 |
| | | TGTGGTGAGTTTTGAGCATATGAGA | 59 | |
| | | AGTACTATCCACATCAACAC | 68 | |
| Proline-rich protein (PRP) | AJ233399 | TGGCTTCCTTAACCTTCCTACTGT | 58 | 64 |
| | | TTGGCAAACCCTTGAGGAAT | 58 | |
| | | ACTCCTTCTTGCTCTTAT | 68 | |
| Mitogen-activated protein kinase (MAPK) | X70703 | CATTCCGCGAATGTTTTGC | 58 | 59 |
| | | TTGGCGTTCAGGAGAAGGTT | 58 | |
| | | AGGGACTTAAAACCC | 70 | |

Table 1. *Cont.*

| Gene Name | GenBank Accession N° | Forward, Reverse Primers and Probes (5′–3′) | $T_m$ (°C) | Amplicon Length |
|---|---|---|---|---|
| **Reactive oxygen species (ROS)** | | | | |
| Superoxide dismutase (SOD) | AB087845 | CCATCATAGGAAGGGCTGTTGT | 59 | 63 |
| | | CGTGACCACCTTTCCCAAGA | 59 | |
| | | CCATGCCGATCCTGAT | 70 | |
| Peroxidase (POX) | AB193816 | ATGCAAGAACAGCAAGCCAAA | 59 | 69 |
| | | GGGTTGCAAGGTCAGATGATG | 59 | |
| | | AACAGTCAAATCCC | 70 | |
| Nitric oxide synthase (NOS) | AY672712 | GGCGGTGGTCAGGGTCTT | 59 | 63 |
| | | CCCTTTGGGACACGCTTTT | 59 | |
| | | TGGAAAGAATGGATCTATT | 68 | |
| Glutathione S-transferase (GST) | AB087837 | GAGAATGCCCTTGGTAAATTTGA | 58 | 70 |
| | | ACGCAATATCCACCAAACTGAAT | 58 | |
| | | CCCCTTCCTTCTTGGTC | 69 | |
| Catalase (Cat) | X60169 | CCAAGTGGTCTCACCACAACAAT | 59 | 69 |
| | | TGACCTCCTCATCCCTGTGAA | 59 | |
| | | CCATGAGGGTTTCATG | 69 | |
| Metallothionein (Metalo) | AB176564 | TCCGGCGAAGATCCAGTTT | 59 | 69 |
| | | CCACACTTGCAGCCACCAT | 59 | |
| | | TGGTGCTGAAATGAGTG | 69 | |
| **Cell wall proteins & Basal defense** | | | | |
| Chitinase (Chit) | L37876 | CCTTCAAGACCGCTTTATGGTT | 58 | 64 |
| | | ACGTCGTGGCAGGATGGTT | 60 | |
| | | ACGCCTCAGTCACCT | 68 | |
| Beta-1,3-glucanase (β 1,3-Glu) | S51479 | TGGAATTGGTTGGGTGAATGT | 58 | 65 |
| | | TTGCAGAGCCTCCATCTGAA | 58 | |
| | | TTGTTTCTGAGAGTGGTTG | 68 | |
| Polygalacturonase inhibiting protein (PGIP) | AB087839 | CAGTGCTTTTCGGGGAGCAA | 59 | 66 |
| | | CAAACGACAGCAAGTTCCTTGA | 59 | |
| | | AAAGGACACAGATACTTGAT | 69 | |
| **JA signaling pathways** | | | | |
| Lipoxygenase (LOX) | X17061 | TGATCCGCGGTCTTCAAGAG | 59 | 60 |
| | | CACCGTATTCTGCGGGATCT | 59 | |
| | | TTCCTCCGAAAAGC | 69 | |
| **Phenylpropanoid & Phytoalexin pathway** | | | | |
| Phenylalanine ammonia lyase (PAL) | D10001 | GCACTTAGAACTTCACCGCAATG | 60 | |
| | | GAAAGTTTCCACCATGCAAAGC | 60 | |
| | | CCCTTTGATTGATGTTTC | 69 | |
| Cinnamate-4-hydroxylase (C4H) | U29243 | GCCATAACCGCCATCACAAT | 59 | 61 |
| | | GGGCCAGGAGGGAGTTTGAA | 59 | |
| | | AACTCCGCGGCAAA | 68 | |
| Chalcone isomerase (CHI) | U03433 | GCTGCAGCATCCTCCATCA | 58 | 56 |
| | | CACCGCTGGGAACTCATGT | 58 | |
| | | CGCAATCCACGTCGAG | 67 | |
| Chalcone synthase (CHS) | D10662 | GACATGGTGGTCGTCGAGGTA | 58 | 70 |
| | | GCCCCCATTCTTTTATAGCTTTC | 58 | |
| | | AGACTAGGGAAAGAGGCT | 70 | |
| Isoflavone reductase (CHR) | S72472 | CTTTTGGCGTTGTACCATTCG | 59 | 68 |
| | | TCTTTGGCAGGGTCAATCTCA | 59 | |
| | | AACAAATAAAGGGAGATGCAG | 70 | |
| **Disease resistance response (DRR)** | | | | |
| DRR230 | AJ308155 | TTGCAGGAACAACGAGCACTT | 60 | 61 |
| | | GCACCAGCAGCGAAAATCAT | 60 | |
| | | CTCAGTGGGAGGTGCA | 69 | |
| DRR276 | M18249 | TGCTGACACTCTTACTCCAAAGGT | 58 | 66 |
| | | CCGTTTCCTTCAACAATTTCG | 58 | |
| | | TTGATGCCATCAAAAGTA | 69 | |
| **Disease resistance response protein (DRR)** | | | | |
| DRR49 | X13383 | GGTGATGCTGCTCCTAGTGAAGA | 58 | 66 |
| | | CTTGAAAAGACCATCCCCCTTA | 58 | |
| | | CAACTCAAGACTGACAAAG | 68 | |
| DRR206 | M18250 | GCTGGAGCTGACCCAATTGT | 59 | 68 |
| | | AAGAAATCTCCAGTACCACCAGTGA | 59 | |
| | | CCAAAACTAGAGATATTTCT | 69 | |

[1] The GAPDH (glyceraldehyde-3-phosphate dehydrogenase) genes were used as an internal reference control for equivalent amplification in the PCR.

*Molecules* **2017**, *22*, 1017

### 4.6.2. Real-Time Reverse Transcription PCR (RT-PCR)

The cDNAs were prepared as follows; 1 µg of total RNA was added to 1.5 µg of oligo(dT)$_{15}$–dNTP (2.5 mM each) and made up to a final volume of 11.5 µL with sterile distilled water. RNA was denatured for 5 min at 70 °C and placed on ice, and then 5 µL of Moloney murine leukemia virus (MMLV) 5× reaction buffer, 300 U of MMLV reverse transcriptase, and 80 U of RNase inhibitor were added. First-strand cDNA was synthesized at 25 °C for 15 min, followed by incubation for 50 min at 42 °C and 2 min at 96 °C. Then, gene-specific fragments were amplified by real-time PCR using the defense gene specific primers and probes listed in Table 1, which were designed using the Primer Express 3 software. The TaqMan assays were carried out as mentioned above. All PCR experiments were carried out using an ABI PRISM 7300 sequence detection system (Applied Biosystems, Foster, CA, USA). The *Mtgap1* gene was used as an internal reference control for equivalent reverse transcription to cDNA and equivalent amplification in the PCR. Expression ratio for each cDNA was calculated relatively to corresponding controls, injected with water, using the $2^{-\Delta\Delta Ct}$ method as described by Livak et al., 2001 [60].

### 4.7. Statistical Analyses

Five technical repetitions were used for each experimental condition and four separate experiments were carried out. For all experiments, significant differences were evaluated using the Tukey test at $p \leq 0.05$.

**Acknowledgments:** This work was supported in part by a grant from the Conseil Régional de Picardie.

**Author Contributions:** S.S. and J.S. carried out the experimental work. S.R. and J.C. prepared the OGs. S.S. analyzed the data and wrote the manuscript.

**Conflicts of Interest:** The authors declare no conflict of interest.

### References

1. Cervone, F.; Hahn, M.G.; DeLorenzo, G.; Darvill, A.; Albersheim, P. Host–pathogen interactions: XXXIII. A plant protein converts a fungal pathogenesis factor into an elicitor of plant defense responses. *Plant Physiol.* **1989**, *90*, 542–548. [CrossRef] [PubMed]
2. Albersheim, P.; Darvill, A.; Augur, C.; Cheong, J.J.; Eberhard, S.; Hahn, M.G.; Marfa, V.; Mohnen, D.; O'Neill, M.A.; Spiro, M.D.; et al. Oligosaccharins: Oligosaccharide regulatory molecules. *Acc. Chem. Res.* **1992**, *25*, 77–83. [CrossRef]
3. Côté, F.; Ham, K.S.; Hahn, M.; Bergmann, C.W.; Côté, F.; Ham, K.S.; Hahn, M.G.; Bergmann, C.W. Oligosaccharide elicitors in host-pathogen interactions. Generation, perception and signal transduction. *Subcell. Biochem.* **1998**, *29*, 385–432. [PubMed]
4. Côté, F.; Hahn, M.G. Oligosaccharins: Structures and signal transduction. *Plant Mol. Biol.* **1994**, *26*, 1379–1411. [CrossRef] [PubMed]
5. Braccini, I.; Perez, S. Molecular basis of Ca(2+)-induced gelation in alginates and pectins: The egg-box model revisited. *Biomacromolecules* **2001**, *2*, 1089–1096. [CrossRef] [PubMed]
6. Cabrera, J.C.; Boland, A.; Messiaen, J.; Cambier, P.; Van Cutsem, P. Egg box conformation of oligogalacturonides: The time-dependent stabilization of the elicitor-active conformation increases its biological activity. *Glycobiology* **2008**, *18*, 473–482. [CrossRef] [PubMed]
7. Farmer, E.E.; Ryan, C.A. Inter plant communication: Airborne methyl jasmonate induces expression of protease inhibitor genes in plant leaves. *Proc. Natl. Acad. Sci. USA* **1990**, *87*, 7713–7716. [CrossRef] [PubMed]
8. Moloshok, T.; Pearce, G.; Ryan, C.A. Oligouronide signaling of proteinase inhibitor genes in plants: Structure-activity relationships of di- and trigalacturonic acids and their derivatives. *Arch. Biochem. Biophys.* **1992**, *294*, 731–734. [CrossRef]
9. Moerschbacher, B.M.; Mierau, M.; Graeßner, B.; Noll, U.; Mor, A.J. Small oligomers of galacturonic acid are endogenous suppressors of disease resistance reactions in wheat leaves. *J. Exp. Bot.* **1999**, *50*, 605–612. [CrossRef]

10. Wydra, K.; Beri, H. Structural changes of homogalacturonan, rhamnogalacturonan I and arabinogalactan protein in xylem cell walls of tomato genotypes in reaction to *Ralstonia solanacearum*. *Physiol. Mol. Plant Pathol.* **2006**, *68*, 41–50. [CrossRef]
11. Wydra, K.; Beri, H. Immunohistochemical changes in methyl-ester distribution of homogalacturonan and side chain composition of rhamnogalacturonan I as possible components of basal resistance in tomato inoculated with *Ralstonia solanacearum*. *Physiol. Mol. Plant Pathol.* **2007**, *70*, 13–24. [CrossRef]
12. Diogo, R.; Wydra, K. Silicon-induced basal resistance in tomato against *Ralstonia solanacearum* is related to modification of pectic cell wall polysaccharide structure. *Physiol. Mol. Plant Pathol.* **2007**, *70*, 120–129. [CrossRef]
13. Osorio, S.; Castillejo, C.; Quesada, M.A.; Medina-Escobar, N.; Brownsey, G.J.; Suau, R.; Heredia, A.; Botella, M.A.; Valpuesta, V. Partial demethylation of oligogalacturonides by pectin methyl esterase 1 is required for eliciting defence responses in wild strawberry (*Fragaria vesca*). *Plant J.* **2008**, *54*, 43–55. [CrossRef] [PubMed]
14. Wiethölter, N.; Graessener, B.; Mierau, M.; Mort, A.J.; Moerschbacher, B.M. Differences in the methyl ester distribution of homogalacturonans from near-isogenic wheat lines resistant and susceptible to the wheat stem rust fungus. *Mol. Plant. Microbe Interact.* **2003**, *16*, 945–952. [CrossRef] [PubMed]
15. Randoux, B.; Renard-Merlier, D.; Mulard, G.; Rossard, S.; Duyme, F.; Sanssené, J.; Courtois, J.; Durand, R.; Reignault, P. Distinct defenses induced in wheat against powdery mildew by acetylated and nonacetylated oligogalacturonides. *Phytopathology* **2010**, *100*, 1352–1363. [CrossRef] [PubMed]
16. Boller, T.; Felix, G. A renaissance of elicitors: Perception of microbe-associated molecular patterns and danger signals by pattern-recognition receptors. *Annu. Rev. Plant Biol.* **2009**, *60*, 379–406. [CrossRef] [PubMed]
17. De Lorenzo, G.; Brutus, A.; Savatin, D.; Sicilia, F.; Gervone, F. Engineering plant resistance by constructing chimeric receptors that recognize damage-associated molecular patterns (DAMPs). *FEBS Lett.* **2011**, *585*, 1521–1528. [CrossRef] [PubMed]
18. Ranf, S.; Eschen-Lippold, L.; Pecher, P.; Lee, J.; Scheel, D. Interplay between calcium signalling and early signalling elements during defence responses to microbe- or damage-associated molecular patterns. *Plant J.* **2011**, *68*, 100–113. [CrossRef] [PubMed]
19. Wydra, K.; Beri, H.; Schacht, H. Polygalacturonase-inhibiting protein (PGIP) and structure and composition of cell wall polysaccharides of tomato in relation to resistance to *Ralstonia solanacearum*. In *Emerging Trends in Plant-Microbe Interactions*; Gananamanickam, S., Balasubramanian, R., Anand, N., Eds.; Centre for Advanced Studies in Botany: Chennai, India, 2005; pp. 217–223.
20. Schacht, T.; Unger, C.; Pich, A.; Wydra, K. Endo- and exopolygalacturonases of *Ralstonia solanacearum* are inhibited by polygalacturonase-inhibiting protein (PGIP) activity in tomato stem extracts. *Plant Physiol. Biochem.* **2011**, *49*, 377–387. [CrossRef] [PubMed]
21. Davis, K.R.; Darvill, A.G.; Alber- sheim, P.; Dell, A. Host–pathogen interactions: XXIX. Oligogalacturonides released from sodium polypectate by endopolygalacturonic acid lyase are elicitors of phytoalexins in soybean. *Plant Physiol.* **1986**, *80*, 568–577. [CrossRef] [PubMed]
22. Davis, K.R.; Hahlbrock, K. Induction of defense responses in cultured parsley cells by plant cell wall fragments. *Plant Physiol.* **1987**, 1286–1290. [CrossRef]
23. Broekaert, W.F.; Pneumas, W.J. Pectic polysaccharides elicit chitinase accumulation in tobacco. *Plant Physiol.* **1988**, *74*, 740–744. [CrossRef]
24. Rasul, S.; Dubreuil-Maurizi, C.; Lamotte, O.; Koen, E.; Poinssot, B.; Alcaraz, G.; Wendehenne, D.; Jeandroz, S. Nitric oxide production mediates oligogalacturonide-triggered immunity and resistance to *Botrytis cinerea* in *Arabidopsis thaliana*. *Plant Cell Environ.* **2012**, *35*, 1483–1499. [CrossRef] [PubMed]
25. Cannesan, M.A.; Gangneux, C.; Lanoue, A.; Giron, D.; Laval, K.; Hawes, M.; Driouich, A.; Vicre-Gibouin, M. Association between border cell responses and localized root infection by pathogenic *Aphanomyces euteiches*. *Ann. Bot.* **2011**, *108*, 459–469. [CrossRef] [PubMed]
26. Shang, H.; Grau, C.R.; Peters, R.D. Oospore germination of *Aphanomyces euteiches* in root exudates and on the rhizoplanes of crop plants. *Plant Dis.* **2000**, *84*, 994–998. [CrossRef]
27. Baldauf, S.L.; Roger, A.J.; Wenk-Siefert, J.; Doolittle, W.F. A kingdom level phylogeny of eukaryotes based on combined protein data. *Science* **2000**, *290*, 972–977. [CrossRef] [PubMed]

28. Limberg, G.; Korner, R.; Buchholt, H.C.; Christensen, T.M.; Roepstorff, P.; Mikkelsen, J.D. Analysis of different de-esterification mechanisms for pectin by enzymatic fingerprinting using endopectin lyase and endopolygalacturonase II from *A. niger. Carbohydr. Res.* **2000**, *327*, 293–307. [CrossRef]

29. Ferrari, S. Oligogalacturonides: Plant damage-associated molecular patterns and regulators of growth and development. *Front. Plant Sci.* **2013**, *4*, 49. [CrossRef] [PubMed]

30. Hadwiger, L.A. Pea-*Fusarium solani* interactions contributions of a system toward understanding disease resistance. *Phytopathology* **2008**, *98*, 372–379. [CrossRef] [PubMed]

31. Rennenberg, H. Glutathione metabolism and possible biological roles in the higher plant. *Phytochemistry* **1982**, *21*, 2771–2781. [CrossRef]

32. Smith, I.K.; Polle, A.; Rennemerg, H. Glutathione. In *Stress Responses in Plants: Adaptation and Acclimation Mechanisms*; Alscher, R.G., Cumming, J.R., Eds.; Wiley & Liss: New York, NY, USA, 1990; pp. 201–215.

33. Liu, N.; Zhang, X.; Sun, Y.; Wang, P.; Li, X.; Pei, Y.; Li, F.; Hou, Y. Molecular evidence for the involvement of a polygalacturonase-inhibiting protein, GhPGIP1, in enhanced resistance to Verticillium and Fusarium wilts in cotton. *Sci. Rep.* **2017**, *7*, 39840. [CrossRef] [PubMed]

34. Ghareeb, H.; Bozsó, Z.; Ott, P.G.; Repenning, C.; Stahl, F.; Wydra, K. Transcriptome of silicon-induced resistance against *Ralstonia solanacearum* in the silicon non accumulator tomato implicates priming effect. *Physiol. Mol. Plant Pathol.* **2011**, *75*, 83–89. [CrossRef]

35. Moiseyev, G.P.; Beintema, J.J.; Fedoreyeve, L.I.; Yakovlev, J.I. High sequence similarity between a ribonuclease from Ginseng calluses and fungus-elicited proteins from parsley indicates that intracellular pathogenesis-related proteins are ribonucleases. *Planta* **1994**, *193*, 470–472. [CrossRef] [PubMed]

36. Riggleman, R.D.; Fristensky, B.; Hadwiger, L.A. The disease resistance response in pea is associated with increased levels of specific mRNAs. *Plant Mol. Biol.* **1985**, *4*, 81–86. [CrossRef] [PubMed]

37. Chang, M.M.; Chiang, C.C.; Martin, M.W.; Hadwiger, L.A. Expression of a pea disease resistance response gene in the potato cultivar shepody. *Am. Potato J.* **1993**, *70*, 635–647. [CrossRef]

38. Breiteneder, H.; Pettenburger, K.; Bito, A.; Valenta, R.; Kraft, D.; Rumpold, H.; Scheiner, O.; Breitenbach, M. The gene coding for the major birch pollen allergen Betv 1 is highly homologous to a pea disease resistance response gene. *EMBO J.* **1989**, *8*, 1935–1938. [PubMed]

39. Badreddine, I.; Lafitte, C.; Heux, L.; Skandalis, N.; Spanou, Z.; Martinez, Y.; Esquerre-Tugaye, M.-T.; Bulone, V.; Dumas, B.; Bottin, A. Cell wall chitosaccharides are essential components and exposed patterns of the phytopathogenic oomycete *Aphanomyces euteiches. Eukaryot. Cell* **2008**, *7*, 1980–1993. [CrossRef] [PubMed]

40. Wang, Y.; Nowak, G.; Culley, D.; Hadwiger, L.A.; Fristensky, B. Constitutive expression of pea defense gene DRR206 confers resistance to blackleg (*Leptosphaeria maculans*) disease in transgenic canola (*Brassica napus*). *Mol. Plant. Microbe Interact.* **1999**, *12*, 410–418. [CrossRef]

41. Seneviratne, H.K.; Dalisay, D.S.; Kim, K.-W.; Moinuddin, S.G.A.; Yang, H.; Hartshorn, C.M.; Davin, L.B.; Lewis, N.G. Non-host disease resistance response in pea (*Pisum sativum*) pods: Biochemical function of DRR206 and phytoalexin pathway localization. *Phytochemistry* **2015**, *113*, 140–148. [CrossRef] [PubMed]

42. Chu, H.D.; Le, Q.N.; Nguyen, H.Q.; Le, D.T. Genome-wide analysis of genes encoding methionine-rich proteins in *Arabidopsis* and Soybean suggesting their roles in the adaptation of plants to abiotic stress. *Int. J. Genomics* **2016**, *2016*, 1–8. [CrossRef] [PubMed]

43. Vassão, D.G.; Kim, K.W.; Davin, L.B.; Lewis, N.G. Lignans (neolignans) and allyl/propenyl phenols: Biogenesis, structural biology, and biological/human health considerations. In *Comprehensive Natural Products Chemistry II, Structural Diversity I*; Townsend, C., Ebizuka, Y., Eds.; Elsevier: Oxford, UK, 2010; Volume 1, pp. 815–928.

44. Chiang, C.C.; Hadwiger, L.A. The *Fusarium solani*-induced expression of a pea gene family encoding high cysteine content proteins. *Mol. Plant. Microbe Interact.* **1991**, *4*, 324–331. [CrossRef] [PubMed]

45. Fristensky, B.; Riggleman, C.R.; Wagoner, W.; Hadwiger, L.A. Gene expression in susceptible and disease resistant interactions of peas induced with *Fusarium solani* pathogens and chitosan. *Physiol. Plant Pathol.* **1985**, *27*, 15–28. [CrossRef]

46. Janisiewicz, W.J.; Pereira, I.B.; Almeida, M.S.; Roberts, D.P.; Wisniewski, M.E.; Kurtenbach, E. Biocontrol activity of recombinant *Pichia pastoris* constitutively expressing pea defensin against blue mold of apple. (Abstr.). *Phytopathology* **2006**, *96*, S53.

47. Selitrennikoff, C.P. Antifungal Proteins. *Appl. Environ. Microbiol.* **2001**, *63*, 834–839. [CrossRef] [PubMed]

48. Lay, F.T.; Anderson, M.A. Defensins–Components of the innate immune system in plants. *Curr. Protein Pept. Sci.* **2005**, *5*, 85–101. [CrossRef]

49. Selim, S.; Negrel, J.; Wendehenne, D.; Ochatt, S.; Gianinazzi, S.; van Tuinen, D. Stimulation of defense reactions in *Medicago truncatula* by antagonistic lipopeptides from *Paenibacillus* sp. strain B2. *Appl. Environ. Microbiol.* **2010**, *76*, 7420–7428. [CrossRef] [PubMed]

50. Lamb, C.; Dixon, R.A. The oxidative burst in plant disease resistance. *Annu. Rev. Plant Physiol. Plant Mol. Biol.* **1997**, *48*, 251–275. [CrossRef] [PubMed]

51. Jeandet, P.; Clément, C.; Courot, E.; Cordelier, S. Modulation of phytoalexin biosynthesis in engineered plants for disease resistance. *Int. J. Mol. Sci.* **2013**, *14*, 14136–14170. [CrossRef] [PubMed]

52. Jeandet, P.; Hébrard, C.; Deville, M.A.; Crouzet, J. Deciphering the role of phytoalexins in plant-microorganism interactions and human health. *Molecules* **2014**, *19*, 18033–18056. [CrossRef] [PubMed]

53. Wu, Q.; Van Etten, H.D. Introduction of plant and fungal genes into pea (*Pisum sativum* L.) hairy roots reduces their ability to produce pisatin and affects their response to a fungal pathogen. *Mol. Plant. Microbe Interact.* **2004**, *17*, 798–804. [CrossRef] [PubMed]

54. Maloney, A.P.; Van Etten, H.D. A gene from the fungal pathogen *Nectria haematococca* that encodes the phytoalexin-detoxifying enzyme pisatin demethylase defines a new cytochrome P450 family. *Mol. Genet. Genomics* **1994**, *243*, 506–514. [CrossRef]

55. Delserone, L.M.; McCluskey, K.D.; Mathews, E.; Van Etten, H.D. Pisatin demethylation by fungal pathogens and nonpathogens of pea: Association with pisatin tolerance and virulence. *Physiol. Mol. Plant Pathol.* **1999**, *55*, 317–326. [CrossRef]

56. Mackintosh, S.F.; Matthews, D.E.; Van Etten, H.D. Two additional genes for pisatin demethylation and their relationship to the pathogenicity of *Nectria haematococca* on pea. *Mol. Plant. Microbe Interact.* **1989**, *2*, 354–362. [CrossRef]

57. Courtois, J.; Seguin, J.P.; Roblot, C.; Heyraud, A.; Gey, C.; Dantas, L.; Barbotin, J.N.; Courtois, B. Exopolysaccharide production by the Rhizobium meliloti M5N1CS strain. Location and quantification of the sites of O-acetylation. *Carbohydr. Polym.* **1994**, *25*, 7–12. [CrossRef]

58. Carman, L.M.; Lockwood, J.L. Factors affecting zoospore production by *Aphanomyces euteiches*. *Phytopathology* **1959**, *49*, 535.

59. Vandemark, G.J.; Kraft, J.M.; Larsen, R.C.; Gritsenko, M.A.; Boge, W.L. A PCR-based assay by sequence-characterized DNA markers for the identification and detection of *Aphanomyces euteiches*. *Phytopathology* **2000**, *90*, 1137–1144. [CrossRef] [PubMed]

60. Livak, K.J.; Schmittgen, T.D. Analysis of relative gene expression data using real-time quantitative PCR and the $2^{-\Delta\Delta CT}$ method. *Methods* **2001**, *25*, 402–408. [CrossRef] [PubMed]

**Sample Availability:** Samples of the compounds are available from the authors.

*molecules*

MDPI

*Article*

# Enhanced Stilbene Production and Excretion in *Vitis vinifera* cv Pinot Noir Hairy Root Cultures

Leo-Paul Tisserant [1,2], Aziz Aziz [2], Nathalie Jullian [1], Philippe Jeandet [2,*], Christophe Clément [2], Eric Courot [2] and Michèle Boitel-Conti [1,*]

[1]  Laboratoire de Biologie des Plantes et Innovation EA 3900, SFR Condorcet FR CNRS 3417, UFR des Sciences, Ilot des Poulies, Université de Picardie Jules Verne, 33 rue Saint Leu, 80039 Amiens Cedex, France; leo-paul.tisserant@univ-reims.fr (L.-P.T.); nathalie.pawlicki@u-picardie.fr (N.J.)
[2]  Unité de Recherche Vignes et Vins de Champagne EA 4707, SFR Condorcet FR CNRS 3417, UFR des Sciences Exactes et Naturelles, Université de Reims Champagne-Ardenne, BP 1039, 51687 Reims Cedex 2, France; aziz.aziz@univ-reims.fr (A.A.); christophe.clement@univ-reims.fr (C.C.); eric.courot@univ-reims.fr (E.C.)
*  Correspondences:philippe.jeandet@univ-reims.fr (P.J.); michele.boitel@u-picardie.fr (M.B.-C.); Tel.: +33-3-913-341 (P.J.); +33-3-2282-7956 (M.B.-C.); Fax: +33-3-913-340 (P.J.); +33-3-227-648 (M.B.-C.)

Academic Editor: Derek J. McPhee
Received: 14 October 2016; Accepted: 7 December 2016; Published: 10 December 2016

**Abstract:** Stilbenes are defense molecules produced by grapevine in response to stresses including various elicitors and signal molecules. Together with their prominent role in planta, stilbenes have been the center of much attention in recent decades due to their pharmaceutical properties. With the aim of setting up a cost-effective and high purity production of resveratrol derivatives, hairy root lines were established from *Vitis vinifera* cv Pinot Noir 40024 to study the organ-specific production of various stilbenes. Biomass increase and stilbene production by roots were monitored during flask experiments. Although there was a constitutive production of stilbenes in roots, an induction of stilbene synthesis by methyl jasmonate (MeJA) after 18 days of growth led to further accumulation of ε-viniferin, δ-viniferin, resveratrol and piceid. The use of 100 μM MeJA after 18 days of culture in the presence of methyl-β-cyclodextrins (MCDs) improved production levels, which reached 1034 μg/g fresh weight (FW) in roots and 165 mg/L in the extracellular medium, corresponding to five-and 570-fold increase in comparison to control. Whereas a low level of stilbene excretion was measured in controls, addition of MeJA induced excretion of up to 37% of total stilbenes. The use of MCDs increased the excretion phenomenon even more, reaching up to 98%. Our results demonstrate the ability of grapevine hairy roots to produce various stilbenes. This production was significantly improved in response to elicitation by methyl jasmonate and/or MCDs. This supports the interest of using hairy roots as a potentially valuable system for producing resveratrol derivatives.

**Keywords:** *Vitis vinifera*; hairy roots; resveratrol; viniferins; methyl jasmonate; cyclodextrins

## 1. Introduction

Stilbenes have been the center of much attention over recent decades due to their valuable biological properties. The most studied stilbene, resveratrol, has been described as an important plant defense molecule, with antifungal and antibacterial activities [1–4]. Although stilbenes are found in grapevines at levels considerably lower than flavonoid-type compounds, they possess outstanding pharmaceutical properties [5] as antioxidants [6] as well as anticancer agents [7,8]. Resveratrol derivatives such as viniferins also show an interesting therapeutic potential, mainly in cancer treatment [9,10]. The main natural sources of resveratrol available on the market result from its extraction from Japanese knotweed (*Falliopa japonica*), which produces large quantities of piceid (a β-D-resveratrol glucoside), which is then de-glycosylated to obtain resveratrol [11,12]. Consumers' preference indeed remains in favor of stilbenes from natural sourcing. Few resveratrol derivatives, except the dehydrodimer ε-(+)-viniferin, are available and the current production of these derivatives does not yet meet the pharmaceutical market standards in terms of quantity and purity. Whereas the biosynthetic pathway of resveratrol has been very well characterized [13], those of many resveratrol derivatives are unknown. Currently, the sustainable production of resveratrol derivatives remains difficult to achieve due to their overall low concentrations in plant extracts and problems in purifying them. Alternative strategies including microorganism or plant cell cultures have been tested as valuable systems for stilbene sourcing [14]. Genetically engineered microorganisms or plants for the production of stilbenes would enable a cost-effective supply of these molecules, but the overall low yields and the still incomplete understanding of the biosynthetic pathways of complex stilbenes make their use of limited interest [15–17]. Moreover, there is a considerable interest in searching for sources of stilbenes without recombinant genetic modification. Grapevine cell suspensions, which can produce and excrete resveratrol into their culture medium, have led to promising results [18,19]. In grapevine cell systems in either flasks or bioreactors, resveratrol production can reach up to 7 g/L in response to elicitation treatments with various compounds such as signal molecules (jasmonate or methyl jasmonate) or chelating agents or drug carriers such as cyclodextrins (CDs) (for a review, see [18,19]).

The use of hairy roots offers the advantage of high genetic and biochemical stabilities of organ culture systems coupled with growth rates similar to those of cell suspensions [20,21], without the need for plant growth regulators. In this context, hairy roots of peanut (*Arachis hypogaea*) and Muscadine grape (*Vitis rotundifolia*) have already been developed [22–24]. Both of these plants have previously been described as naturally producing resveratrol and resveratrol derivatives [25]. Although stilbenes have been described in various taxonomically unrelated species, the highest molecular diversity within this chemical group has been found in the *Vitis* spp. genus [26]. Use of the sequenced cultivar *Vitis vinifera* cv Pinot Noir 40024 hairy roots thus appears a good tool for a comprehensive analysis of the organ-specific biosynthesis of stilbenes. Moreover, Pinot Noir has been reported as naturally producing high levels of stilbenes in canes [27] and berries [28].

In this study, we describe the establishment of *Vitis vinifera* cv Pinot Noir PN 40024 hairy root lines from in vitro plantlets. Various culture conditions were tested and the growth kinetics were defined in those yielding the highest growth rate. The ability of *Vitis vinifera* hairy roots to produce and excrete various stilbenes was monitored during growth kinetics experiments in response to methyl jasmonate (MeJA) or CD treatment alone or in combination.

## 2. Results and Discussion

### 2.1. Establishment of Hairy Root Lines (HRs)

As suggested by some previous results [29], the transformation of *Vitis* spp. by *R. rhizogenes* is very dependent on species, on the bacterial strain used as well as the explants tested. In this study, only the sequenced grapevine cultivar PN 40024 [30] was tested, as it represents particular interest for research purpose. Two different bacterial strains were tested to infect various explants from in vitro plantlets. The phytohormone 2,4-D was used to make plant cells more competent, as previously reported in *Arabidopsis thaliana* and *Beta vulgaris* L. [31,32].

Leaves with petioles and stem explants of plantlets of *V. vinifera* PN 40024 transformed using *Rhizobium rhizogenes* ICPBTR7, led to a rapid loss of viability of the explants and thus yielded no roots either by immersion or wounding of the explants or by direct pricking of the plantlets. The plantlets pricked with an infected needle showed no development of roots regardless of the strain used.

Only the Pinot Noir leaves and stem explants transformed with *R. rhizogenes* ATCC15834 by immersion in the bacterial suspension or by wounding with an infected scalpel yielded a particularly large amount of transformed roots. Over the 200 explants tested, several hundreds of roots developed after two weeks, mostly along the petioles and leaf veins.

Once isolated on solid B5 medium supplemented with 3% sucrose, these lines showed a slow growth rate during the first weeks of culture and grew thinner over time until death after two months (Figure 1a). These roots were probably adventitious roots due to the very well-known rooting ability of grapevine and as a result of the addition of 2,4-dichlorophenoxyacetic acid in the bacterial suspension used to infect the explants. A second series of roots appeared on the same Pinot Noir explants four weeks after infection with *R. rhizogenes* ATCC 15834 using the two previous methods. Once they were 1–2 cm long, roots were isolated and cultivated on solid B5 medium supplemented with 3% sucrose (B5-30). These roots showed a completely different phenotype from the first ones, with a diameter ranging from 1 to 3 mm and a higher branching rate (Figure 1a,b). This phenotype showed some similarity to that described on Muscadine grape, with thick roots, very short root hairs, and a higher branching rate (10–15 branches per 5 cm) compared to the 4–5 branches per 5 cm for Muscadine HRs, as reported in a previous work [24]. In the second series of roots that appeared, 80 different lines were isolated and cultivated on B5-30 solid medium. The 60 lines displaying the best growth rates were then transferred to a B5-30 liquid medium (Figure 1c) to be further selected on growth capacity. Nine lines were thus conserved and their transformation was verified by confirmation of the presence of transferred *rolC* gene and the absence of *virD2* bacterial virulence gene (Figure S1). As few differences in terms of phenotype, growth rate and phytochemical profile were observed between these lines (data not shown), the root line 19A was chosen as representative and used further.

**Figure 1.** Observations of the different phenotypes of roots isolated:(**a**) roots from the first series isolated after the second subculture (28 days); (**b**) roots from the second series isolated after the second subculture (21 days of culture); and (**c**) roots from the second series after 40 days growth in 100-mL shaken flasks with 20 mL culture medium. Bars represent 1 cm.

## 2.2. Effects of Culture Medium and Sucrose Concentration on HR Growth

The composition of the medium used is known to have an important impact on the growth capacity and secondary metabolite production of in vitro cultures. The observed slow growth of the selected lines strongly suggested that the medium composition could be further improved.

To search for better culture conditions, four commonly used culture media, Gamborg (B5), McCown (MC), Murashige and Skoog (MS), Schenk and Hildebrandt (SH), and their half-strength dilutions, were compared. In addition, six different sucrose concentrations ranging from 1% to 6% were tested (Figure 2).

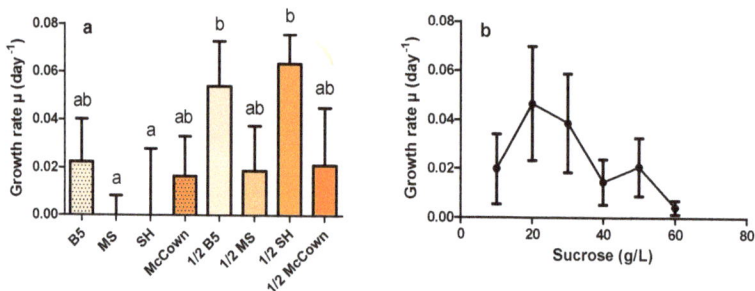

**Figure 2.** Effects of media and sucrose concentration on HR growth rate: (**a**) effect of various standard media and 1/2 dilutions, with 3% sucrose on the growth rate of line 16C; and (**b**) effect of sucrose concentrations in B5 medium on the growth of line 19A. Average growth rate μ and its standard error were calculated from weight measurements made over 35 days with renewal of culture media every three or four days. Each point is composed of three biological replicates. Statistical differences were tested using ANOVA ($\alpha$ = 5%). Letters above dataset point out statistically different groups.

A preference for 1/2 media seems to occur regardless of the medium tested (Figure 2a). Only the B5 and MC media enabled growth when non-diluted. Amongst the 1/2 media, it would seem that the SH medium yielded the best growth rate (μ), reaching μ = 0.063 day$^{-1}$ (doubling time, Td = 10.9 days), along with B5. Determination of the growth rate using different sucrose concentrations in the B5 medium (Figure 2b) showed that the range of 1% to 3% sucrose concentrations in the medium seems to be the most effective, reaching a growth rate of 0.046 day$^{-1}$ (doubling time Td = 14.9 days) for 2% sucrose. As well as showing a low growth rate, the roots grown on medium with 5% and 6% sucrose started browning after three weeks, suggesting significant stress under these conditions. This effect could be caused by an osmotic stress due to high sugar concentrations [33]. Thus, 1/2 SH medium with 2% sucrose was used for further experiments.

*2.3. Growth and Stilbene Production Kinetics*

2.3.1. Biomass Production

The growth kinetics were evaluated using line 19A in the improved growth conditions of 1/2 SH medium with 2% sucrose (Figure 3).

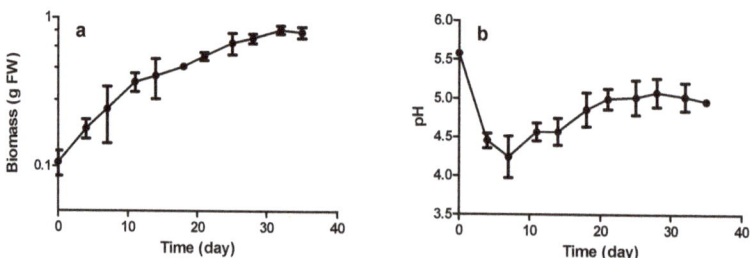

**Figure 3.** Increase in: biomass (fresh weight) (**a**); and pH (**b**) during growth kinetics. Each point represents the mean with standard deviation of three biological replicates.

The semi-log representation of the growth curve (Figure 3a) of grapevine HRs shows no significant lag phase, starting directly with an exponential growth phase up to Day 11, followed by a slowing phase until Day 31 and a stationary phase up to Day 35. The rapid diminution in growth could be due to the relatively low amounts of sucrose in the culture medium. Figure 3b displays a rapid fall in the pH during the first days of culture up to 4–7 days. The pH then rises to 5 at day 20 before stabilizing, in the phases corresponding to both the slower growth and the stationary phases. This suggests that pH can be used as a rapid, non-destructive tool to monitor growth in grapevine HR cultures.

Growth rate was calculated between Day 0 and Day 11, giving a $\mu$ value of 0.11 day$^{-1}$ (Td = 6.2 days). In comparison, peanut and Muscadine HRs have been reported to have a doubling time of 10.2 and 10.7 days, respectively, without any addition of growth regulators [23,24].

### 2.3.2. Constitutive Stilbene Production in Roots

The analysis of stilbenes during the time course experiments shows that a basal production takes place in roots (Figure 4), where they probably play a role as phytoanticipins as already mentioned in natural grapevine roots as well as in the perennial parts of the plant [34].

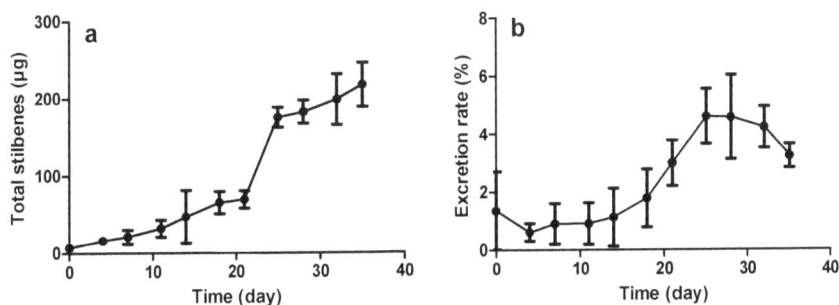

**Figure 4.** Stilbene production during growth kinetics: (**a**) total production of stilbenes (µg of stilbenes per culture flask); and (**b**) excreted stilbenes as a proportion of total stilbenes. Each point represents the mean with standard deviation of three biological replicates.

Here, only the resveratrol glucoside, piceid, resveratrol, and the two dehydrodimers, $\varepsilon$-viniferin and $\delta$-viniferin, were quantified using external standards. Further use of the term "total stilbenes" refers to the sum of all these molecules. A close relationship between the growth phase and the production of secondary metabolites in roots can be observed, especially when the metabolism tends to switch from primary to secondary at the end of the exponential growth phase. As shown in Figure 4a, the basal production of stilbenes per flask rises slowly until the end of the exponential phase at Day 21, before rapidly increasing up to 175 µg at Day 25 and 217 µg after 35 days. It is interesting to note the same trend in the proportion of excreted stilbenes (Figure 4b). The low proportion of 1%–5% excreted stilbenes vs. total stilbenes is consistent with those measured in similar systems without elicitation [23,24].

### 2.4. Induction of Stilbene Production in HRs in Response to Various Elicitors

### 2.4.1. Elicitation with Methyl Jasmonate

Stilbene production, and particularly that of resveratrol, is known to be strongly induced by various environmental stresses, signaling molecules and drug transporters [19,35]. Such compounds have been used in various plant cell systems to mimic a stress and thus induce the production of numerous metabolites of interest [14,36,37]. Amongst them, the jasmonic acid ester, methyl jasmonate (MeJA), has been shown to induce stilbene production mainly in grapevine cell suspensions [38–47].

Figure 5 shows the response in terms of stilbene production in roots and in the culture medium following treatment with 100 μM MeJA up to 10 days after induction. Addition of 0.1% EtOH ($v/v$) (corresponding to the alcoholic content upon addition of MeJA) to the control cultures shows no significant effect of the solvent on root or medium stilbene contents. All stilbene production values in roots were expressed here as micrograms per gram of fresh weight. In order to relate the obtained data with root dry weight, we determined during replicate experiments the ratio dry root weight over fresh root weight which was in the order of 16% ± 7%. As a result, stilbene production in μg/g DW can be obtained by multiplying stilbene production in μg/g fresh weight (FW) by a factor of around 7.

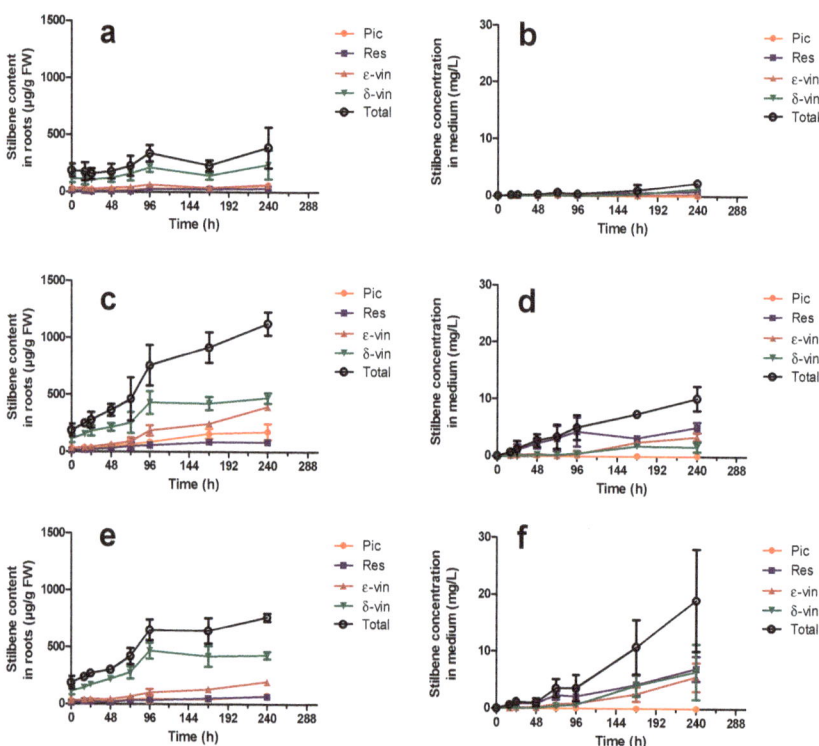

**Figure 5.** Amounts of stilbenes produced after 18 days of culture following MeJA elicitation in: HRs (**a,c,e**); and in the culture medium (**b,d,f**). (**a,b**) EtOH at 0.1% was used as the control; (**c,d**) 100 μM MeJA; and (**e,f**) 200 μM MeJA. Pic: piceid (●); Res: Resveratrol (■); ε-vin: ε-viniferin (▲); δ-vin: δ-viniferin (▼); Total (○) (sum of previous stilbenes). Each point represents the mean with standard deviation of three biological replicates.

Figure 5a displays a basal stilbene production in roots of around 200 μg/g FW, reaching up to 390 μg/g FW 10 days after elicitation. The low stilbene concentrations measured in the extracellular medium, reaching a maximum of 2.3 mg/L at Day 10, are consistent with those previously described [23,24]. However, application of MeJA at doses of 100 μM and 200 μM has a clear inducing effect on the stilbene contents of both the roots and the extracellular medium. Unlike described in peanut hairy roots treated with MeJA, the stilbene concentration in the culture medium did not reach its maximum after 12–48 h and no fall in concentration was observed during the tested period [22]. One hundred micromoles of MeJA (Figure 5c) gives rise to a quick increase in the stilbene root content between two and four days after induction, reaching 757 μg/g FW and followed by a steady increase up to 1126 µg/g FW after 10 days, corresponding to 2.2- and 2.9-fold increases respectively in comparison to control.

In the extracellular medium (Figure 5d), 100 µM MeJA causes a similar effect, reaching stilbene contents of 5 and 10 mg/L after four and 10 days of treatment, corresponding to 12.6- and 4.3-fold increases, respectively, in comparison to control. Application of 200 µM MeJA shows a comparable effect on the total stilbene root content up to Day 4, reaching 650 µg/g FW (Figure 5e), although stabilizing afterwards. The browning of the roots observed upon treatment with 200 µM MeJA suggests a possible effect of this compound on root viability and biomass, as previously reported in *Brassica* hairy root cultures [48]. However, only the treatment with 100 µM MeJA displays statistically significant differences in biomass in comparison to the control 10 days after elicitation. Quantification of the corresponding excreted stilbenes again shows a response similar to the use of 100 µM MeJA until Day 4, reaching 3.5 mg/L. The stilbene concentration then rises to 19 mg/L at Day 10 (Figure 5f), which corresponds to 8.8- and 8.1-fold increases, respectively, in comparison to control. This observation is probably due to a release of intracellular stilbenes into the medium. All the data indicate that four days following MeJA treatment is the best timing, i.e. when induction is at a maximum and to minimize a possible detrimental effect of MeJA on root viability. For further experiments, stilbene measurements were thus carried out four days after MeJA induction. Excretion rates of stilbenes are also significantly affected by MeJA treatments (Figure 6).

While the control excretion rate rises slowly from 1% to 11% of total stilbenes in 10 days, the cultures with MeJA display excretion rates of 31% and 37% for 100 µM and 200 µM MeJA, respectively, after 10 days. No piceid was found in the medium of MeJA-treated roots and less than 10% of δ-viniferin was excreted while almost 30% ε-viniferin and 60% of resveratrol produced were found in the culture medium (Figure S2).

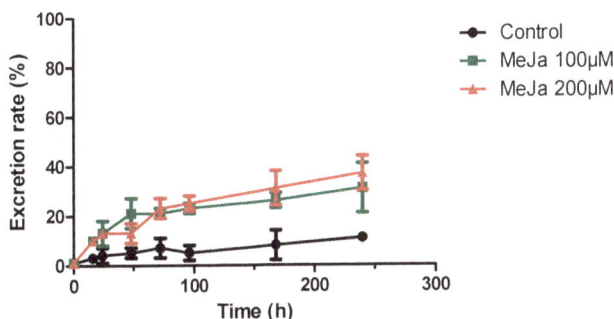

**Figure 6.** Proportion of stilbenes in the culture medium over time in control cultures (●) and cultures treated with 100 µM (■) or 200 µM MeJA (▲). Each point represents the mean with standard deviation of three biological replicates.

2.4.2. Elicitation Experiments with Cyclodextrins and/or Methyl Jasmonate

Cyclodextrins (CDs) and particularly methyl-β-CDs (MCDs) have also been described as strong inducers of stilbene production in grapevine cell suspensions [49]. They have been reported to display a synergistic effect when used in combination with MeJA [35,41,50–53]. However, their effect on the production of stilbenes in grapevine HRs is unknown as yet. The effect of CAVASOL® W-7 MCDs alone or in combination with MeJA was evaluated on both root growth (Figure 7) and stilbene content (Figure 8). UPLC chromatograms are displayed in Figure S3.

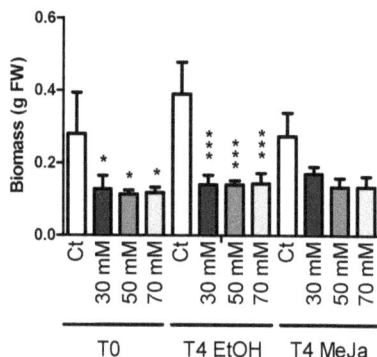

**Figure 7.** Effect of eliciting treatments on the biomass. The tested MCD concentrations were 0, 30, 50 and 70 mM. One hundred micromoles of MeJA and EtOH 0.1% were applied after 18 days of culture and for four days. Each point represents the mean with standard deviation of three biological replicates. Each value was compared to its corresponding control using ANOVA test (* $p < 0.5$, *** $p < 0.001$). Biomasses of different controls were not statistically different.

CD concentrations used were chosen according to previous reports showing high stilbene production levels using grapevine cell suspensions [51,54,55]. Due to the high MCD concentrations applied, these were prepared directly in the culture medium before autoclaving. The HRs were thus cultivated with 0, 30, 50 or 70 mM MCD for 18 days before treatment with MeJA. Measurements were made four days after the application of MeJA. Figure 7 shows a strong effect of MCDs on HR growth at concentrations of 30, 50 and 70 mM, with almost no growth during the first culture phase of 18 days. MeJA treatment inhibited the growth of roots without MCDs but had no additional effect when applied on cultures containing MCDs. Previous results obtained for grapevine cell suspensions treated with 50 mM MCDs described a detrimental effect of the latter alone on cell growth [51]. MeJA treatment is known to have an even stronger inhibitory effect on the growth of grapevine cell suspension cultures [18,19,39,42,56]. Similar observations were made on HR cultures of *Scutellaria lateriflora* [57]. The use of MCDs on HRs has been reported to enhance the production of other phenolic compounds in *S. lateriflora* and to induce the production of stilbenes including resveratrol and derivatives such as prenylated arachidin-1 and arachidin-3 in peanut hairy root cultures [22].

Different elicitors were tested on those cultures, including sodium acetate, $H_2O_2$ as well as MeJA, CD or the combination of both. The best stilbene production in this system was achieved with 100 µM MeJA in combination with 9 g/L MCD, reaching 249 mg/L stilbenes excreted in the culture medium with a majority of arachidin-1 and arachidin-3. The use of MCDs on grapevine HR cultures displays a large induction effect on stilbenes accumulated in roots and excreted into the medium (Figure 8). After 18 days of culture, a clear increase in all quantified stilbenes in roots treated with MCDs was observed: control roots showed intracellular concentrations of 175 µg/g FW. In contrast, MCD-treated roots contained 999, 762 and 580 µg stilbenes/g FW before MeJA treatment for 30, 50 and 70 mM MCDs, respectively. The concentrations of stilbenes measured in the culture medium were 0.1 mg/L for the control and 71, 106, and 98 mg/L for the same CD-treated cultures. These concentrations represent an increase of stilbene production in comparison to control of 5.7, 4.4 and 3.3 folds in the roots and of 711, 1060 and 980 folds in the culture medium (Figure S4).

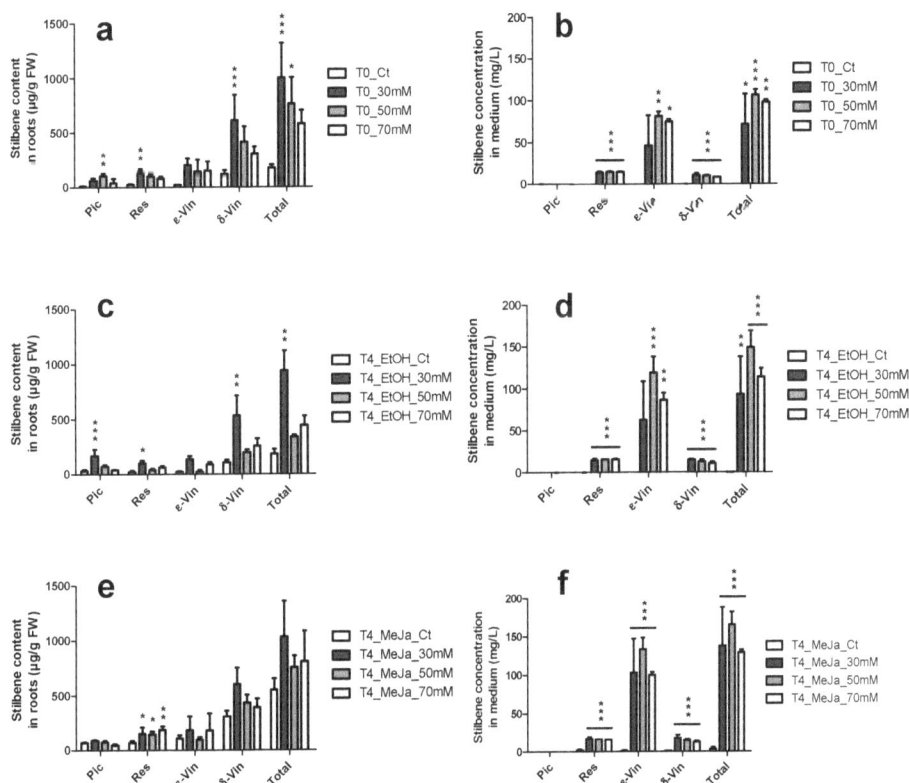

**Figure 8.** Amounts of stilbenes produced in response to treatments with MCDs in the culture media over time. Stilbene measured in HRs and in the culture medium, respectively: (**a,b**) after 18 days of culture and before applying 100 μM MeJA or EtOH; (**c,d**) four days after applying EtOH; and (**e,f**) four days after applying MeJA at 100 μM. Each point represents the mean with standard deviation of three biological replicates. Pic: piceid; Res: resveratrol; ε-vin: ε-viniferin; δ-vin: δ-viniferin. Each value was compared to its corresponding control using ANOVA test (* $p < 0.5$, ** $p < 0.01$, *** $p < 0.001$).

The MeJA treatment was applied after 18 days of culture and its effect was measured four days afterwards. Addition of 100 μM MeJA induced further accumulation of stilbenes in the roots and the culture medium (Figure 8), as previously reported in grapevine cell suspensions [18,19]. Control roots and medium contained 183 μg/g FW and 0.3 mg/L stilbenes, respectively, after four days. A maximum production of stilbenes in roots was observed for roots treated with 30 mM MCDs and 100 μM MeJA, reaching up to 1034 μg/g FW total stilbenes after four days, which represents a 5.6-fold increase in production. In the medium, the highest stilbene concentrations were observed in the cultures treated with 50 mM MCDs and 100 μM MeJA, which yielded up to 165 mg/L total stilbenes after four days, corresponding to an increase of 571 folds in excreted stilbenes. The measured rates of excreted stilbenes show extremely high values of more than 90% of total stilbenes for all three tested MCD concentrations (Figure 9). δ-viniferin was slightly less excreted, with 80% found in the extracellular medium and piceid which remains mostly intracellular (Figure S3b).

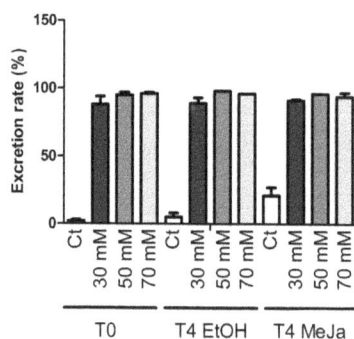

**Figure 9.** Excreted stilbenes as a proportion of total stilbenes in response to MeJA, MCDs and in combination. Each point represents the mean with standard deviation of three biological replicates.

These data support the fact that, besides their well-known eliciting effects on stilbene production, MCDs facilitate the outflow of stilbenes by increasing their solubility in the medium [58]. Although 30 mM MCDs leads to the highest stilbene content in roots, a dose of 50 mM is more suitable due the high quantity of excreted stilbenes obtained. In fact, high yields of excreted stilbenes facilitate their extraction from the cultures. Production of resveratrol reaching up to 5 g/L has been reported using MCDs in grapevine cell suspensions [59]. Although the use of HRs presents unique advantages, this system still needs further optimization in terms of growth conditions to reach similar stilbene concentrations. For example, the biomass reported here with HRs was 8 g FW/L after 18 days of culture with MCDs. In comparison, grapevine cell suspensions can reach up to 500 g FW/L [19]. Various approaches could be used to improve growth and thus biomass production before elicitation to reduce this limitation. A thorough optimization of the culture medium, by identifying limiting nutrients and potential growth factors, could lead to an improvement of growth. In the same way, growth conditions can be modified to optimize gas transfers, by shaking or changing the volume of the cultures. Comparing specific stilbene production in grapevine cell suspensions shows a total stilbene production of 13.4 mg/g FW [53], whereas it reached 18.1 mg/g FW in grapevine HRs, demonstrating that this system could represent a promising tool for stilbene production.

## 3. Materials and Methods

### 3.1. Plant and Bacterial Materials

Six-week-old in vitro plantlets from *Vitis vinifera* cv Pinot Noir 40024 were used for the transformation experiments. They were cultivated in a 16 h/8 h day/night photoperiod on McCown Woody Plant Medium (Dushefa, Haarlem, The Netherlands). All plant cultures were kept at 23 °C.

Two bacterial strains were tested: *Rhizobium rhizogenes* strain ATCC 15834 and *Rhizobium rhizogenes* ICPB TR7. The cultures were initiated in 10 mL Yeast Mannitol Broth YMB medium [60] for 24 h at 28 °C before being pelleted and resuspended in 20 mL in the corresponding plant medium supplemented with 1 mg/L 2,4-dichlorophenoxyacetic acid (2,4-D). 2,4-D was used to induce a short dedifferentiation to facilitate transformation [31,32]. The explants tested consisted of entire leaves with petioles or 1-cm-long sections of the stem. For plant infection, three different methods were compared: (1) wounding explants with a scalpel previously infected with the bacterial suspension; (2) immersing wounded explants in the bacterial suspension for 5 min; and (3) directly pricking 24 in vitro plantlets with an infected needle on 3 spots of the stem.

After infection, explants were transferred onto solid B5 medium [61] and co-cultivated in darkness. After 48 h, the explants were rinsed in B5 + 600 mg/L cefotaxime and cultured on solid B5 medium 3% sucrose supplemented with 300 mg/L cefotaxime. When the first roots appeared on the explants, and measured 1 cm or more, they were isolated and cultured on B5 medium + 300 mg/L cefotaxime. They were grown for 3 months on solid B5 medium + 300 mg/L cefotaxime and subcultured every 3 weeks, then transferred to liquid B5 medium + 300 mg/L cefotaxime. Each line was routinely subcultured every 40 days and the medium was changed every 10 days. The antibiotic pressure was removed 4 months after isolation, once the absence of *Rhizobacteria* was confirmed by PCR.

### 3.2. Confirmation of the Genetic Transformation

The status of genetically transformed roots was verified by PCR, by searching for the *rolC* transgene and the absence of bacterial *virD2*. Primer characteristics are described in Table 1 and sequences were taken from data available in the literature [62,63]. Specific amplification was achieved using Firepol® polymerase (Solis Biodyne) with a thermal cycler (Progene, Thechne): 12 min activation at 95 °C, followed by 35 cycles of 30 s at 95 °C, 30 s at 56 °C and 45 s at 72 °C, and a final 2 min extension time at 72 °C. The bacterial strain was used as a double positive control, a wild type *Escherichia coli* strain was used as a double negative control and *Brassica rapa* hairy roots were used as a transformation-positive control [64].

**Table 1.** Sequences of primers used for confirmation of the genetic transformation [61,63].

| Primer | Direction | Sequence |
|--------|-----------|----------|
| *rolC* | Forward | 5′-ATGGCTGAAGACGACCTGTG-3′ |
|        | Reverse | 5′-TAGCCGATTGCAAACTTGCAC-3′ |
| *virD2* | Forward | 5′-ATGCCCGATCGAGCTCAAG-3′ |
|         | Reverse | 5′-GACCCAAACATCTCGGCTG-3′ |

### 3.3. Improvement of Growth Conditions

The best medium and sucrose concentrations tested were determined in non-limiting conditions. Each culture was started in 100-mL flasks, containing 20 mL medium and one primary root tip of approximately 2 cm. They were cultivated at 110 rpm in darkness for 35 days. Weight measurements and medium renewal, to avoid excessive medium loss over time, were carried out every 3 or 4 days. All tests were done in triplicate.

The tested media were MS [65], B5 [61], MC which is used to cultivate *Vitis vinifera* cv Pinot Noir in vitro plantlets [66] and SH [67] as well as 1/2 MS, 1/2 B5, 1/2 MC and 1/2 SH containing 3% sucrose. The different sucrose concentrations tested were 1%, 2%, 3%, 4%, 5% and 6% in B5 medium.

The growth rate μ was determined from the exponential trend line of the mean growth curve, with biomass equation $X = X_0 \, e^{\,\mu(t-T0)}$. The doubling time Td was calculated using $Td = \frac{\ln(2)}{\mu}$ [68].

### 3.4. Growth Kinetics

The growth kinetic parameters were calculated from cultures in 100-mL flasks, containing 20 mL 1/2 SH medium with 2% sucrose, inoculated with two primary root tips of approximately 2 cm. Three flasks were used for each point to determine the fresh weight. Roots and media were stored at −20 °C for stilbene quantification. The growth rate μ was calculated using the slope of the curve given by mean values of triplicates.

### 3.5. Stilbene Extraction

Stilbenes were extracted from the medium as described in [69,70]. Liquid–liquid extractions using 10 mL ethyl acetate for 10 mL medium were carried out. Ethyl acetate phases were then collected, evaporated in a rotavapor (Laborota 4000-Efficient, Heidolph with PC3001 VARIO vacuum pump, Vacuubrand, Schwabach, Germany) and the extracts were resuspended in 1 mL methanol (MeOH) for UPLC quantification.

Stilbene extraction from roots was adapted from [69,70] using roots ground in liquid nitrogen. One hundred micrograms was weighed and extracted with 1mL 85% MeOH for 2 h in a thermoshaker at 850 rpm, then centrifuged. The supernatant was collected in a glass tube and the pellet was resuspended in 1 mL 100% MeOH for 1 h with a thermomixer (Eppendorf) at 850 rpm, and then centrifuged. The supernatants were pooled, evaporated in a Vacuum Concentrator 5301 (Eppendorf) and resuspended in 1 mL MeOH for UPLC quantification.

*3.6. UPLC Analysis*

UPLC analyses were done using an Acquity UPLC (Waters, Guyancourt, France) system, with a C18 Cortecs® (Waters) column (2.1 × 100 mm, particles: 1.6 µm, pore size: 90 Å), maintained at 30 °C. The separation conditions were taken from [68]. Briefly, a gradient at 0.5 mL/min starting from 90% $H_2O$ +0.1% formic acid (A) and 10% acetonitrile +0.1% formic acid(B) to 40% A and 60% B, followed by 5 min rinsing was used. Stilbene fluorescence was measured with an Aqcuity fluorometer (Waters) with an excitation wavelength of 330 nm and an emission wavelength at 375 nm. Piceid (3-*O*-β-D-resveratrol glucoside), resveratrol (3,5,4′-trihydroxystilbene), δ- and ε-viniferins (resveratrol dehydrodimers), in both their *cis* and *trans* forms, were identified and quantified as described previously and used as external standards [71,72]. Resveratrol, piceid and ε-viniferin were purchased from Sigma at the time of the experiments. Supplementary data concerning stilbene identification are available in [73]. Quantification of piceid and resveratrol was expressed as *trans*-resveratrol equivalents and that of δ and ε-viniferins as *trans*-ε-viniferin equivalents. All extractions were done in subdue light to limit isomerization of *trans* stilbenes to the *cis* forms. The concentrations shown are the sum of *cis* and *trans* forms measured. Total stilbenes represent the sum of these stilbenes.

*3.7. Induction Treatments*

Methyl jasmonate (MeJA) (Sigma Aldrich, St. Quentin Fallavier, France) was added to the culture medium (1/2 SH supplemented with 2% sucrose) from an ethanol (EtOH) solution after the exponential growth phase, corresponding to 18 days of culture. Final concentrations at 100 µM and 200 µM were chosen, with a maximum concentration of 0.1% EtOH in the medium. Controls were obtained by addition of EtOH to a final concentration of 0.1%. All experiments were conducted in triplicate.

For the MeJA and CD experiments, CAVASOL® W-7 MCDs (WackerChemie AG, Burghausen, Germany) were dissolved in the culture medium (1/2 SH supplemented with 2% sucrose) to final concentrations of 30 mM, 50 mM and 70 mM before autoclaving and MeJA was added after 18 days at a final concentration of 100 µM.

# 4. Conclusions

Hairy root systems are particularly attractive as study and bioproduction tools due to their specific properties such as high genetic and phytochemical stability, good growth rates and their highly controlled environment.

In the present work, the growth and phenotype of grapevine cv Pinot Noir HRs has been reported. Moreover, their ability to respond to elicitation treatments that lead to the significant production and excretion of stilbenes into the culture medium has been demonstrated. The results show that grapevine hairy root cultures are able to provide high yields of secreted stilbenes, particularly ε-viniferin, when treated with methyl jasmonate and MCDs. This makes them a good candidate for the production of highly pure stilbenes. The up-scaling of such systems has already been proven feasible [74] and achieved by different companies (10,000 L *Panax* bioreactor) [75].

**Supplementary Materials:** Supplementary materials can be accessed at: http://www.mdpi.com/1420-3049/21/12/1703/s1.

**Acknowledgments:** This research was supported by the SFR Condorcet and funding was provided by the "Région Champagne Ardenne". We thank Carol Robins for English corrections to the manuscript.

**Author Contributions:** Leo-Paul Tisserant conducted experiments and wrote the manuscript;Aziz Aziz helped in performing experiments with stilbene extraction and analysis by UPLC;Nathalie Jullian helped to obtain HR and analyzed the data about the HR cultures;Philippe Jeandet analyzed the data and edited the whole manuscript;Christophe Clément and Eric Courot designed research about *V. vinifera* and elicitation;Michèle Boitel-Conti designed research and analyzed data about HR cultures of *V. vinifera* (HR obtained and characterized stilbenoid production in in vitro cultures). All authors read and approved the manuscript.

**Conflicts of Interest:** The authors declare no conflicts of interest.

## Abbreviations

The following abbreviations are used in this manuscript:

| | |
|---|---|
| B5 | Gamborg medium |
| CD | Cyclodextrin |
| HR | Hairy roots |
| JA | Jasmonate |
| MeJA | Methyl jasmonate |
| MS | Murashige and Skoog medium |
| SH | Schenk and Hildebrandt medium |

## References

1. Adrian, M.; Jeandet, P. Effects of resveratrol on the ultrastructure of *Botrytis cinerea* conidia and biological significance in plant/pathogen interactions. *Fitoterapia* **2012**, *83*, 1345–1350. [PubMed]
2. Adrian, M.; Jeandet, P.; Veneau, J.; Weston, L.A.; Bessis, R. Biological activity of resveratrol a stilbenic compound from grapevines against *Botrytis* cinerea, the causal agent for gray mold. *J. Chem. Ecol.* **1997**, *23*, 1689–1702. [CrossRef]
3. Chalal, M.; Klinguer, A.; Echairi, A.; Meunier, P.; Vervandier-Fasseur, D.; Adrian, M. Antimicrobial activity of resveratrol analogues. *Molecules* **2014**, *19*, 7679–7688. [CrossRef] [PubMed]
4. Paulo, L.; Oleastro, M.; Eugenia Gallardo, E.; Queiroz, J.A.; Domingues, F. Antimicrobial properties of resveratrol: A review. In *Science against Microbial Pathogens: Communicating Current Research and Technological Advances*; Méndez-Vilas, A., Ed.; Research Center: Badajoz, Spain, 2015; pp. 1225–1237.
5. Pezzuto, J.M. The phenomenon of resveratrol: Redefining the virtues of promiscuity. *Ann. N. Y. Acad. Sci.* **2001**, *1215*, 123–130. [CrossRef]
6. Santos, J.A.; de Carvaho, G.S.G.; Oliveira, V.; Raposo, N.R.B.; da Silva, A.D. Resveratrol and analogues: A review of antioxidant activity and applications to human health. *Recent Pat. Food Nutr. Agric.* **2013**, *5*, 144–153. [CrossRef] [PubMed]
7. Aluyen, J.K.; Ton, Q.N.; Tran, T.; Yang, A.E.; Gottlieb, H.B.; Bellanger, R.A. Resveratrol: Potential as anticancer agent. *J. Diet. Suppl.* **2012**, *9*, 45–56. [PubMed]
8. Pervaiz, S.; Holme, A.L. Resveratrol: Its biologic targets and functional activity. *Antioxid. Redox Signal.* **2009**, *11*, 2851–2897. [CrossRef] [PubMed]
9. Xue, Y.Q.; Di, J.M.; Luo, Y.; Cheng, K.J.; Wei, X.; Shi, Z.; Xue, Y.Q.; Di, J.M.; Luo, Y.; Cheng, K.J.; et al. Resveratrol oligomers for the prevention and treatment of cancers resveratrol oligomers for the prevention and treatment of cancers. *Oxidative Med. Cell. Longev.* **2014**, *2014*, 765832. [CrossRef]
10. Giovannelli, L.; Innocenti, M.; Santamaria, A.R.; Bigagli, E.; Pasqua, G.; Mulinacci, N. Antitumoural activity of viniferin-enriched extracts from *Vitis vinifera* L. cell cultures. *Nat. Prod. Res.* **2014**, *28*, 2006–2016. [CrossRef] [PubMed]
11. Vastano, B.C.; Chen, Y.; Zhu, N.; Ho, C.T.; Zhou, Z.; Rosen, R.T. Isolation and identification of stilbenes in two varieties of *Polygonum cuspidatum*. *J. Agric. Food Chem.* **2000**, *48*, 253–256. [CrossRef] [PubMed]
12. Waterhouse, A.L.; Lamuela-Raventós, R.M. The occurrence of piceid, a stilbene glucoside in grape berries. *Phytochemistry* **1994**, *37*, 571–573. [CrossRef]
13. Lu, D. Relevant enzymes, genes and regulation mechanisms in biosynthesis pathway of stilbenes. *Open J. Med. Chem.* **2012**, *2*, 15–23. [CrossRef]
14. Donnez, D.; Jeandet, P.; Clément, C.; Courot, E. Bioproduction of resveratrol and stilbene derivatives by plant cells and microorganisms. *Trends Biotechnol.* **2009**, *27*, 706–713. [CrossRef] [PubMed]

15. Delaunois, B.; Cordelier, S.; Conreux, A.; Clément, C.; Jeandet, P. Molecular engineering of resveratrol in plants. *Plant Biotechnol. J.* **2009**, *7*, 2–12. [CrossRef] [PubMed]

16. Jeandet, P.; Delaunois, B.; Aziz, A.; Donnez, D.; Vasserot, Y.; Cordelier, S.; Courot, E. Metabolic engineering of yeast and plants for the production of the biologically active hydroxystilbene resveratrol. *J. Biomed. Biotechnol.* **2012**, *2012*, 579089. [CrossRef] [PubMed]

17. Jeandet, P.; Clément, C.; Courot, E.; Cordelier, S. Modulation of phytoalexin biosynthesis in engineered plants for disease resistance. *Int. J. Mol. Sci.* **2013**, *14*, 14136–14170. [PubMed]

18. Jeandet, P.; Clément, C.; Courot, E. Resveratrol production at large scale using plant cell suspensions. *Eng. Life Sci.* **2014**, *14*, 622–632. [CrossRef]

19. Jeandet, P.; Clément, C.; Tisserant, L.P.; Crouzet, J.; Courot, E. Use of grapevine cell cultures for the production of phytostilbenes of cosmetic interest. *C. R. Chim.* **2016**, *19*, 1062–1070.

20. Georgiev, M.I.; Pavlov, A.I.; Bley, T. Hairy root type plant in vitro systems as sources of bioactive substances. *Appl. Microbiol. Biotechnol.* **2007**, *74*, 1175–1185. [CrossRef] [PubMed]

21. Giri, A.; Narasu, M.L. Transgenic hairy roots: Recent trends and applications. *Biotechnol. Adv.* **2000**, *18*, 1–22. [CrossRef]

22. Yang, T.; Fang, L.; Nopo-Olazabal, C.; Condori, J.; Nopo-Olazabal, L.; Balmaceda, C.; Medina-Bolivar, F. Enhanced production of resveratrol, piceatannol, arachidin-1, and arachidin-3 in hairy root cultures of peanut co-treated with methyl jasmonate and cyclodextrin. *J. Agric. Food Chem.* **2015**, *63*, 3942–3950. [CrossRef] [PubMed]

23. Medina-Bolivar, F.; Condori, J.; Rimando, A.M.; Hubstenberger, J.; Shelton, K.; O'Keefe, S.F.; Bennett, S.; Dolan, M.C. Production and secretion of resveratrol in hairy root cultures of peanut. *Phytochemistry* **2007**, *68*, 1992–2003. [CrossRef] [PubMed]

24. Nopo-Olazabal, C.; Hubstenberger, J.; Nopo-Olazabal, L.; Medina-Bolivar, F. Antioxidant activity of selected stilbenoids and their bioproduction in hairy root cultures of Muscadine grape (*Vitis rotundifolia* Michx.). *J. Agric. Food Chem.* **2013**, *61*, 11744–11758. [CrossRef] [PubMed]

25. Parage, C.; Tavares, R.; Réty, S.; Baltenweck-Guyot, R.; Poutaraud, A.; Renault, L.; Heintz, D.; Lugan, R.; Marais, G.A.B.; Aubourg, S.; et al. Structural, functional and evolutionary analysis of the unusually large stilbene synthase gene family in grapevine. *Plant Physiol.* **2012**, *160*, 1407–1419. [CrossRef] [PubMed]

26. Pawlus, A.D.; Sahli, R.; Bisson, J.; Rivière, C.; Delaunay, J.C.; Richard, T.; Gomès, E.; Bordenave, L.; Waffo-Téguo, P.; Mérillon, J.M. Stilbenoid profiles of canes from *Vitis* and *Muscadinia* species. *J. Agric. Food Chem.* **2013**, *61*, 501–511. [CrossRef] [PubMed]

27. Lambert, C.; Richard, T.; Renouf, E.; Bisson, J.; Waffo-Téguo, P.; Bordenave, L.; Ollat, N.; Mérillon, J.M.; Cluzet, S. Comparative analyses of stilbenoids in canes of major *Vitis vinifera* L. cultivars. *J. Agric. Food Chem.* **2013**, *61*, 11392–11399. [CrossRef] [PubMed]

28. Jeandet, P.; Bessis, R.; Gautheron, B. The production of resveratrol (3,5,4'-trihydroxystilbene) by grape berries in different developmental stages. *Am. J. Enol. Vitic.* **1991**, *42*, 41–46.

29. Jittayasothorn, Y.; Yang, Y.; Chen, S.; Wang, X.; Zhong, Y. Influences of *Agrobacterium rhizogenes* strains, plant genotypes, and tissue types on the induction of transgenic hairy roots in *Vitis* species. *Vitis* **2011**, *50*, 107–114.

30. Jaillon, O.; Aury, J.M.; Noel, B.; Policriti, A.; Clepet, C.; Casagrande, A.; Choisne, N.; Aubourg, S.; Vitulo, N.; Jubin, C.; et al. The grapevine genome sequence suggests ancestral hexaploidization in major angiosperm phyla. *Nature* **2007**, *449*, 463–467. [CrossRef] [PubMed]

31. Sangwan, R.S.; Bourgeois, Y.; Brown, S.; Vasseur, G.; Sangwan-Norreel, B. Characterization of competent cells and early events of *Agrobacterium*-mediated genetic transformation in *Arabidopsis thaliana*. *Planta* **1992**, *188*, 439–456. [CrossRef] [PubMed]

32. Krens, F.A.; Trifonova, A.; Keizer, L.C.P.; Hall, R.D. The effect of exogenously-applied phytohormones on gene transfer efficiency in sugarbeet (*Beta vulgaris* L.). *Plant Sci.* **1996**, *116*, 97–106. [CrossRef]

33. Do, C.B.; Cormier, F. Accumulation of anthocyanins enhanced by a high osmotic potential in grape (*Vitis vinifera* L.) cell suspensions. *Plant Cell Rep.* **1990**, *9*, 143–146. [CrossRef] [PubMed]

34. Wang, W.; Tang, K.; Yang, H.R.; Wen, P.F.; Zhang, P.; Wang, H.L.; Huang, W.D. Distribution of resveratrol and stilbene synthase in young grape plants (*Vitis vinifera* L. cv. Cabernet Sauvignon) and the effect of UV-C on its accumulation. *Plant Physiol. Biochem.* **2010**, *48*, 142–152. [CrossRef] [PubMed]

35. Almagro, L.; Belchí-Navarro, S.; Sabater-Jara, A.B.; Vera-Urbina, J.C.; Selles-Marchart, S.; Bru, R.; Pedreno, M.A. *Handbook of Natural Products*; Ramawat, K.G., Merillon, J.M., Eds.; Springer: Berlin, Germany, 2013; pp. 1683–1713.

36. Namdeo, A.G. Plant cell elicitation for production of secondary metabolites: A review. *Pharmacogn. Rev.* **2007**, *1*, 69–79.

37. Ramachandra Rao, S.; Ravishankar, G.A. Plant cell cultures: Chemical factories of secondary metabolites. *Biotechnol. Adv.* **2002**, *20*, 101–153. [CrossRef]

38. Belhadj, A.; Telef, N.; Saigne, C.; Cluzet, S.; Barrieu, F.; Hamdi, S.; Mérillon, J.M. Effect of methyl jasmonate in combination with carbohydrates on gene expression of PR proteinsstilbene and anthocyanin accumulation in grapevine cell cultures. *Plant Physiol. Biochem.* **2008**, *46*, 493–499. [CrossRef] [PubMed]

39. Donnez, D.; Kim, K.H.; Antoine, S.; Conreux, A.; de Luca, V.; Jeandet, P.; Clément, C.; Courot, E. Bioproduction of resveratrol and viniferins by an elicited grapevine cell culture in a 2 L stirred bioreactor. *Process Biochem.* **2011**, *46*, 1056–1062. [CrossRef]

40. Krisa, S.; Larronde, F.; Budzinski, H.; Decendit, A.; Deffieux, G.; Mérillon, J.M. Stilbene production by *Vitis vinifera* cell suspension cultures: Methyl jasmonate induction and $^{13}$C biolabeling. *J. Nat. Prod.* **1999**, *62*, 1688–1690. [CrossRef]

41. Martinez-Esteso, M.J.; Sellés-Marchart, S.; Vera-Urbina, J.C.; Pedreño, M.A.; Bru-Martinez, R. Changes of defense proteins in the extracellular proteome of grapevine (*Vitis vinifera* cv Gamay) cell cultures in response to elicitors. *J. Proteom.* **2009**, *73*, 331–341. [CrossRef] [PubMed]

42. Santamaria, A.R.; Mulinacci, N.; Valletta, A.; Innocenti, M.; Pasqua, G. Effects of elicitors on the production of resveratrol and viniferins in cell cultures of *Vitis vinifera* L. cv Italia. *J. Agric. Food Chem.* **2011**, *59*, 9094–9101. [CrossRef] [PubMed]

43. Santamaria, A.R.; Innocenti, M.; Mulinacci, N.; Melani, F.; Valletta, A.; Sciandra, I.; Pasqua, G. Enhancement of viniferin production in *Vitis vinifera* L. cv. Alphonse Lavallée cell suspensions by low-energy ultrasound alone and in combination with methyl jasmonate. *J. Agric. Food Chem.* **2012**, *60*, 11135–11142. [CrossRef] [PubMed]

44. Tassoni, A.; Fornalè, S.; Franceschetti, M.; Musiani, F.; Michael, A.J.; Perry, B.; Bagni, N. Jasmonates and Na-orthovanadate promote resveratrol production in *Vitis vinifera* cv. Barbera cell cultures. *New Phytol.* **2005**, *166*, 895–905. [CrossRef] [PubMed]

45. Xu, A.; Zhan, J.C.; Huang, W.D. Effects of ultraviolet C, methyl jasmonate and salicylic acid alone or in combination on stilbene biosynthesis in cell suspension cultures of *Vitis vinifera* L. cv. Cabernet Sauvignon. *Plant Cell Tissue Organ Cult.* **2015**, *122*, 197–211. [CrossRef]

46. Yue, X.; Zhang, W.; Deng, M. Hyper-production of $^{13}$C-labeled trans-resveratrol in *Vitis vinifera* suspension cell culture by elicitation and in situ adsorption. *Biochem. Eng. J.* **2011**, *53*, 292–296. [CrossRef]

47. Santamaria, A.R.; Antonacci, D.; Caruso, G.; Cavaliere, C.; Gubbiotti, R.; Laganà, A.; Valletta, A.; Pasqua, G. Stilbene production in cell cultures of *Vitis vinifera* L. cvs Globe and Michele Palieri elicited by methyl jasmonate. *Nat. Prod. Res.* **2010**, *24*, 1488–1498. [CrossRef] [PubMed]

48. Kastell, A.; Smetanska, I.; Ulrichs, C.; Cai, Z.; Mewis, I. Effects of phytohormones and jasmonic acid on glucosinolate content in hairy root cultures of *Sinapis alba* and *Brassica rapa*. *Appl. Biochem. Biotechnol.* **2013**, *169*, 624–635. [CrossRef] [PubMed]

49. Bru, R.; Sellés, S.; Casado-Vela, J.; Belchí-Navarro, S.; Pedreño, M.A. Modified cyclodextrins are chemically defined glucan inducers of defense responses in grapevine cell cultures. *J. Agric. Food Chem.* **2006**, *54*, 65–71. [CrossRef] [PubMed]

50. Almagro, L.; Belchí-Navarro, S.; Martínez-Márquez, A.; Bru, R.; Pedreño, M.A. Enhanced extracellular production of trans-resveratrol in *Vitis vinifera* suspension cultured cells by using cyclodextrins and coronatine. *Plant Physiol. Biochem.* **2015**, *97*, 361–367. [CrossRef] [PubMed]

51. Belchí-Navarro, S.; Almagro, L.; Lijavetzky, D.; Bru, R.; Pedreño, M.A. Enhanced extracellular production of trans-resveratrol in *Vitis vinifera* suspension cultured cells by using cyclodextrins and methyljasmonate. *Plant Cell Rep.* **2012**, *31*, 81–89. [CrossRef] [PubMed]

52. Lijavetzky, D.; Almagro, L.; Belchi-Navarro, S.; Martinez-Zapater, J.M.; Bru, R.; Pedreno, M.A. Synergistic effect of methyljasmonate and cyclodextrin on stilbene biosynthesis pathway gene expression and resveratrol production in Monastrell grapevine cell cultures. *BMC Res. Notes* **2008**, *1*, 132. [CrossRef] [PubMed]

53. Vera-Urbina, J.C.; Selles-Marchart, S.; Martinez-Esteso, M.J.; Pedreno, M.A.; Bru-Martinez, R. *Resveratrol: Source Production and Health Benefits*; Delmas, D., Ed.; Nova Science Publishers Inc.: New York, NY, USA, 2013; pp. 19–40.

54. Zamboni, A.; Vrhovsek, U.; Kassemeyer, H.H.; Mattivi, F.; Velasco, R. Elicitor-induced resveratrol production in cell cultures of different grape genotypes (*Vitis* spp.). *Vitis* **2006**, *45*, 63–68.

55. Belchí-Navarro, S.; Almagro, L.; Sabater-Jara, A.B.; Fernández-Pérez, F.; Bru, R.; Pedreño, M.A. Induction of trans-resveratrol and extracellular pathogenesis-related proteins in elicited suspension cultured cells of *Vitis vinifera* cv Monastrell. *J. Plant Physiol.* **2013**, *170*, 258–264. [CrossRef] [PubMed]

56. Repka, V.; Fischerová, I.; Šilhárová, K. Methyl jasmonate is a potent elicitor of multiple defense responses in grapevine leaves and cell-suspension cultures. *Biol. Plant.* **2004**, *48*, 273–283. [CrossRef]

57. Marsh, Z.; Yang, T.; Nopo-Olazabal, L.; Wu, S.; Ingle, T.; Joshee, N.; Medina-Bolivar, F. Effect of light, methyl jasmonate and cyclodextrin on production of phenolic compounds in hairy root cultures of *Scutellaria lateriflora*. *Phytochemistry* **2014**, *107*, 50–60. [CrossRef] [PubMed]

58. Silva, F.; Figueiras, A.; Gallardo, E.; Nerín, C.; Domingues, F.C. Strategies to improve the solubility and stability of stilbene antioxidants: A comparative study between cyclodextrins and bile acids. *Food Chem.* **2014**, *145*, 115–125. [CrossRef] [PubMed]

59. Bru-Martinez, R.; Pedreno, M.A. Method for the Production of Resveratrol in Cell Cultures. U.S. Patent 20060205049 A1, 14 September 2006.

60. Hooykaas, P.J.J.; Klapwijk, P.M.; Nuti, M.P.; Schilperoort, R.A.; Rörsch, A. Transfer of the *Agrobacterium tumefaciens* TI plasmid to avirulent *Agrobacteria* and to *Rhizobium ex planta*. *Microbiology* **1977**, *98*, 477–484. [CrossRef]

61. Gamborg, O.L.; Miller, R.A.; Ojima, K. Nutrient requirements of suspension cultures of soybean root cells. *Exp. Cell Res.* **1968**, *50*, 151–158. [CrossRef]

62. Ruslan, K.; Selfitri, A.D.; Bulan, S.A.; Rukayadi, Y.; Elfahmi, T. Effect of *Agrobacterium rhizogenes* and elicitation on the asiaticoside production in cell cultures of *Centella asiatica*. *Pharmacogn. Mag.* **2012**, *8*, 111–115. [PubMed]

63. Yang, D.C.; Choi, Y.E. Production of transgenic plants via *Agrobacterium rhizogenes*-mediated transformation of *Panax ginseng*. *Plant Cell Rep.* **2000**, *19*, 491–496. [CrossRef]

64. Huet, Y.; Ekouna, J.P.E.; Caron, A.; Mezreb, K.; Boitel-Conti, M.; Guerineau, F. Production and secretion of a heterologous protein by turnip hairy roots with superiority over tobacco hairy roots. *Biotechnol. Lett.* **2013**, *36*, 181–190. [CrossRef] [PubMed]

65. Murashige, T.; Skoog, F. A revised medium for rapid growth and bio assays with tobacco tissue cultures. *Physiol. Plant.* **1962**, *15*, 473–497. [CrossRef]

66. Lloyd, G.; McCown, B. Commercially-feasible micropropagation of mountain laurel *Kalmia latifolia* by use of shoot-tip culture. *Comb. Proc. Int. Plant Propag. Soc. USA* **1980**, *30*, 421–426.

67. Schenk, R.U.; Hildebrandt, A.C. Medium and techniques for induction and growth of monocotyledonous and dicotyledonous plant cell cultures. *Can. J. Bot.* **1972**, *50*, 199–204. [CrossRef]

68. Mairet, F.; Sierra, J.; Glorian, V.; Villon, P.; Shakourzadeh, K.; Boitel-Conti, M. A new approach to define optimized range of medium composition for enhancement of hairy root production in fed-batch process. *Bioprocess Biosyst. Eng.* **2008**, *3*, 2257–2265. [CrossRef] [PubMed]

69. Gruau, C.; Trotel-Aziz, P.; Villaume, S.; Rabenoelina, F.; Clément, C.; Baillieul, F.; Aziz, A. *Pseudomonas fluorescens* PTA-CT2 triggers local and systemic immune response against *Botrytis cinerea* in grapevine. *Mol. Plant-Microbe Interact.* **2015**, *28*, 1117–1129. [CrossRef] [PubMed]

70. Hatmi, S.; Trotel-Aziz, P.; Villaume, S.; Couderchet, M.; Clément, C.; Aziz, A. Osmotic stress-induced polyamine oxidation mediates defence responses and reduces stress-enhanced grapevine susceptibility to *Botrytis cinerea*. *J. Exp. Bot.* **2014**, *65*, 75–88. [CrossRef] [PubMed]

71. Aziz, A.; Trotel-Aziz, P.; Dhuicq, L.; Jeandet, P.; Couderchet, M.; Vernet, G. Chitosan oligomers and copper sulfate induce grapevine defense reactions and resistance to gray mold and downy mildew. *Phytopathology* **2006**, *96*, 1188–1194. [CrossRef] [PubMed]

72. Verhagen, B.; Trotel-Aziz, P.; Jeandet, P.; Baillieul, F.; Aziz, A. Improved resistance against *Botrytis cinerea* by grapevine-associated bacteria that induce a prime oxidative burst and phytoalexin production. *Phytopathology* **2011**, *101*, 768–777. [CrossRef] [PubMed]

73. Tisserant, L.P.; Hubert, J.; Lequart, M.; Borie, N.; Maurin, N.; Pilard, S.; Jeandet, P.; Aziz, A.; Renault, J.H.; Nuzilllard, J.M.; et al. $^{13}$C RNM and LC-MS Chemical profiling of major stilbenes of pharmaceutical significance produced by elicited *Vitis vinifera* cv. Pinot noir hairy root cultures. *J. Nat. Prod.* **2016**, *79*, 2846–2855. [CrossRef]

74. Georgiev, M.I.; Agostini, E.; Ludwig-Müller, J.; Xu, J. Genetically transformed roots: From plant disease to biotechnological resource. *Trends Biotechnol.* **2012**, *30*, 528–537. [CrossRef] [PubMed]

75. Sivakumar, G.; Yu, K.W.; Paek, K.Y. Production of biomass and ginsenosides from adventitious roots of *Panax ginseng* in bioreactor cultures. *Eng. Life Sci.* **2005**, *5*, 333–342. [CrossRef]

**Sample Availability:** Not available.

*Article*

# A Focused Multiple Reaction Monitoring (MRM) Quantitative Method for Bioactive Grapevine Stilbenes by Ultra-High-Performance Liquid Chromatography Coupled to Triple-Quadrupole Mass Spectrometry (UHPLC-QqQ)

**Elías Hurtado-Gaitán [1], Susana Sellés-Marchart [1,2], Ascensión Martínez-Márquez [1], Antonio Samper-Herrero [1] and Roque Bru-Martínez [1,3,]***

[1]   Departamento Agroquímica y Bioquímica, Facultad de Ciencias, Universidad de Alicante, 03690 Alicante, Spain; ehurtadog7@gmail.com (E.H.-G.); susana.selles@ua.es (S.S.-M.); asun.martinez@ua.es (A.M.-M.); antonio.samper@ua.es (A.S.-H.)
[2]   Genomics and Proteomics Unit, SSTTI Universidad de Alicante, 03690 Alicante, Spain
[3]   Instituto de Investigación Sanitaria y Biomédica de Alicante ISABIAL-Fundación para el Fomento de la Investigación Sanitaria y Biomédica de la Comunitat Valenciana FISABIO, 03010 Alicante, Spain
*   Correspondence: roque.bru@ua.es; Tel.: +34-96-590-3400

Academic Editor: Philippe Jeandet
Received: 23 January 2017; Accepted: 1 March 2017; Published: 7 March 2017

**Abstract:** Grapevine stilbenes are a family of polyphenols which derive from *trans*-resveratrol having antifungal and antimicrobial properties, thus being considered as phytoalexins. In addition to their diverse bioactive properties in animal models, they highlight a strong potential in human health maintenance and promotion. Due to this relevance, highly-specific qualitative and quantitative methods of analysis are necessary to accurately analyze stilbenes in different matrices derived from grapevine. Here, we developed a rapid, sensitive, and specific analysis method using ultra-high-performance liquid chromatography coupled to triple-quadrupole mass spectrometry (UHPLC-QqQ) in MRM mode to detect and quantify five grapevine stilbenes, *trans*-resveratrol, *trans*-piceid, *trans*-piceatannol, *trans*-pterostilbene, and *trans*-ε-viniferin, whose interest in relation to human health is continuously growing. The method was optimized to minimize in-source fragmentation of piceid and to avoid co-elution of *cis*-piceid and *trans*-resveratrol, as both are detected with resveratrol transitions. The applicability of the developed method of stilbene analysis was tested successfully in different complex matrices including cellular extracts of *Vitis vinifera* cell cultures, reaction media of biotransformation assays, and red wine.

**Keywords:** UHPLC; MRM; resveratrol; piceid; piceatannol; epsilon-viniferin; pterostilbene; cell culture; bioconversion; wine

---

## 1. Introduction

Grapevine (*Vitis vinifera* L.) stilbenes are a family of polyphenols which derive from *trans*-Resveratrol (*trans*-R: 3,5,4′-trihydroxystilbene), found in different plant organs, including berries and, consequently, also in wines. *Trans*-R is synthesized as an end-product of a branch of the phenylpropanoid pathway; further *trans*-R may undergo biochemical modifications, such as hydroxylation, glycosylation, methylation, or oligomerization, giving rise to several bioactive derivatives (Figure 1). The synthesis and accumulation of stilbenes is considered to be part of the general defense mechanism as these are considered as the phytoalexins of *Vitis* species since they display strong antifungal and antimicrobial activities [1–3]. In fact, *trans*-R is accumulated in grapevine vegetative tissues and berries, as well as in cell cultures in response to abiotic and biotic stress [1,4–9]. Stilbenes have been demonstrated to have a large number of bioactive properties in relation to major human diseases such as cancer, atherosclerosis, or neurodegeneration [10]. Many studies have reported the potential of *trans*-R in human health as an antioxidant in model systems of cardiovascular diseases [11], as a cytotoxic agent for several cancer lines [12–15], as an anti-inflammatory, anti-aging, and antioxidant agent [16,17], having a beneficial role on neurological system [18], as well as health maintenance and survival in adverse scenarios related with obesity and aging [19,20]. Moreover, other studies have demonstrated that some *trans*-R polyhydroxy and polymethoxy derivatives also exhibit enhanced pharmacological activity and bioavailability [21]. Such potential of *trans*-R and other stilbenes have called the attention of numerous companies producing a large variety of dietary supplements and cosmetics. Stilbenes are presently in natural foods, such as grapes, berries of *Vacinum* species, cocoa, and peanuts, and their derived products [22–26]. In this sense, stilbenes, especially resveratrol, play a key role in wine health-promoting properties (see [27] for a review). Once the potential health benefits of stilbenes have been demonstrated, it is necessary to have highly specific qualitative and quantitative methods to accurately analyze stilbenes in different matrices derived from grapevine.

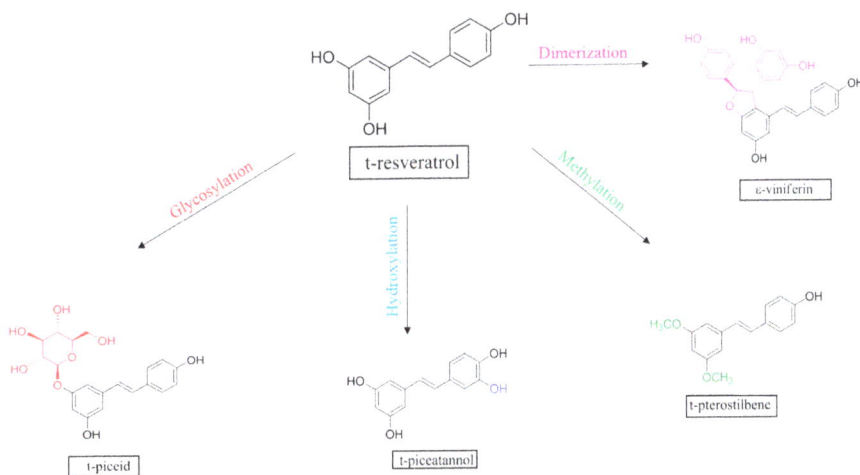

**Figure 1.** Grapevine stilbenes: t-Resveratrol and derivatives produced by different biochemical reactions.

Stilbenes have been detected by different analytical techniques coupled to chromatographic separation. In wine and grape samples, several research groups report the separation and detection of phenolic compounds using HPLC coupled to detection by either UV-VIS and fluorescence [28], or MS and MS/MS [29–33]. However, the specificity of detection of these techniques is limited because different compounds, having similar mass or spectroscopic properties may co-elute in a chromatographic run. The incorporation of triple quadrupole mass spectrometers (HPLC-QqQ) to analyze phenolic compounds significantly improved the detection and allowed the simultaneous quantitation of multiple compounds in complex matrices in a reduced time. Jaitz et al. quantified 11 polyphenolic compounds from red wines in less than 13 min [34]. In other plant matrices, HPLC-QqQ MS was applied to quantify 28 polyphenols in cocoa extracts [35] and, more recently, 20 polyphenols from carob (*Ceratonia siliqua* L.) flour by improving the chromatography separation and speed by using ultra-high-performance liquid chromatography (UHPLC) [36]. UHPLC allows faster separation and enhanced chromatographic resolution as compared to classical HPLC, due to reduced particle size and increased column pressure (up 1000 bar). UHPLC coupled to triple quadrupole mass spectrometry using multiple reaction monitoring (MRM) mode has been proposed as an alternative for fast analysis of phenolic compounds in grapes and wine. UHPLC-QqQ in MRM mode has been used to detect and quantify 152 polyphenols in rose wine, in 30 min [37], which is quite a noticeable improvement for the analysis of general wine phenolics. The large increase in the number of compounds analyzed is achieved at expense of longer chromatographic runs and possibly, some loss of specificity. and/or sensitivity when different co-eluting compounds share transitions or are present at dramatically different abundances.

Thus, when focusing on particular compound families (i.e., stilbenes), improved methods that provide effective compound separation and quantitation in different kinds of matrices in short chromatographic runs are necessary.

Our study aimed at developing an analytical method for efficient separation and accurate quantification of stilbenes produced by *Vitis vinifera*, of application to different matrices, such as red wine and plant material extracts, among others. Five important bioactive stilbenes were chosen for analysis by UHPLC-QqQ MS operated in MRM detection mode, in order to improve the detection and quantitation of stilbenes in a short time of analysis.

## 2. Results

### 2.1. Optimization of Ion Source Parameters

To carry out this work we utilized an Agilent 6490 QQQ instrument equipped with a JetStream® ion source. As this instrument uses a fixed 380 V fragmentor voltage this parameter cannot be optimized. Glycosylated polyphenols, such as piceid, are prone to in-source fragmentation, thus, it was necessary to optimize some parameters of the ion source to minimize this effect. To that, a standard solution of piceid was directly infused into the instrument and several parameters were tested to maximize the signal intensity. The cell acceleration voltage, tested at 2 and 5 V, was set at 2 V for subsequent analysis. Next, the gas curtain temperature was tested at five values in the range of 250 °C to 400 °C, with 300 °C being selected as a compromise between signal intensity and compound in-source fragmentation. The gas flow had almost no effect on fragmentation, but a clear improvement in the signal intensity was observed at 12 L/min. Eventually, we tested several capillary voltages in a range of 2500–4500 V, getting the best result at 3000 V. As the other stilbene compounds do not undergo in-source fragmentation, the source parameters optimized for piceid were used throughout in subsequent analysis.

### 2.2. Selection of MRM Transitions and Optimization of Chromatographic Separation

Five stilbenes found in *Vitis* species, namely t-Piceid, t-Piceatannol, t-R, t-ε-Viniferin and t-Pterostilbene were used in this study. MRM conditions were first optimized in positive detection mode. Each compound was individually infused and, while operating the MS instrument in product ion scan mode, four discrete collision energies were applied (0, 10, 20, and 30 eV) in order to detect fragment ions and automatically select the four most intense ones. Then, the collision energy for each transition was optimized by applying a 0–40 eV ramp. The transitions selected for each compound, optimal collision energies and intensities are listed in Table 1. Since the transitions selected for piceid gave rise to non-specific peaks in the chromatogram, the fragment ions were changed for those of *trans*-R, thus producing clean chromatographic traces. Only the product ion $m/z$ 229.1 was kept as being the most abundant transition for piceid. Due to in-source fragmentation, piceid generates resveratrol parent ions that are detected through the resveratrol transitions; thus, peak overlapping of resveratrol and piceid must be avoided.

**Table 1.** MRM parameters, chromatographic attributes, and quantitative response of stilbene compounds in standard samples.

| Compound | Formula | Mass (Da) | Precursor Ion ($m/z$) | Product Ion ($m/z$) | Collision Energy (eV) | % Intensity | Retention Time (Min–Max) *trans/cis* | LOD (mg/L) | LOQ (mg/L) | Linearity [b] |
|---|---|---|---|---|---|---|---|---|---|---|
| **Piceid** | $C_{20}H_{22}O_8$ | 390.13 | 391.1 | 114.8 308.8 349.9 229.1 [a] | 20 8 0 8 | 86.5 67.5 100 24.4 | 0.93–0.99/1.52–1.63 | 0.04 | 0.07 | 640 |
| **Resveratrol** | $C_{14}H_{12}O_3$ | 228.08 | 229.09 | 107.1 [a] 91 [a] 135 [a] 165 [a] | 24 24 8 28 | 100 45.7 80.9 35.6 | 1.85–1.99/2.55–2.65 | 0.12 | 0.22 | 800 |
| **Piceatannol** | $C_{14}H_{12}O_4$ | 244.07 | 245.08 | 107.1 135.1 152 181.1 | 20 12 36 24 | 95.5 100 11.6 38.7 | 1.25–1.35/1.90–2.00 | 0.12 | 0.15 | 1600 |
| **ε-Viniferin** | $C_{28}H_{22}O_6$ | 454.14 | 455.15 | 107.1 215.1 349.1 199.1 | 32 20 16 24 | 100 75.0 38.5 32.0 | 3.00–3.17/2.80–2.90 | 0.07 | 0.09 | 1777 |
| **Pteroestilbene** | $C_{16}H_{16}O_3$ | 256,11 | 257.12 | 181 133.1 91 165.1 | 40 12 28 40 | 90.1 100 58.5 72.2 | 5.18–5.28/5.26–5.35 | 0.06 | 0.08 | 1000 |

[a] Transitions used for piceid analysis. [b] Ratio between the highest and lowest concentrations in linear range.

To optimize the separation conditions of these five stilbenes in a short chromatographic run, a mix of standard was injected in a UHPLC-QQQ from Agilent (Santa Clara, CA, USA). Several trials were conducted with different gradients in the mobile phase, containing water as solvent A and acetonitrile or acetonitrile:methanol as solvent B. Using several binary gradients starting at 5% acetonitrile resulted in a poor resolution and late elution of peaks and, in addition, peaks corresponding to *cis*-Piceid and *trans*-R appeared overlapping. To improve the poor separation observed, ternary gradients of water:acetonitrile:methanol were assayed starting at 35% B for an earlier peak elution. A 10 min gradient was tested using from 40:60 up to 70:30 acetonitrile:methanol as solvent B, thus achieving a significant improvement in the chromatographic separation, 70:30 being the proportion that showed the best separation of all the stilbenes. A key improvement of this chromatographic optimization has been the *cis*-Piceid peak neat separation from the *trans*-R peak, otherwise *trans*-R quantification would be overestimated as the sum of the signal of both compounds. As can be seen in Figure 2, the gradient was finally shortened to 6 min (the total run after equilibration was 8 min) without compromising resolution. Minor peaks, supposed to be the respective *cis*-isomers as judged by the co-elution of all compound-specific transitions, were also detected. As mainly the *trans* isomers are commercialized as standards, we exposed the standard mixture to UV light to promote photoisomerization in order to characterize the peaks corresponding to the *cis* isomers. Under the optimized chromatographic conditions, all stilbenes both, as *trans* and *cis* isomers were detectable and eluted clearly differentiated (Figure 2). The only co-eluting compounds were *trans*-R and *cis*-piceatannol, but such overlapping is unambiguously resolved by their specific transitions.

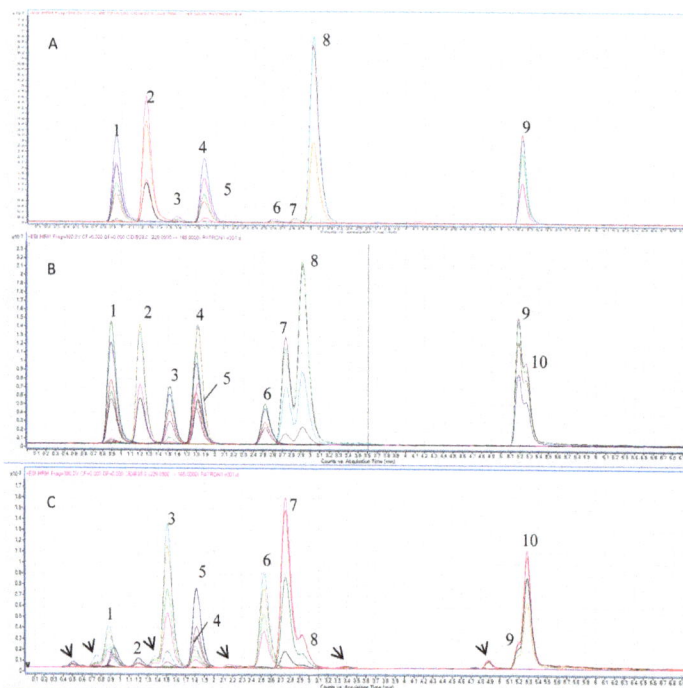

**Figure 2.** Chromatogram of MRM transitions of stilbenes in the standard mix. 1: *trans*-Piced; 2: *trans*-piceatannol; 3: *cis*-Piceid; 4: *trans*-Resveratrol; 5: *cis*-Piceatannol; 6: *cis*-Resveratrol; 7: *cis*-Viniferin; 8: *trans*-Viniferin; 9: *trans*-Pterostilbene; 10: *cis*-Pterostilbene. (**A**): Without UV light exposition; (**B**) short exposition to UV light; and (**C**) prolonged exposition to UV light. Arrows indicate the presence of new compounds originated by the prolonged UV exposition.

## 2.3. MRM Quantitation of Stilbenes in a Standard Mixture

After MRM has been successfully optimized to detect target stilbenes produced by *V. vinifera*, performance of the method to quantify these compounds was assessed. For quantitative MRM, the quantifier transition chromatogram of each targeted precursor $m/z$ was used, and the rest of transitions provided the evidence of the presence or absence of the stilbene in the standard mixture.

The quantifier transitions used for each compound were $m/z$ 391.1→229.1 for piceid, $m/z$ 229.09→107.1 for resveratrol, $m/z$ 245.08→135.1 for piceatannol, $m/z$ 455.15→107.1 for viniferin, and $m/z$ 257.12→133.1 for pterostilbene. Detailed data of metabolite fragmentation is provided as Supplementary Material in file S1. The optimized transition for each specific compound constitute a robust assay that can be used to quantify the target stilbenes in complex mixtures by MRM analysis. The estimated concentration of standard stilbene solutions were calculated, results are shown in Supplementary Material file S1.

In order to explore the linear dynamic range of the instrument's response and the limits of detection (LOD) and quantification (LOQ) for each stilbene, serial dilutions of the standard mixture covering several orders of magnitude were analyzed with the MRM method and results are summarized in Table 1 and Tables S1–S5 in Supplementary Material, file S2. Linear relationship between peak area and concentration was calculated for all the stilbenes, considering only the peaks corresponding to *trans* isomers. At high concentrations of stilbenoids a loss of linearity was apparent, holding linearity up to 40 mg/L for piceid and pterostilbene, 50 mg/L for Resveratrol and 160 mg/L for piceatannol and ε-viniferin. The response of the detector was linear over ca. three orders of magnitude on average for the set of stilbenes analyzed. To determine LOD and LOQ the signal-to-noise ratio was used establishing rather conservative cut-off values. The LOD ranged from 0.04 to 0.12 mg/L and the LOQ from 0.07 to 0.22 mg/L.

## 2.4. Examples of Applications of the Optimized Method

In order to validate the applicability of the MRM method developed we carried out stilbene analysis in different experimental scenarios. On the one hand, we applied the method to the analysis of stilbenes extracted from grapevine cell cultures. *V. vinifera* cell cultures might produce some of these five stilbenes incubated either in normal growth medium or under elicitation conditions [38]. To that, several samples of methanolic extracts of elicited cell cultures of *V. vinifera* were used for the applications of the method in complex mixtures. Use of external calibration curves with three levels of standards enable the detection and quantification of all stilbenes assayed with the exception of pterostilbene, which was absent in this sample. The quantifier transition chromatogram for each stilbene detected is shown in Figure 3 and the calculated concentrations are in Table 2. As expected, *trans*-R was the most abundant stilbene produced by *V. vinifera* under elicitation, but others including piceid, piceatannol, and ε-viniferin were also present at significant levels.

**Table 2.** Quantitation of stilbenes in *V. vinifera* cell culture extracts upon elicitation.

| Compound | Concentration (mg/L) in | | |
| --- | --- | --- | --- |
| | Extract #1 | Extract #2 | Extract #3 |
| Piceid | 10.92 | 14.07 | 16.35 |
| Resveratrol | 46.93 | 62.24 | 50.82 |
| Piceatannol | 1.44 | 1.46 | 1.38 |
| ε-Viniferin | 1.48 | 0.93 | 0.11 |
| Pteroestilbene | n.d. | n.d. | n.d. |

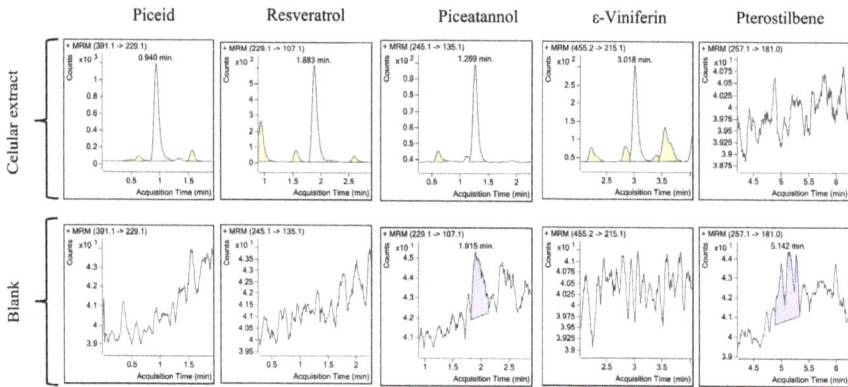

**Figure 3.** Quantitation of stilbenes from *V. vinifera* cell cultures upon elicitation. Transition chromatograms for each compound are shown. Blank is methanol, 80%.

On the other hand, the method was applied to follow the enzymatic assay of bioconversion of *trans*-R through an hydroxylation reaction into a tetrahydroxystilbene [39]. A solution of *trans*-R standard and NADPH substrates was incubated in the presence of a protein crude extract from grapevine cell cultures. Stilbene reaction products after one hour incubation, as well as residual *trans*-R substrate, was extracted and analyzed with the developed method. As shown in Figure 4A, *trans*-R peak decreased in abundance at the time a new peak at the same retention time as *trans*-Piceatannol appeared, thus confirming that grapevine cell extracts have resveratrol hydroxylation activity.

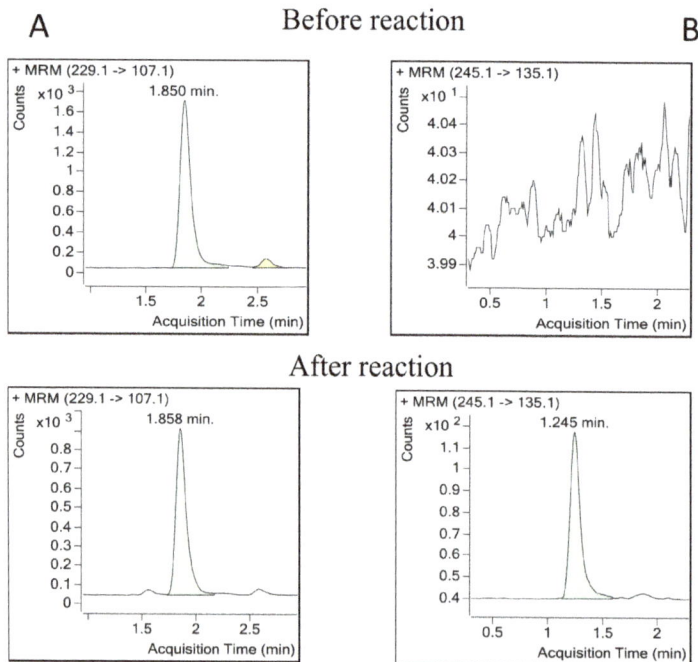

**Figure 4.** Bioconversion of *trans*-Resveratrol to *trans*-Piceatannol. (**A**) Resveratrol transition chromatogram; and (**B**) piceatannol transition chromatogram.

Eventually, red wine samples were analyzed using the MRM method in order to detect and quantify the absolute amount of the five targeted stilbenes in this highly complex matrix. To reduce interferences and increase sensitivity of the assay, stilbenes from both spiked and non-spiked samples were recovered by solid phase extraction [40]. However, the resulting sample is still highly complex as evidenced by an HPLC-UV analysis (Figure S2) in conditions described by Martinez-Esteso et al. [5]. As shown in Figure 5, out of the five stilbenes analyzed, piceid, resveratrol and ε-viniferin were present, but only their respective *trans* isomers were quantified (Table 3). In addition, their *cis* isomers as well as other viniferins were also present and, presumably, in higher amounts as judged by the intensity of their corresponding transition peaks, but the lack of standards precluded their absolute quantification.

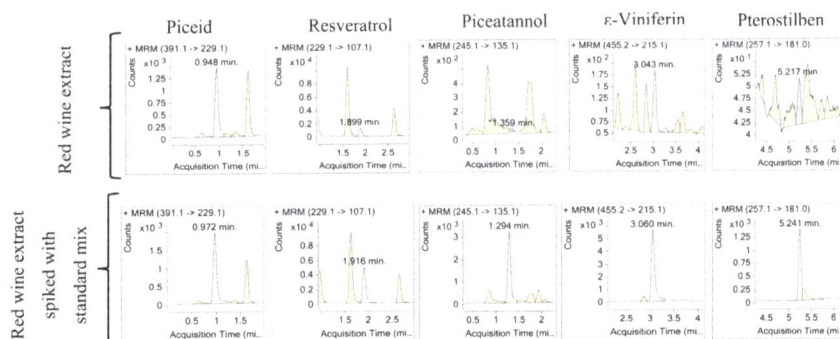

**Figure 5.** Quantitation of stilbenes in red wine extract and in the same red wine extract spiked with a standard stilbene mixture. Transition chromatograms for each compound are shown.

**Table 3.** Quantitation of stilbenes in red wine extract.

|  | t-Piceid (mg/L) | t-Resveratrol (mg/L) | t-Piceatannol (mg/L) | ε-Viniferin (mg/L) | t-Pterostilbene (mg/L) |
|---|---|---|---|---|---|
| wine | 0.826 | 0.281 | n.d. | 0.014 | n.d. |
| S/N | 140.24 | 60.86 | 0.93 | 15.6 | 1.67 |

n.d. Not detected; S/N Signal-to-Noise.

Although the quantitation limits of the stilbenes in this method can be above the existing amount in wines, the protocol used for sample preparation combining solid phase extraction, leading to 33-fold concentration, and several calibration points of standard spiking allows for the confident analysis of concentrations in wine well below LOQ, as is the case for ε-viniferin.

## 3. Discussion

In the context of the occurrence of bioactive stilbenes in natural sources, a liquid chromatography-based effective methodology to analyze both qualitatively and quantitatively the content of these compounds in different matrices is necessary. The way towards efficient analytical methods of polyphenols has been focused in making both a more efficient chromatographic separation and a more specific detection system. Thus, we have currently arrived to UHPLC, that lead to sharp peak resolution in short runs, and mass detectors, from which the triple quadrupole (QqQ) operated in SRM/MRM mode offers the highest specificity, excellent sensitivity and an extreme multiplexing capacity. Overall benefits involve a significant reduction of the analysis time and, thus, of costs. Using HPLC-MS/MS stilbenes from elicited grapevine cell cultures were qualitatively analyzed, detecting up to 20 different compounds in ca. 30 min chromatographic runs [33]. There also currently exist UHPLC-MRM methods focused on stilbenes [41], however these were specifically designed to analyze resveratrol and piceid metabolites in urine bearing negatively-charged groups such as glucuronides and sulfates derived from phase II reactions. Recent examples of the application of UHPLC-QqQ to the analysis of polyphenols in specific matrices, as rose wine [37] or several fruits and beverages [42], have extensively demonstrated these capacities. Trying to exploit these capacities to extreme has led to develop the simultaneous assay of 152 different polyphenols in non-processed samples of rose wine [37], but this remarkable achievement was done at the cost of longer time of analysis, introduction of quantitative uncertainty and loss of specificity for some compounds which include stilbenes. In particular, the *trans* and *cis* configurations of resveratrol, piceid and two viniferin isomers of unknown configuration included in such analysis eluted between 12 and 18 min. The short availability of polyphenolic standards and, in particular, most of the *cis* isomers of stilbenes, precludes the direct quantification of these configurations. The similarity between *cis* and *trans* fragmentation spectrum allows their easy detection with the major transitions of the *trans* standards, as seen in UV-exposed standard mix (Figure 2), but the different relative abundance of fragment ions introduces uncertainty in their absolute quantitation. By means of an extensive UV exposition we intended to quantitatively generate the corresponding *cis* standards but the appearance of new peaks from almost every compound revealed that such *trans*/*cis* photoisomerization is not quantitative as the reaction may proceed to generate other isomers [43], so we have not performed *cis*-isomer quantitation. The same is true for the diversity of viniferins resulting from the coupling of two resveratrol units producing several dehydrodimers [44] that, in turn, may also have *trans* and *cis* configurations. In-source fragmentation is frequently occurring in glucoside and galactoside derivatives of phenolic compounds as syringetin [42] or piceid seen here. This phenomenon lead to detect aglycone at the retention time of the glycosylated form, in addition to the aglycone at its own retention time. Although we tried to avoid this phenomenon through source parameters optimization, only an improvement of the piceid signal was achieved, perhaps because a key parameter of the ion source to attain such goal, i.e., the fragmentor voltage [45], is fixed for this instrument. For MRM detection, co-elution of analytes is not generally a problem unless these are detected by the same transitions, as is the case for the glycosylated *cis*-Piceid and the aglycone *trans*-Resveratrol due to the in-source fragmentation of the former, therefore co-elution has to be avoided. Quantitation of *trans*-Resveratrol, which has been extensively performed in plant extracts, juices and especially wines during the last two decades for its relevance as bioactive compound [46], might therefore be deeply affected by this issue. Although these compounds co-eluted in most of the chromatographic conditions assayed here, we were eventually able to establish the right conditions for a neat separation, becoming the first UHPLC-MRM method to solve this problem, thus assigning each transition peak area to a unique stilbene compound. The use of methanol as solvent in the mobile phase has been reported to increase considerably the response when the instrument is operating in positive ionization mode (ESI+) [47]. In the most of methods to determine and quantify polyphenols based on LC-MS or MS/MS [34,37,42,48–50], solvent B consisted of acetonitrile or methanol, but here we achieved the best chromatographic resolution using a mixture 70:30 acetonitrile:methanol as solvent B.

As mentioned above, resveratrol dimers exist in diverse isomeric forms but only the ε-*trans*-viniferin is commercialized as standard. The *cis* isomer was produced by UV exposition and, conversely to the rest of *cis* isomers for the stilbenes included in this study, it eluted earlier than the *trans* (Figure 2). This has some implications on the tentative assignment of structural variants of these compounds in complex samples i.e., wine, when no standards are available. In previous studies using HPLC coupled to an ion trap mass spectrometer we found several viniferin dimers eluting between resveratrol and pterostilbene standards [5,32], thus, the gap between ε-*trans*-viniferin and pterostilbene in the current method is not worthy to shorten as complex samples may contain these isomers, as can be seen in Figure 5.

MRM has proven to be a successful technique for detection and quantitation of target metabolites. The set of transitions selected for each stilbene show a high specificity, and allow both their determination and quantitation in an efficient manner in the standard mixture. Although we monitored all of them only the most intense transition was used as the quantifier, and the peak area of it and the nominal concentration was used to construct the calibration curves for each compound. The limit of quantitation (LOQ) and limit of detection (LOD) were based on the S/N of standards and the values, in the tens of μg/L range, were well below the levels typically described for these compounds, above the hundreds of μg/L range. Only pterostilbene can be a scarcer compound in biological samples as its accumulation in grapevine tissues occurs locally and in response to pathogen infection [51], and even in engineered cell lines its cytotoxicity and fast degradation precludes its high accumulation [38]. The response of the detector was linear over ca. three orders of magnitude, thus covering the levels found in the majority of matrices. In this way, target stilbenes could be quantitatively analyzed in different types of samples, including grapevine cell extracts, bioconversion reaction media, and wine.

In conclusion, major achievements of the analytical method developed include a higher analysis throughput by means of a significant shortage of chromatographic runs and a higher specificity and accuracy provided by a good separation of compounds detected with the same transitions. Focusing on the particular family of stilbene compounds provides the analytical advantages over broad compound coverage methods. Many studies were done in the past twenty years on resveratrol and piceid content in natural sources, such as fruits, juices, and beverages, including wine, as well as cultures of grapevine cells. There is a renewed interest in other stilbenes such as piceatannol and pterostilbene, which show enhanced pharmacological and pharmacodynamical properties with respect to its relative resveratrol and, oligomers of resveratrol begin to being explored as neuroprotecting compounds [52]. The number of compounds quantifiable with this method may increase as new stilbenes become available as standards.

## 4. Materials and Methods

### 4.1. MRM Method Optimization

#### 4.1.1. Chemicals and Reagents

*Trans*-Resveratrol, *trans*-Piceid, and *trans*-Pterostilbene standards were supplied by Chromadex (Irvine, CA, USA); *trans*-ε-Viniferin from Actichem (Mountaban, France) and *trans*-Piceatannol were from Cayman Chemical Company (Ann Arbor, MI, USA). Thermo Fisher scientific (Waltham, MA, USA) supplied formic acid, acetonitrile, water, methanol, and trifluoroacetic acid of LC-MS grade.

#### 4.1.2. Preparation of Standard Solutions

Stock solutions of each individual standard stilbene were prepared in 80% methanol, resveratrol (500 mg/L), piceid (400 mg/mL), piceatannol (160 mg/mL), ε-viniferin (800 mg/mL), and pterostilbene (400 mg/mL). These stock solutions were used to prepare a mixture containing the following concentration of standards: resveratrol (50 mg/L), piceid (40 mg/mL), piceatannol (90 mg/mL), ε-viniferin (90 mg/mL), and pterostilbene (40 mg/mL). Dilutions of either pure of mixed standards were carried out in 80% methanol. To generate *cis*-isomers of the standards, the stilbene mixture was exposed to UV light using a germicide lamp for 3 h (short exposition) or 24 h (prolonged exposition).

#### 4.1.3. Chromatography and Mass Conditions

LC-MS analysis were performed in an Agilent 1290 Infinity UHPLC coupled to an Agilent 6490 triple quadrupole mass spectrometer through an Agilent Jet Stream® ion source (Agilent, Santa Clara, CA, USA). Separation of analytes was performed on a Zorbax Extended C18 column (2.1 microns × 50 mm, 1.8 μm; 1200 bar maximal pressure; Agilent, Santa Clara, CA, USA). In optimized conditions the mobile phase consisted in solvent A (0.05% trifluoroacetic acid) and solvent B (acetonitrile:methanol 70:30 with 0.05% trifluoroacetic acid) using the following gradient: 0 min 35% B. 2 min 35% B, 2.3 min 36.4% B, 3 min 37% B, 4 min 40% B, 5 min 65% B, 6 min 65% B at a constant flow rate of 0.4 mL/min. Unless otherwise stated, injection volume was 1 μL.

Multiple reaction monitoring (MRM) analysis mode was used to monitor the transitions from precursor ions to dominant product ions. Several specific transitions were used to determine each compound, and the most intense transition was used for the quantitation (quantifier transition). The optimized source parameters were: capilar voltage 3 kV, gas curtain temperature 300 °C, gas flow 12 L/min, cell acceleration voltage 2 V, a nebulizer (40 psi). For each transition, the collision energy applied was optimized in order to detect the greatest possible intensity. Dwell time 20 ms. Data acquisition was performed with the software Mass Hunter Workstation Quantitative Analysis version B06 SP01 (Agilent, Santa Clara, CA, USA).

### 4.1.4. Calibration Curves and Lower Limit of Quantitation

Method validation was performed by studying the linear dynamic range, precision of the analysis, and limit of detection (LOD) and quantitation (LOQ) for the standard compounds. The linear dynamic range was evaluated using serial dilutions of standard stock solutions and plotting the peak area of the quantifier transition of each stilbene vs. the nominal concentration of each corresponding standard compound. The upper linear response was determined by fitting the data starting at a moderately low concentration (from 0.2 to 1.6 mg/L depending on the compound) to a straight line by linear regression analysis method. High concentration data were removed until $R^2 \geq 0.99$. The LODs and LOQs were established for each compound as the concentration of standard providing a signal to noise ratio $\geq 5$ and $\geq 10$, respectively. For the construction of calibration curves of individual stilbenes, serial dilutions from the standard mix solution were performed. The volume injected is the same as the one mentioned above.

### 4.2. Method Applications

### 4.2.1. Analysis and Quantitation of Stilbenes Produced by *V. vinifera* Cell Culture Upon Elicitation

*Vitis vinifera* cv. Gamay cell cultures were handled and subjected to elicitation as described in [2]. Samples of stilbenes from the extracellular medium and from cells were prepared as described in [5]. One µL aliquot was injected for LC-MS MRM analysis.

### 4.2.2. Biotransformation Assays

Crude protein extracts were prepared from elicited *V. vinifera* cv. Gamay cell cultures as described in [6]. Resveratrol hydroxylation reaction was assayed as described in [39,53]. The reaction mixture contained 0.2 mM of *trans*-R delivered in DMSO (Fluka Chemika-Sigma Aldrich, St. Louis, MO, USA), 1 mM ascorbic acid, 1 mM NADPH, and 100 mM potassium phosphate buffer, pH 7.4, and started by adding 50 µL of protein extract to complete a final volume of 1 mL. The reaction mixture was incubated for 1 h at 37 °C in the dark and terminated by the addition of 0.5 mL ethyl acetate to extract stilbenoids twice. The solvent was evaporated in a Speed-vac (Eppendorff, Hamburg, Germany) and the solid residue resuspended in 0.2 mL 80% methanol. One µL of reaction medium extract was injected for UHPLC-MS MRM analysis under the optimized conditions.

Absolute concentration of stilbenes in real samples of cell extracts and enzymatic reaction extract was determined using an external calibration curve performed with three levels of standards in duplicate.

### 4.2.3. Analysis and Quantitation of Stilbenes in Red Wine

Red wine (80% Cabernet Sauvignon and 20% Alicante Bouschet), used in this study was elaborated during the 2012 vintage in a local wine cellar and the analyzed sample was taken right after the alcoholic fermentation and then clarified by centrifugation at $10,000 \times g$ for 20 min. Wine samples of 2 mL were extracted by solid phase according to *Kallithraka* et al. [40]. After that, extracted stilbenes were evaporated at 30 °C in a Speed-Vac centrifuge (Eppendorff, Hamburg, Germany), resuspended in 60 µL methanol 80% and stored at −20 °C until analysis. One µL of sample extract was injected for UHPLC-MS MRM analysis under the optimized conditions. To minimize the effects of sample preparation on analysis, a sample of wine was spiked with the standard mix, four serial dilutions were prepared by mixing the spiked sample with fresh wine before proceeding to solid phase extraction. A non-spiked wine aliquot was extracted in parallel. Then, concentration of the standard added to the wine sample was plotted against the measured peak area of the quantifier transition of the extracted sample. In this way the percentage of stilbene recovery from wine was obviated and the real concentration of each stilbene compound is obtained from the intercept of the fitting straight line with the standard concentration axis (A = Ao + F.C, where A is the peak area of any sample; Ao is the peak area of the non-spiked sample, F is the response factor and C standard concentration), as shown in Figure S1.

**Supplementary Materials:** The supplementary materials are available online.

**Acknowledgments:** A.M.-M. acknowledges a grant from Conselleria d'Educacio, Cultura I Sport de la Generalitat Valenciana (FPA/2013/A/074). This work has been supported by grants from University of Alicante (VIGROB-105), the Spanish Ministry of Economy and Competitiveness (BIO2014-51861-R), and European funds for Regional development (FEDER). Technical assistance of Mario Antonazzo is acknowledged.

**Author Contributions:** R.B.-M. and S.S.-M. conceived and designed the experiments; E.H.-G., S.S.-M., A.M.-M., and A.S.-H. performed the experiments and analyzed the data; E.H.-G. and R.B.-M. wrote the paper.

**Conflicts of Interest:** The authors declare no conflict of interest.

## References

1. Pezet, R.; Gindro, K.; Viret, O.; Richter, H. Effects of resveratrol and pterostilbene on *Plasmopara viticola* zoospore mobility and disease development. *Vitis* **2004**, *43*, 145–148.
2. Bru, R.; Selles, S.; Casado-Vela, J.; Belchi-Navarro, S.; Pedreño, M.A. Modified cyclodextrins are chemically defined glucan inducers of defense response in grapevine cell cultures. *J. Agric. Food Chem.* **2006**, *54*, 65–71. [CrossRef] [PubMed]
3. Adrian, M.; Jeandet, P. Effects of resveratrol on the ultrastructure of Botrytis cinerea conidia and biological significance in plant/pathogen interactions. *Fitoterapia* **2012**, *83*, 1345–1350. [CrossRef] [PubMed]
4. Cantos, E.; Espin, J.C.; Fernandez, M.J.; Oliva, J.; Tomas-Barberan, F.A. Post-harvest UV-C irradiated grapes as potential source for producing stilbene-enriched red wines. *J. Agric. Food Chem.* **2003**, *51*, 1208–1241. [CrossRef] [PubMed]
5. Martinez-Esteso, M.J.; Selles, S.; Vera-Urbina, J.C.; Pedreño, M.A.; Bru, R. Changes of defense proteins in the extracelular proteome of grapevine (*Vitis vinifera* cv. Gamay) cell cultures in response to elicitors. *J. Proteom.* **2009**, *73*, 331–341. [CrossRef] [PubMed]
6. Martinez-Esteso, M.J.; Selles, S.; Vera-Urbina, J.C.; Pedreño, M.A.; Bru, R. DIGE analysis of proteome changes accompanying large resveratrol production by grapevine (*Vitis vinifera* cv. Gamay) cell cultures in response to methyl-β-cyclodextrin and methyl jasmonate elicitors. *J. Proteom.* **2011**, *74*, 1421–1436. [CrossRef] [PubMed]
7. Wang, W.; Tang, K.; Yang, H.R.; Wen, P.F.; Zhang, P.; Wang, H.L.; Huang, W.D. Distribution of resveratrol and stilbene synthase in yong grape plants (*Vitis vinifera* L. cv. Cabernet sauvignon) and the effect of UV-C on its accumulation. *Plant Physiol. Biochem.* **2010**, *48*, 142–152. [CrossRef] [PubMed]
8. Jeandet, P.; Clément, C.; Courot, E.; Cordelier, S. Modulation of phytoalexin biosynthesis in engineered plants for disease resistance. *Int. J. Mol. Sci.* **2013**, *14*, 14136–14170. [CrossRef] [PubMed]
9. Jeandet, P.; Hébrard, C.; Deville, M.A.; Cordelier, S.; Dorey, S.; Aziz, A.; Crouzet, J. Deciphering the role of phytoalexins in plant-microorganism interactions and human health. *Molecules* **2014**, *19*, 18033–18056. [CrossRef] [PubMed]
10. Tsai, H.-Y.; Ho, C.-T.; Chen, Y.-K. Biological actions and molecular effects of resveratrol, pterostilbene and 3′-hydroxypterostilbene. *J. Food Drug Anal.* **2016**, *25*, 134–147. [CrossRef]
11. Bradamante, S.; Barenghi, L.; Villa, A. Cardiovascular protective effects of resveratrol. *Cardiovasc. Drug Rev.* **2014**, *22*, 169–188. [CrossRef]
12. Le Corre, L.; Chalabi, N.; Delort, L.; Bignon, Y.J.; Bernard-Gallon, D.J. Resveratrol and breast cancer chemoprevention: Molecular mechanism. *Mol. Nutr. Food Res.* **2005**, *49*, 462–471. [CrossRef] [PubMed]

13. Wolter, F.; Ulrich, S.; Stein, J. Molecular mechanism of the chemopreventive effects of resveratrol and its analogs in colorectal cancer: Key role of polyamines. *J. Nutr.* **2004**, *134*, 3219–3222. [PubMed]
14. Ratan, H.L.; Steward, W.P.; Gescher, A.J.; Mellon, J.K. Resveratrol—A prostate cancer chemopreventive agent? *Urol. Oncol.* **2002**, *7*, 223–227. [CrossRef]
15. Fernández-Pérez, F.; Belchí-Navarro, S.; Almagro, L.; Bru, R.; Pedreño, M.A.; Gómez-Ros, L.V. Cytotoxic effect of natural *trans*-Resveratrol obtained from elicted *Vitis vinifera* cell cultures on three cancer lines. *Plant Foods Hum. Nutr.* **2012**, *67*, 422–429. [CrossRef] [PubMed]
16. De la Lastra, C.A.; Villegas, I. Resveratrol as an anti-inflamatory and antiaging agent: Mechanism and clinical implications. *Mol. Nutr. Food Res.* **2005**, *49*, 405–430. [CrossRef] [PubMed]
17. De la Lastra, C.A.; Villegas, I. Resveratrol as an antioxidant and pro-oxidant agent: Mechanism and clinical implications. *Biochem. Soc. Trans.* **2007**, *35*, 1156–1160. [CrossRef] [PubMed]
18. Okawara, M.; Katsuki, H.; Kurimoto, E.; Chibata, H.; Kume, T.; Akaike, K. Resveratrol protects dopaminergics neurons in midbrains slice culture from multiple insults. *Biochem. Pharmacol.* **2007**, *73*, 550–560. [CrossRef] [PubMed]
19. Baur, J.A.; Sinclair, D.A. Therapeutic potential of resveratrol: The in vivo evidence. *Nat. Rev. Drug Discov.* **2006**, *5*, 493–506. [CrossRef] [PubMed]
20. Reinisalo, M.; Kårlund, A.; Koskela, A.; Kaarniranta, K.; Karjalainen, R.O. Polyphenol Stilbenes: Molecular Mechanisms of Defence against Oxidative Stress and Aging-Related Diseases. *Oxid. Med. Cell. Longev.* **2015**. [CrossRef] [PubMed]
21. Szekeres, T.; Saiko, P.; Fritzer-Szekeres, M.; Djavan, B.; Jäger, W. Chemopreventive effects of resveratrol and resveratrol derivatives. *Ann. N. Y. Acad. Sci.* **2011**, *1215*, 89–98. [CrossRef] [PubMed]
22. Langcake, P.; Pryce, R.J. A new class of phytoalexins from grapevines. *Experientia* **1977**, *33*, 151–152. [CrossRef] [PubMed]
23. Siemann, E.H.; Creasy, L.L. Concentration of the phytoalexin resveratrol in wine. *Am. J. Enol. Vitic.* **1992**, *43*, 49–52.
24. Sanders, T.H.; McMichael, R.W., Jr.; Hendrix, K.W. Occurrence of resveratrol in edible peanuts. *J. Agric. Food Chem.* **2000**, *48*, 1243–1246. [CrossRef] [PubMed]
25. Rimando, A.M.; Kalt, W.; Magee, J.B.; Dewey, J.; Ballington, J.R. Resveratrol, pterostilbene, and piceatannol in vaccinium berries. *J. Agric. Food Chem.* **2004**, *52*, 4713–4719. [CrossRef] [PubMed]
26. Hurst, W.J.; Glinski, J.A.; Miller, K.B.; Apgar, J.; Davey, M.H.; Stuart, D.A. Survey of the *trans*-R and *trans*-piceid content of cocoa-containing and chocolate products. *J. Agric. Food Chem.* **2008**, *56*, 8374–8378. [CrossRef] [PubMed]
27. Guilford, J.M.; Pezzuto, J.M. Wine and health: A review. *Am. J. Enol. Vitic.* **2011**, *62*, 471–486. [CrossRef]
28. Jeandet, P.; Breuil, A.C.; Adrian, M.; Weston, L.A.; Debord, S.; Meunier, P.; Maume, B.; Bessis, R. HPLC analysis of grapevine phytoalexins coupling photodiode array detection and fluorometry. *Anal. Chem.* **1997**, *69*, 5172–5177. [CrossRef]
29. Kallithraka, S.; Tsoutsouras, R.; Tzourou, E.; Lanaridis, P. Principal phenolic compounds in Greek red wines. *Food Chem.* **2006**, *99*, 784–793. [CrossRef]
30. Prosen, H.; Strilic, M.; Kocar, D.; Rusjan, D. In vino veritas: LC-MS in wine analysis. *LC-GC Eur.* **2007**, *20*, 617–621.
31. Püssa, T.; Janar, F.; Paul, K.; Ain, R. Survey of grapevine *Vitis vinifera* stem polyphenols by liquid chromatography-diode array detection-tamdem mass spectrometry. *J. Agric. Food Chem.* **2006**, *54*, 7488–7494. [CrossRef] [PubMed]
32. Sun, B.; Leandro, M.C.; de Freitas, V.; Spranger, M.I. Fractionation of red wine polyphenols by solid-phase extraction and liquid chromatography. *J. Chromatogr. A* **2006**, *1128*, 27–38. [CrossRef] [PubMed]
33. Mulinacci, N.; Innocenti, M.; Santamaria, A.R.; la Marca, G.; Pasqua, G. High-performance liquid chromatography/electrospray ionization tandem mass spectrometric investigation of stilbenoids in cell cultures of *Vitis vinifera* L., cv. Malvasia. *Rapid Commun. Mass Spectrom.* **2010**, *24*, 2065–2073. [CrossRef] [PubMed]
34. Jaitz, L.; Siegl, K.; Eder, R.; Rak, G.; Abranko, L.; Koellensperger, G.; Hann, S. LC-MS/MS analysis of phenols for classification of red wine according to geographic origin, grape variety and vintage. *Food Chem.* **2010**, *122*, 366–372. [CrossRef]

35. Ortega, N.; Romero, M.P.; Macià, A.; Reguant, J.; Anglès, N.; Morelló, J.R.; Motilva, M.J. Obtention and characterization of phenolic extracts from differents cocoa sources. *J. Agric. Food Chem.* **2008**, *65*, 9621–9627. [CrossRef] [PubMed]

36. Ortega, N.; Macià, A.; Romero, M.P.; Trullols, E.; Morelló, J.R.; Anglès, N.; Motilva, M.J. Rapid determination of phenolic compound and alkaloids of carob flour by improved liquid chromatography tamdem mass spectrometry. *J. Agric. Food Chem.* **2009**, *57*, 7239–7244. [CrossRef] [PubMed]

37. Lambert, M.; Meudec, E.; Verbaere, A.; Mazerolles, G.; Wirth, J.; Masson, G.; Cheynier, V.; Sommerer, N. A High-Troughput UHPLC-QqQ-Ms Method for polyphenol profiling in Rosé wines. *Molecules* **2015**, *20*, 7890–7914. [CrossRef] [PubMed]

38. Martínez-Márquez, A.; Morante-Carriel, J.A.; Ramírez-Estrada, K.; Cusidó, R.M.; Palazon, J.; Bru-Martínez, R. Production of highly bioactive resveratrol analogues pterosilbene and piceatannol in metabolically engineered grapevine cell cultures. *Plant Biotech. J.* **2016**, *14*, 1813–1825. [CrossRef] [PubMed]

39. Piver, B.; Fer, M.; Vitrac, X.; Merillon, J.M.; Dreano, Y.; Berthou, F.; Lucas, D. Involvement of cytochrome P450 1A2 in the biotransformation of *trans*-resveratrol in human liver microsomes. *Biochem. Pharmacol.* **2004**, *68*, 773–782. [CrossRef] [PubMed]

40. Kallithraka, S.; Arvanitoyannis, I.; El-Zajouli, A.; Kelafas, P. The application of an improved method for *trans*-R to determine the origin of Greek red wines. *Food Chem.* **2001**, *75*, 355–363. [CrossRef]

41. Rotches-Ribalta, M.; Urpi-Sarda, M.; Llorach, R.; Boto-Ordoñez, M.; Jauregui, O.; Chiva-Blanch, G.; Perez-Garcia, L.; Jaeger, W.; Guillen, M.; Corella, D.; et al. Gut and microbial resveratrol metabolite profiling after moderate long-term consumption of red wine versus dealcoholized red wine in humans by an optimized ultra-high-pressure liquid chromatography tandem mass spectrometry method. *J. Chromatogr. A* **2012**, *1265*, 105–113. [CrossRef] [PubMed]

42. Vrhovsek, U.; Masuero, D.; Gasperotti, M.; Franceschi, P.; Caputi, L.; Viola, R.; Mattivi, F. A versatile targeted metabolomics method for the rapid quantifaction of multiple class of phenolics in fruits and beverages. *J. Agric. Food Chem.* **2012**, *60*, 8831–8840. [CrossRef] [PubMed]

43. Yang, I.; Kim, E.; Kang, J.; Han, H.; Sul, S.; Park, S.B.; Kim, S.K. Photochemical generation of a new, highly fluorescent compound from non-fluorescent resveratrol. *Chem. Commun.* **2012**, *48*, 3839–3841. [CrossRef] [PubMed]

44. Morales, M.; Ros Barcelo, A.; Pedreno, M.A.; Hemantaranja, A. Plant stilbenes: Recent advances in their chemistry and biology. *Adv. Plant Physiol.* **2000**, *3*, 39–70.

45. Abrankó, L.; García-Reyes, J.F.; Molina-Diaz, A. In-source fragmentation and accurate mass analysis of multiclass flavonoid conjugates by electrospray ionization time-of-flight mass spectrometry. *J. Mass Spectrom.* **2011**, *46*, 478–488. [CrossRef] [PubMed]

46. Stervbo, U.; Vang, O.; Bonnesen, C. A review of the content of the putative chemopreventive phytoalexin resveratrol in red wine. *Food Chem.* **2007**, *101*, 449–457. [CrossRef]

47. Jemal, M.; Hawthrone, D.J. Effect of High Performance Liquid Chromatography mobile phase (Methanol versus Acetonitrile) on the positive and negative ion Electrospray Response of a compound that contains both an unsaturated lactone and a methyl sulfone group. *Rapid Commun. Mass Spectrom.* **1999**, *13*, 61–66. [CrossRef]

48. Omar, J.M.; Yang, H.; Li, S.; Marquardt, R.R.; Jones Peter, J.H. Development of an Improved Reverse-Phase High-Performance Liquid Chromatogrpahy Method for the simultaneous Analyses of *trans-/cis*-Resveratrol, Quercitin, and Emodin in commercial Reveratrol supplements. *J. Agric. Food Chem.* **2014**, *62*, 5812–5817. [CrossRef] [PubMed]

49. Pinasseau, L.; Verbaere, A.; Roques, M.; Meudec, E.; Vallverdu-Queralt, A.; Terrier, N.; Boulet, J.C.; Cheyner, V.; Sommerer, N. A Fast and robus UHPLC-MRM-Ms method to characterize and Quantify Grape Skin Tannins after Chemical Depolymerization. *Molecules* **2016**, *21*, 1409. [CrossRef] [PubMed]

50. Buiarelli, F.; Coccioli, F.; Jasionowska, R.; Merolle, M.; Terracciano, A. Analysis of some stilbenes in Italian wines by liquid chromatography/tándem mass spectrometry. *Rapid Commun. Mass Spectrom.* **2007**, *21*, 2955–2964. [CrossRef] [PubMed]

51. Schmidlin, L.; Poutaraud, A.; Claudel, P.; Mestre, P.; Prado, E.; Santos-Rosa, M.; Wiedemann-Merdinoglu, S.; Karst, F.; Merdinoglu, D.; Hugueney, P. A stress-inducible resveratrol *O*-methyltransferase involved in the biosynthesis of pterostilbene in grapevine. *Plant Physiol.* **2008**, *148*, 1630–1639. [CrossRef] [PubMed]

52. Richard, T.; Poupard, P.; Nassra, M.; Papastamoulis, Y.; Iglesias, M.L.; Krisa, S.; Waffo-Teguo, P.; Mérillon, J.M.; Monti, J.P. Protective effect of ε-viniferin on β-amyloid peptide aggregation investigated by electrospray ionization mass spectrometry. *Bioorganic Med. Chem.* **2011**, *19*, 3152–3155. [CrossRef] [PubMed]
53. Kim, D.H.; Ahn, T.; Jung, H.C.; Pan, J.G.; Yun, C.H. Generation of the human metabolite piceatannol from the anticancer preventive agent resveratrol by bacterial cytochrome P450 BM3. *Drug Metab. Dispos.* **2009**, *37*, 932–936. [CrossRef] [PubMed]

**Sample Availability:** Not Available.

*molecules*

MDPI

*Article*

# Simultaneous Ultra Performance Liquid Chromatography Determination and Antioxidant Activity of Linarin, Luteolin, Chlorogenic Acid and Apigenin in Different Parts of Compositae Species

Seung Hwan Hwang [1], Ji Hun Paek [1] and Soon Sung Lim [1,2,*]

[1] Department of Food Science and Nutrition, Hallym University, 1 Hallymdeahak-gil, Chuncheon 24252, Korea; isohsh@gmail.com (S.H.H.); hun6678@gmail.com (J.H.P.)
[2] Institute of Natural Medicine, Hallym University, 1 Hallymdeahak-gil, Chuncheon 24252, Korea
* Correspondence: limss@hallym.ac.kr; Tel.: +82-33-248-2133; Fax: +82-33-251-0663

Academic Editor: Philippe Jeandet
Received: 30 September 2016; Accepted: 17 November 2016; Published: 23 November 2016

**Abstract:** Linarin (LA), luteolin (LE), chlorogenic acid (CA) and apigenin (AP) are four major flavonoids with various promising bioactivities found in Compositae (COP) species. A reliable, reproducible and accurate method for the simultaneous and quantitative determination of these four major flavonoids by Ultra Performance Liquid Chromatography (UPLC) analysis was developed. This method should be appropriate for the quality assurance of COP. The UPLC separation was carried out using an octadecylsilane (ODS) Hypersil (2.1 mm × 250 mm, 1.9 µm) and a mobile phase composed of acetonitrile and 0.1% formic acid in water at a flow rate 0.44 mL/min and ultraviolet (UV) detection 254 nm. Gradient elution was employed. The method was precise, with relative standard deviation below 3.0% and showed excellent linearity ($R^2 > 0.999$). The recoveries for the four flavonoids in COP were between 95.49%–106.23%. The average contents of LA, LE, CA and AP in different parts (flower, leave and stem) of COP were between 0.64–1.47 g/100 g, 0.66–0.89 g/100 g, 0.32–0.52 g/100 g and 0.16–0.18 g/100 g, respectively. The method was accurate and reproducible and it can provide a quantitative basis for quality control of COP.

**Keywords:** UPLC; linarin; luteolin; chlorogenic; apigenin; Compositae

## 1. Introduction

There is considerable recent evidence showing that free radicals induce oxidative damage to cause pathological effects on humans, including DNA damage, aging, and cancer [1,2]. Recently, there has been a global trend towards the using phenolic compounds extracted from fruits, vegetables, oilseeds, and herbal plants [3,4].

Dietary foods contain a variety of free radical scavenging antioxidants, such as phenolic compounds (tocopherols, flavonoids, and phenolic acids) [5]. Plant phenols have free radical scavenging properties due to their redox potential [6]. Phenolic compounds are commonly found in both edible and inedible plants, and have been reported to have multiple biological effects, including antioxidant activity. The antioxidant activity of phenolic compounds is mainly caused by their redox properties, which play an important role in adsorbing and neutralizing free radicals, by quenching singlet and triplet oxygen or by forming peroxides. Antioxidants are also of immense interest to health professionals, as they may help to protect the body against damage caused by reactive oxygen species (ROS) [7].

Dried Compositae (COP) flowers have traditionally been used in Korea for their anti-inflammatory and antioxidant activity. Plants from the family COP have been reported to have anti-inflammatory, antimicrobial, and antitumor properties [8–11]. Antioxidants are generally abundant in polyphenolic substances. The known major flavones present in COP plants are apigenin, acacetin, luteolin, diosmetin, eriodictyol, chlorogenic acid, and linarin [12,13]. Therefore, in the current study linarin, luteolin, chlorogenic acid, and apigenin were selected as the key compounds for extraction.

Method validation of analytical tests is conducted to ensure that the methodology is accurate, specific, reproducible, and robust over the specified range of analysis. Method validation provides an assurance of reliability during normal use, sometimes referred to as "the process of providing documented evidence that the method does what it is intended to do" [14]. Many validation methods including high-performance liquid chromatography (HPLC) [15], mass spectrometry (MS) [16], and capillary electrophoresis (CE) [17] have been used in the qualitative or quantitative analysis of natural extract, drug candidates and processed food. Despite the availability of various analytical techniques, currently there is an increasing demand for the fast and sensitive analysis of samples which could reduce costs and achieve high sample throughput. Among them, Ultra Performance Liquid Chromatography (UPLC) systems allow the use of small particle-packed columns with small diameter. The particles are designed to be able to resist high pressures, in contrast to conventional HPLC [18]. For this reason, UPLC can give improvements in speed, resolution, rapid and sensitivity of analysis, time savings, and solvent consumption compared to the previously used HPLC method [19]. Most of these advantages may be attributed to moving from HPLC to UPLC.

On the basis of the International Conference on Harmonization guidelines [20], an UPLC analytical method requires validation to confirm its linearity, recovery, and precision. In this study, a UPLC method for the quantification of flavonoid compounds from different parts of COP plant products was developed and validated for the first time. We show not only the validation of a reliable, fast, and easy methodology for quantification of flavonoid compounds, but in addition, the antioxidant capacity of the extracts was determined by 1,1-Diphenyl-2-picrylhydrazyl (DPPH) free radical-scavenging activity, and the contents of total polyphenols and total flavonoids were determined for different parts of COP plant products. Finally, the relationships between polyphenol and flavonoid contents and antioxidant activity were explored.

## 2. Results

### 2.1. Structural Determination of Isolate Compounds

The four flavonoids were separated from a 1 g sample of a primrose small Compositae (Kugya-sunjong, Table 1, Entry 13) methanol leaves extract by Sephadex LH-20 column chromatography to obtain chlorogenic acid (CA) (10.3 mg), luteolin (LE) (11.8 mg) apigenin (AP) (8.2 mg) and linarin (LA) (9.2 mg) (Figure 1). These compounds were identified by comparing their [1]H- and [13]C-NMR spectra and NMR correlation spectra such as correlation spectroscopy (COSY), heteronuclear multiple bond correlation (HMBC) and heteronuclear multiple quantum correlation (HMQC) with previously reported data [21,22].

**Table 1.** List of Compositae family samples.

| Entry | Plant Species | |
|---|---|---|
| | Name | Scientific Name |
| 1 | Double flower Compositae (Kugya-baekcheon) | *Chrysanthemum morifolium* variants |
| 2 | Red Korean Compositae (Kugya-seonnyeo) | *Chrysanthemum morifolium* variants |
| 3 | Dark red Compositae (Kugya-jinju) | *Chrysanthemum morifolium* variants |
| 4 | Pink Compositae (Kugya-dowon) | *Chrysanthemum morifolium* variants |
| 5 | Yellow short Compositae (Kugya-hana) | *Chrysanthemum morifolium* variants |
| 6 | Red Compositae (Kugya-myungseong) | *Chrysanthemum morifolium* variants |
| 7 | Yellow small Compositae (Kugya-baram) | *Chrysanthemum morifolium* variants |
| 8 | White double Compositae (Kugya-cheonsa) | *Chrysanthemum morifolium* variants |

Table 1. *Cont.*

| Entry | Plant Species | |
| --- | --- | --- |
| | Name | Scientific Name |
| 9 | Dark yellow Compositae (Kugya-gyeongjin) | *Chrysanthemum morifolium* variants |
| 10 | White short Compositae (Kugya-somang) | *Chrysanthemum morifolium* variants |
| 11 | Yellow Compositae (Kugya-kughang) | *Chrysanthemum morifolium* variants |
| 12 | White Compositae (Kugya-sinsun) | *Chrysanthemum indicum* variants |
| 13 | Primrose small Compositae (Kugya-sunjong) | *Chrysanthemum indicum* variants |
| 14 | Primrose short Compositae (Kugya-gamhea) | *Chrysanthemum indicum* variants |
| 15 | Yellow large Compositae (Kugya-gamthae) | *Chrysanthemum indicum* variants |
| 16 | White small Compositae (Kugya-sulwha) | *Chrysanthemum indicum* variants |
| 17 | Primrose Compositae (Kugya-sunjeong) | *Chrysanthemum indicum* variants |
| 18 | Yellow lanugo Compositae (Kugya-gumi) | *Chrysanthemum zawadskii* var. *latilobum* |
| 19 | White lanugo Compositae (Kugya-baekhae) | *Aster sphathulifolius Maxim* |
| 20 | Makino Compositae (Kugya-makino) | *Dendranthema makinoi (Matsum)* |

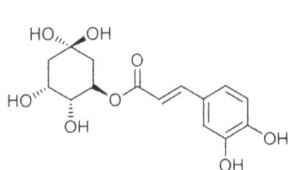

Chlorogenic acid

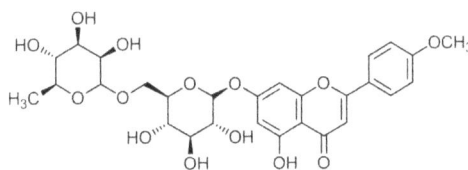

Linarin

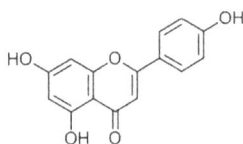

Apigenin

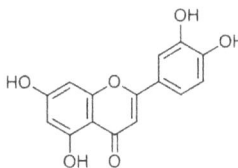

Luteolin

**Figure 1.** Chemical structures of chlorogenic acid, linarin, apigenin and luteolin.

## 2.2. Optimization of UPLC Conditions

The effectiveness of the UPLC separation was tested using the four flavonoids isolated from Entry 13. The gradient elution profile was optimized to obtain the highest resolution of four flavonoids and the shortest time of the analysis. Two solvents in a mobile phase 0.1% formic acid (FA) in water (solvent A) and acetonitrile (ACN) (solvent B), were selected and run according to the programmed gradient elution.

Under the chromatographic conditions of the current experiment, the addition of 0.1% FA in water increased the resolution of the peaks. Unfortunately, neither isocratic elution nor gradient elution resulted in good chromatographic separation of LA and AP. Temperature was then used in this study. It was found that when the column temperature was maintained at 45 °C, it produced a good chromatographic peak. The wavelength for detection was tested at 254, 280, and 360 nm. The wavelength for detection was then set at 254 nm, which is where the four flavonoids showed the maximum absorption, as measured by a diode array detector (DAD). Therefore, the best resolution of all the peaks was obtained using a gradient of the mobile phase consisting of ACN and 0.1% FA in water within 14 min. The retention times of CA, LA, AP, and LE were 2.8, 7.7, 8.7, and 10.3 min (Figure 2).

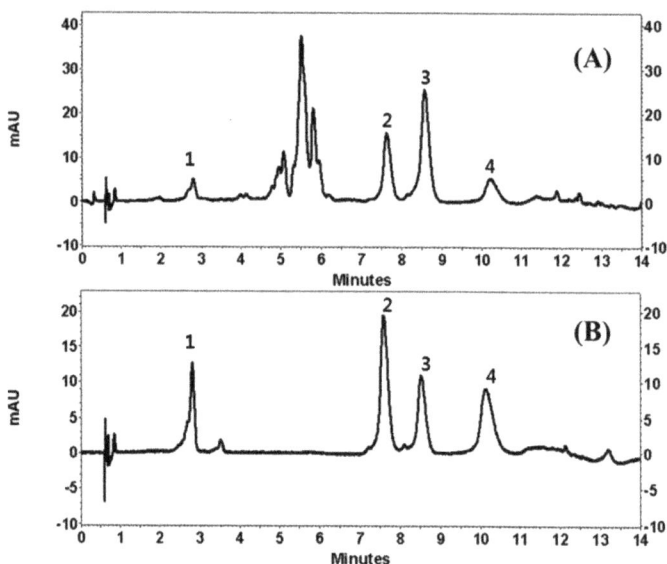

**Figure 2.** Ultra Performance Liquid Chromatography (UPLC) chromatogram of Compositae (Entry 13 in Table 1, (**A**)) and standard mixtures of the isolated compounds (1: chlorogenic acid; 2: linarin; 3: apigenin and 4: luteolin, (**B**)). mAU: miliabsorbance units.

## 2.3. Linearity

Serially diluted solutions of the four flavonoids prepared in the range of 1, 10, 25, 50 and 100 µg/mL were injected into the UPLC, and calibration curve equations were calculated. As shown in Table 2, Linearity ($R^2$) showed a correlation higher than of 0.999, with the linear ranges also being determined. The detection and quantification limits for CA, LA, AP, and LT at the signal-to-noise ratio of the four flavonoids were 0.38, 0.13, 0.11, 0.26 µg/mL$^{-1}$ and 1.15, 0.41, 0.35, 0.79 µg/mL$^{-1}$, respectively.

**Table 2.** Statistical analysis for the calibration curves of four flavonoids ($n = 3$).

| Compound | Slope | Intercept | $R^{2\,a}$ | LOD [b] | LOQ [c] |
|---|---|---|---|---|---|
| Linarin | 4600.44 ± 73.86 | −9.71 ± 1.89 | 0.999 | 0.13 | 0.41 |
| Luteolin | 6829.50 ± 185.08 | −107.54 ± 54.17 | 0.999 | 0.26 | 0.79 |
| Chlorogenic acid | 3400.03 ± 51.41 | −5.01 ± 3.92 | 0.999 | 0.38 | 1.15 |
| Apigenin | 5614.93 ± 56.57 | 16.26 ± 2.01 | 0.999 | 0.11 | 0.35 |

[a] $R^2$, correlation coefficient for the 5 data points in the calibration curves ($n = 3$); [b] LOD, limit of detection (µg/mL$^{-1}$, Signal to noise (S/N) = 3); [c] LOQ, limit of quantification (µg/mL$^{-1}$, S/N = 10).

## 2.4. Precision

Intra-day and inter-day variability was used to validate the precision of the UPLC method. To assess repeatability, four flavonoid standard solutions were injected three times at concentrations of 2.5, 10, and 100 µg/mL onto the UPLC system, and relative standard deviation (RSD) values were calculated for the retention time and peak area.

As shown in Tables 3 and 4, the intra-day and inter-day peak area of the RSD were LA < 1.68%, LE < 1.87%, CA < 2.23%, and AP < 1.46%. The intra-day and inter-day retention time of the RSDs for the different compounds were <0.65% for LA, <1.12% for LE, <0.35% for CA, and <0.1% for AP, with an RSD of less than 3.0%.

**Table 3.** Intra-day precision data for the retention time and peak area of four flavonoids.

| Concentration (µg/mL) | Intra-day Precision (*n* = 3) | | | | | | | |
|---|---|---|---|---|---|---|---|---|
| | Linarin | | Luteolin | | Chlorogenic Acid | | Apigenin | |
| | Rt [a] | Area [b] | Rt | Area | Rt | Area | Rt | Area |
| 100 | 0.07 | 1.56 | 0.20 | 0.60 | 0.25 | 1.70 | 0.08 | 0.94 |
| 10 | 0.23 | 1.47 | 0.13 | 2.15 | 0.24 | 0.45 | 0.04 | 1.53 |
| 2.5 | 0.01 | 1.68 | 0.24 | 1.87 | 0.38 | 2.23 | 0.05 | 1.46 |

[a] Relative standard deviation of retention time (Rt) (% RSD); [b] Relative standard deviation of peak area (% RSD).

**Table 4.** Inter-day precision data for the retention times and peak area of four flavonoids.

| Concentration (µg/mL) | Inter-Day Precision (*n* = 3) | | | | | | | |
|---|---|---|---|---|---|---|---|---|
| | Linarin | | Luteolin | | Chlorogenic Acid | | Apigenin | |
| | Rt [a] | Area [b] | Rt | Area | Rt | Area | Rt | Area |
| 100 | 0.65 | 0.29 | 1.12 | 0.68 | 0.35 | 0.86 | 0.10 | 1.77 |
| 10 | 0.61 | 1.47 | 0.12 | 2.21 | 0.29 | 1.20 | 0.03 | 1.38 |
| 2.5 | 0.06 | 1.38 | 0.33 | 2.34 | 0.09 | 2.04 | 0.08 | 2.05 |

[a] Relative standard deviation of retention time (% RSD); [b] Relative standard deviation of peak area (% RSD).

## 2.5. Accuracy (Recovery)

A recovery test was used to determine accuracy. The methanol extract was spiked with four flavonoids to observe changes in the recovery rate (%). Accuracy was evaluated by measuring the mean recovery (%) of four flavonoids from the spiked extract solution versus the non-spiked extract sample. Each sample was analyzed three times, and the recovery rate was calculated by using the calibration curve obtained from the results of linearity test. The accuracy was determined for the different compounds, where LA had an accuracy of 99.12% ± 0.90%, LE 95.49% ± 0.23, CA 103.07% ± 0.36%, and AP 106.23% ± 0.33% (Table 5).

**Table 5.** Recovery data of four flavonoids.

| Compounds | Original Amount (µg) | Spiked Amount (µg) | Determined Amount (µg) | Recovery (%, Mean ± RSD, *n* = 3) |
|---|---|---|---|---|
| Linarin | 13.91 | 9.80 | 23.49 | 99.12 ± 0.90 |
| Luteolin | 6.29 | 9.80 | 15.37 | 95.49 ± 0.23 |
| Chlorogenic acid | 4.19 | 9.50 | 14.12 | 103.07 ± 0.36 |
| Apigenin | 3.02 | 9.70 | 13.49 | 106.23 ± 0.33 |

## 2.6. Quantification of the Four Compounds, Total Polyphenol Content, Total Flavonoid Content and Antioxidant Activity

Relatively high contents of LA, LE, and CA was observed from (Table 6): flowers of 10 and 18 COP (1.37 and 1.38 g/100 g), leaves of 6 and 11 COP (3.52 and 3.51 g/100 g), stems of 6 and 16 COP (1.53 and 1.27 g/100 g) of LA and flowers of 6, 9 and 10 COP (1.49, 1.55 and 1.64 g/100 g), leaves of 8, 11, 13 and 17 COP (1.36, 1.17, 1.31 and 1.39 g/100 g), of LE and flowers of 14 and 20 COP (1.02 and 1.02 g/100 g), leaves of 17 COP (2.55 g/100 g), of CA, respectively.

**Table 6.** Linarin, luteolin, chlorogenic acid, and apigenin content in different parts of Compositae (g/100 g).

| Entry | Flower (n = 3) | | | | Leave (n = 3) | | | | Stem (n = 3) | | | |
|---|---|---|---|---|---|---|---|---|---|---|---|---|
| | Linarin | Luteolin | Chlorogenic Acid | Apigenin | Linarin | Luteolin | Chlorogenic Acid | Apigenin | Linarin | Luteolin | Chlorogenic Acid | Apigenin |
| 1 | 0.24 ± 0.008 | 0.25 ± 0.011 | 0.29 ± 0.059 | - | 1.32 ± 0.051 | 0.34 ± 0.001 | 0.37 ± 0.022 | 0.01 ± 0.010 | 0.77 ± 0.027 | - | 0.25 ± 0.012 | - |
| 2 | 0.67 ± 0.118 | - | - | - | 1.87 ± 0.002 | 0.45 ± 0.013 | 0.26 ± 0.044 | 0.26 ± 0.016 | 0.66 ± 0.119 | - | - | - |
| 3 | 0.32 ± 0.035 | - | 0.62 ± 0.081 | - | 0.55 ± 0.092 | 0.78 ± 0.012 | 0.24 ± 0.008 | - | 0.40 ± 0.146 | - | 0.15 ± 0.007 | - |
| 4 | - | 0.54 ± 0.051 | 0.14 ± 0.033 | 0.02 ± 0.012 | 0.10 ± 0.004 | 0.32 ± 0.002 | 0.49 ± 0.140 | 0.13 ± 0.034 | - | - | 0.23 ± 0.001 | - |
| 5 | 0.08 ± 0.007 | 0.52 ± 0.027 | 0.58 ± 0.096 | - | 0.29 ± 0.029 | 0.86 ± 0.110 | 0.16 ± 0.013 | - | - | - | - | - |
| 6 | 1.10 ± 0.086 | 1.49 ± 0.010 | 0.78 ± 0.018 | - | 3.52 ± 0.077 | 0.34 ± 0.005 | 0.30 ± 0.022 | - | 1.53 ± 0.641 | - | 0.19 ± 0.002 | - |
| 7 | 0.48 ± 0.077 | 0.39 ± 0.041 | 0.27 ± 0.011 | - | 3.33 ± 0.297 | 0.34 ± 0.012 | 0.77 ± 0.017 | 0.11 ± 0.008 | 0.64 ± 0.027 | - | 0.42 ± 0.005 | - |
| 8 | 0.17 ± 0.001 | 0.43 ± 0.013 | - | - | 2.00 ± 0.061 | 1.36 ± 0.087 | 0.43 ± 0.020 | - | 0.23 ± 0.038 | - | - | - |
| 9 | - | 1.55 ± 0.057 | - | - | 0.17 ± 0.007 | 0.32 ± 0.001 | - | - | - | - | - | - |
| 10 | 1.37 ± 0.002 | 1.64 ± 0.069 | - | 0.21 ± 0.001 | 1.92 ± 0.119 | 0.73 ± 0.045 | - | 0.16 ± 0.004 | 0.56 ± 0.109 | - | - | - |
| 11 | 0.59 ± 0.065 | 0.95 ± 0.204 | - | - | 3.51 ± 0.061 | 1.17 ± 0.088 | 0.20 ± 0.005 | 0.15 ± 0.006 | 0.74 ± 0.028 | - | - | - |
| 12 | 0.39 ± 0.071 | 0.82 ± 0.100 | 0.52 ± 0.057 | - | 0.54 ± 0.030 | 0.30 ± 0.004 | 0.25 ± 0.002 | - | 0.23 ± 0.141 | - | 0.13 ± 0.014 | - |
| 13 | 0.25 ± 0.042 | 0.40 ± 0.031 | 0.19 ± 0.010 | - | 0.80 ± 0.031 | 1.31 ± 0.051 | 0.29 ± 0.035 | 0.46 ± 0.002 | 0.44 ± 0.078 | - | 0.33 ± 0.005 | - |
| 14 | 1.34 ± 0.220 | 0.41 ± 0.010 | 1.02 ± 0.072 | - | 2.10 ± 0.043 | - | 2.95 ± 0.014 | - | 0.89 ± 0.263 | - | - | - |
| 15 | 0.16 ± 0.025 | - | - | - | 0.58 ± 0.045 | - | - | - | 0.15 ± 0.033 | - | - | - |
| 16 | 1.29 ± 0.076 | 0.78 ± 0.026 | - | 0.12 ± 0.013 | 0.89 ± 0.039 | 0.23 ± 0.003 | 0.69 ± 0.038 | - | 1.27 ± 0.252 | - | 0.35 ± 0.020 | - |
| 17 | 0.46 ± 0.036 | 0.37 ± 0.005 | - | 0.27 ± 0.099 | 2.62 ± 0.102 | 1.39 ± 0.034 | 2.55 ± 0.076 | - | 0.73 ± 0.185 | - | 0.87 ± 0.023 | - |
| 18 | 1.38 ± 0.277 | 0.89 ± 0.171 | 0.34 ± 0.092 | 0.18 ± 0.014 | 0.31 ± 0.012 | 0.35 ± 0.011 | 0.28 ± 0.009 | - | 0.49 ± 0.076 | - | 0.13 ± 0.001 | - |
| 19 | - | - | 0.58 ± 0.049 | - | - | - | 0.41 ± 0.023 | - | - | - | 0.22 ± 0.002 | - |
| 20 | - | 0.73 ± 0.100 | 1.02 ± 0.072 | - | - | - | 1.13 ± 0.057 | - | - | - | 0.55 ± 0.004 | - |
| Average | 0.64 | 0.89 | 0.52 | 0.16 | 1.47 | 0.66 | 0.69 | 0.18 | 0.65 | - | 0.32 | - |

The leaf extracts contained more LA than the other plant parts. The following order of average LA content was observed: leaf (1.47 g/100 g) > stem (0.65 g/100 g) > flower (0.64 g/100 g). LE content showed the following order: flower (0.89 g/100 g) > leaf (0.66 g/100 g). LE was not observed in the stems. The average CA content had the following order: leaf (0.69 g/100 g) > stem (0.52 g/100 g) > flower (0.32 g/100 g). The average AP content was similar in both the leaves (0.18 g/100 g) and the flowers (0.16 g/100 g), but AP was not observed in the stem.

The content of the various compounds did not vary according to cultivar/breed. Total polyphenol content (TPC) and total flavonoid content (TFC) of different plant parts of the 20 COP are shown in Table 7. The TPC of 20 COP was determined using a linear gallic acid standard curve. The TPC of the 20 COP ranged from 0.31 to 6.78 g gallic acid equivalents (GAE)/100 g, with lower values being obtained from the stems, while the higher values were obtained from the flowers and leaves. The TFC of 20 COP was determined using a linear catechin standard. The TFC of the 20 COP ranged from 0.25 to 10.45 g catechin equivalents (CE)/100 g. The TFC of some leaves was higher than the content of flowers; however, most leaves had a lower content than the flowers. It has been previously reported that the TPC and TFC of *Chrysanthemum indicum* L. flower was 2.80 g/100 g and 1.89 g/100 g [23].

The different plant parts of the 20 COP had different antioxidant activity levels, as shown in Table 7. The methanol extract of the different plant parts of the 20 COP were initially evaluated for antioxidant activity at concentration of 1 mg/mL using the DPPH free radical scavenging test system. The plant extracts were derived from the flower, leaf and stem parts, which inhibited antioxidant activity by 8.17%–92.47%, 15.12%–91.17%, and 9.61%–88.2%. The flower parts of Red COP (Kugya-myungseong, Entry 6) exhibited the highest antioxidant activity: 92.47%. Antioxidant activity was significantly correlated with TPC and TFC ($R^2$ = 0.681 and 0.781, respectively; Figure 3). Previous studies have reported strong relationships for antioxidant activity with TPC and TFC in several fruits, vegetables, and grain products [12]. The antioxidant activity was shown to be significantly correlated with TPC and TFC ($R^2$ = 0.681 and 0.781, respectively).

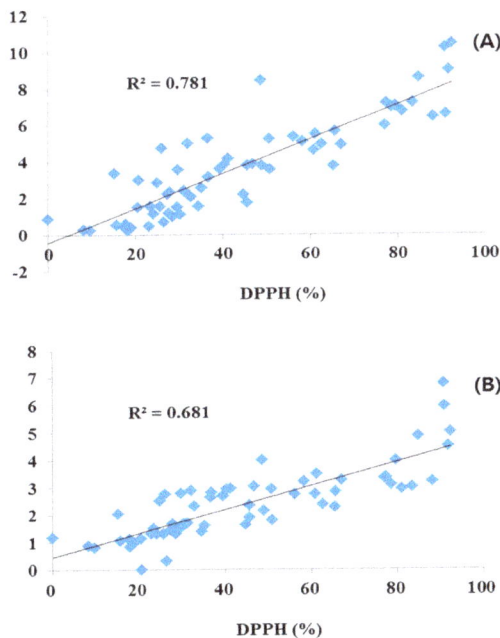

**Figure 3.** Correlation coefficients of antioxidant activity with total polyphenol (**A**) and total flavonoid content (**B**) in different parts of Compositae.

**Table 7.** Total polyphenol content (TPC, gallic acid equivalents (GAE) g/100 g), total flavonoid content (TFC, catechin equivalents (CE) g/100 g), and the 1,1-Diphenyl-2-picrylhydrazyl (DPPH) scavenging activity (%, 1 mg/mL) in different parts of Compositae.

| Entry | Flower (n = 3) | | | Leave (n = 3) | | | Stem (n = 3) | | |
|---|---|---|---|---|---|---|---|---|---|
| | DPPH (%) | TFC | TPC | DPPH (%) | TFC | TPC | DPPH (%) | TFC | TPC |
| 1 | 20.47 ± 2.66 | 1.52 ± 0.06 | 1.13 ± 0.08 | 39.47 ± 0.42 | 3.58 ± 0.23 | 2.68 ± 0.18 | 67.07 ± 2.00 | 4.89 ± 0.04 | 3.25 ± 0.29 |
| 2 | 8.17 ± 0.75 | 0.31 ± 0.04 | 0.88 ± 0.06 | 25.04 ± 1.91 | 2.83 ± 0.16 | 2.51 ± 0.15 | 23.04 ± 2.52 | 0.47 ± 0.10 | 1.32 ± 0.09 |
| 3 | 81.17 ± 0.17 | 6.73 ± 0.13 | 2.93 ± 0.04 | 36.82 ± 1.83 | 3.13 ± 0.04 | 2.81 ± 0.26 | 30.32 ± 0.83 | 1.10 ± 0.14 | 1.67 ± 0.13 |
| 4 | 48.88 ± 1.58 | 3.78 ± 0.31 | 2.14 ± 0.08 | 50.65 ± 0.08 | 5.21 ± 0.47 | 2.94 ± 0.24 | 25.72 ± 1.12 | 1.55 ± 0.07 | 1.32 ± 0.15 |
| 5 | 62.64 ± 3.58 | 4.91 ± 0.30 | 2.36 ± 0.19 | 29.82 ± 2.75 | 3.55 ± 0.04 | 2.78 ± 0.30 | 17.67 ± 1.78 | 0.60 ± 0.03 | 1.11 ± 0.12 |
| 6 | 92.47 ± 0.67 | 10.45 ± 0.41 | 5.00 ± 0.97 | 36.65 ± 1.25 | 5.26 ± 0.45 | 2.64 ± 0.26 | 45.65 ± 2.77 | 1.75 ± 0.04 | 1.91 ± 0.04 |
| 7 | 35.29 ± 3.66 | 2.54 ± 0.34 | 1.60 ± 0.17 | 48.70 ± 0.83 | 4.42 ± 0.34 | 3.99 ± 0.02 | 65.44 ± 0.32 | 3.72 ± 0.01 | 2.27 ± 0.25 |
| 8 | 31.29 ± 2.00 | 2.42 ± 0.23 | 1.70 ± 0.24 | 46.76 ± 1.41 | 3.85 ± 0.21 | 3.04 ± 0.22 | 18.94 ± 2.32 | 0.40 ± 0.08 | 0.99 ± 0.08 |
| 9 | 56.17 ± 0.58 | 5.31 ± 0.21 | 2.75 ± 0.44 | 32.76 ± 0.58 | 2.08 ± 0.03 | 2.31 ± 0.18 | 15.62 ± 0.25 | 0.56 ± 0.03 | 1.05 ± 0.13 |
| 10 | 77.47 ± 1.25 | 7.16 ± 0.34 | 3.35 ± 0.27 | 26.11 ± 1.00 | 4.72 ± 0.37 | 2.76 ± 0.29 | 26.50 ± 1.27 | 0.71 ± 0.04 | 0.31 ± 0.34 |
| 11 | 45.58 ± 0.25 | 3.78 ± 0.37 | 2.35 ± 0.37 | 40.35 ± 1.00 | 3.80 ± 0.11 | 2.93 ± 0.27 | 24.17 ± 0.59 | 1.14 ± 0.03 | 1.32 ± 0.14 |
| 12 | 60.94 ± 1.00 | 4.58 ± 0.11 | 2.73 ± 0.25 | 15.12 ± 1.08 | 3.38 ± 0.08 | 2.04 ± 0.17 | 18.31 ± 1.00 | 0.27 ± 0.04 | 1.06 ± 0.07 |
| 13 | 27.88 ± 0.83 | 2.32 ± 0.11 | 1.69 ± 0.15 | 41.23 ± 0.08 | 4.13 ± 0.24 | 2.96 ± 0.32 | 28.69 ± 2.12 | 1.01 ± 0.13 | 1.30 ± 0.13 |
| 14 | 78.71 ± 2.99 | 6.94 ± 0.28 | 3.08 ± 0.34 | 91.17 ± 0.10 | 6.59 ± 0.01 | 5.95 ± 0.33 | 29.61 ± 1.84 | 1.53 ± 0.10 | 1.50 ± 0.11 |
| 15 | 23.47 ± 2.25 | 1.54 ± 0.25 | 1.51 ± 0.15 | 19.64 ± 0.72 | 0.87 ± 0.07 | 1.19 ± 0.16 | 9.61 ± 3.11 | 0.25 ± 0.04 | 0.82 ± 0.08 |
| 16 | 83.52 ± 0.83 | 7.23 ± 0.40 | 3.02 ± 0.52 | 61.29 ± 0.67 | 5.49 ± 0.52 | 3.47 ± 0.24 | 34.63 ± 2.16 | 1.55 ± 0.13 | 1.37 ± 0.07 |
| 17 | 27.47 ± 0.75 | 2.22 ± 0.34 | 1.46 ± 0.12 | 90.94 ± 0.56 | 10.22 ± 0.20 | 6.78 ± 0.69 | 88.20 ± 0.39 | 6.45 ± 0.47 | 3.21 ± 0.18 |
| 18 | 65.64 ± 0.83 | 5.62 ± 0.31 | 2.84 ± 0.35 | 32.05 ± 0.75 | 4.99 ± 0.10 | 2.88 ± 0.01 | 27.77 ± 1.81 | 1.17 ± 0.13 | 1.44 ± 0.11 |
| 19 | 77.11 ± 1.41 | 5.94 ± 0.28 | 3.34 ± 0.39 | 20.76 ± 3.08 | 3.01 ± 0.24 | 2.51 ± 0.08 | 17.81 ± 1.92 | 0.38 ± 0.08 | 0.83 ± 0.23 |
| 20 | 79.64 ± 1.83 | 7.02 ± 0.23 | 3.96 ± 0.33 | 58.35 ± 2.33 | 5.06 ± 0.03 | 3.21 ± 0.25 | 44.73 ± 1.80 | 2.17 ± 0.07 | 1.64 ± 0.16 |
| Average | 54.20 | 4.61 | 2.49 | 42.18 | 4.51 | 3.12 | 32.97 | 1.58 | 1.48 |

## 3. Discussion

Quality assurance (QA) and quality control (QC) are very important and a suitable analytical method is needed to enable producers to check raw materials, processed foods and plant extracts. The developed UPLC method offers advantages over those previously reported using conventional liquid chromatography by showing a faster chromatographic total run time and better chromatographic performance. Since a high number of samples are needed for QA/QC analysis, the very fast UPLC separation combined with an adequate efficiency described in this work allows the application of the method for routine analysis. The applicability of our method was finally verified by using different parts of 20 COP. Extract stock solutions (10 µg/mL) were analyzed in order to evaluate peak separation and method suitability. Figure 2 showed the resulting chromatograms, where separation was highly satisfactory in all the samples.

The UPLC-DAD method developed in this study was applied, for the first time, to the quantitative analysis of the flavonoids identified in the different parts of COP methanol extracts. A weighed amount (10.0 mg) of dried methanol extract of the different parts of COP plant products was dissolved in 1.0 mL of methanol and filtered through a 0.45 µm filter before analysis. The external standard calibration curve of four flavonoids was generated by using five data points and injections were performed in triplicate for each concentration level (injection volume: 10 µL). In addition, the calibration curve was obtained by plotting the peak area of the compound at each level versus the concentration of the samples and the amount of the four flavonoids in sample extracts was determined by using the calibration curves of the compounds. Our analytical method was applied to simultaneous determination of the four flavonoids in the sample extracts. In Table 6, the flavonoid contents were expressed as g/100 g (dried material). The contents of LA, LE, CA and AP were in the following ranges: LA of 0.16–1.37 (flower), 0.10–3.52 (leave), 0.23–1.53 (stem) g/100 g; LE of 0.25–1.64 (flower), 0.23–1.39 (leave) g/100 g; CA of 0.14–1.02 (flower), 0.20–2.55 (leave), 0.13–0.55 (stem) g/100 g; AP of 0.02–0.27 (flower), 0.01–0.46 (leave) g/100 g dried material, respectively.

The limit of detection (LOD) values of LA, LT, CA and AP were 0.13, 0.26, 0.38 and 0.11 µg/mL$^{-1}$, respectively, and the limit of quantification (LOQ) values were 0.41 (LA), 0.79 (LT), 1.15 (CA) and 0.35 (AP) µg/mL$^{-1}$ (Table 2). The LOD and LOQ values obtained for the standard compounds indicate a suitable sensitivity of the UPLC-DAD method proposed for the analysis of sample extracts. However, LE and AP contents were under the LOQ value in stem.

Leaves had a generally higher content of the four flavonoids (0.18–1.47 g/100 g) than flowers (0.16–0.89 g/100 g), whereas, stems had the lower contents of the four flavonoids (0.32–0.65 g/100 g) and LE and AP were not detected. On the other hand, the COP samples showed considerable differences in antioxidant activity over a 32.97%–54.20% range. We suggest that antioxidant activity was changed by the TPC and TFC. Among them, Entry 6 flowers showed strong antioxidant activity of 92.47% (TFC: 10.45 g/100 g; TPC: 5.00 g/100 g), but leaves and stems of Entry 6 showed lower activities with 36.65 (TFC: 5.26 g/100 g; TPC: 2.64 g/100 g) and 45.65% (TFC: 1.75 g/100 g; TPC: 1.91 g/100 g). In addition, Entry 17 leaves and stems had strong inhibition of 90.94% (TFC: 10.22 g/100 g; TPC: 6.78 g/100 g) and 88.20% (TFC: 6.45 g/100 g; TPC: 3.21 g/100 g), more than flowers 27.47% (TFC: 2.22 g/100 g; TPC: 1.46 g/100 g). Similarly, Entry 3, 10, 16 and 20 showed relatively high antioxidant activity; all of these samples also contained above average levels of TFC and TPC. The TFC-rich samples generally showed very strong activities with high TPC. Antioxidant activity was significantly correlated with TPC and TFC.

HPLC including the MS method allows the separation of most constituents, but there are several problems in applying this method to natural extracts: resolution is often incomplete, the analysis time too long and costly, and chromatographic performance quickly deteriorates [20]. Previously Zhang et al. reported on a rapid quantitative determination of andrographolide, neoandrographolide, 14-deoxyandrographolide and 14-deoxyl-11,12-didehydroandrographolide in *Andrographis paniculata* extract by reverse phase UPLC and gradient elution using ACN-water as mobile phase (0–2 min, 20%–25% ACN (A); 2–5 min, 25%–35% A; 5–7 min, 35% A; 7–10 min, 35%–55% A) at a flow rate of

0.5 min/mL, detecting wavelength at 220 nm [24]. Li et al. reported the development and validation of a UPLC method for the simultaneous quantification of five flavonoids (baicalin, wogonoside, baicalein, wogonin and oroxylin) in *Scutellariae Radix* [25]. Wang et al. also reported a UPLC quantification of LE, rutin, quercetin and betulinic acid in an extract of *Disporopsis pernyi* (Hua) Diels with more than 30 min of run time and a gradient elution with a mobile phase of ACN and water containing 0.1% FA in water and with a flow rate of 0.2 mL/min, with detection at 210, 254, and 280 nm [26]. In addition, Miranda et al. were to develop and validate a novel UPLC-DAD method for simultaneously quantifying chloroquine and primaquine in tablet formulations [27] and UPLC was completely validated and successfully applied to the pharmacokinetic and bioavailability study after sublingual vein and oral administration of corilagin to rats by Zheng et al. [19].

Our UPLC method represents an excellent technique for the simultaneous determination of LA, LE, CA and AP in COP with good sensitivity, accuracy, precision, and linearity. The method gives a good resolution of the four flavonoids with a short analysis run time (14 min). The UPLC method can be used as quality control of polyphenolic constituents in COP and can serve as a reference role for the determination of constituents in natural sources, drug candidate and pharmaceutical preparations.

## 4. Materials and Methods

### 4.1. General Experimental Procedures

UPLC was performed using an Accela UPLC system (Accela 1250, Thermo, Boston, MA, USA). $^1$H- and $^{13}$C-NMR spectra and correlation NMR spectra such as COSY, HMBC, HMQC, and distortionless enhancement by polarization transfer (DEPT) were obtained from an Avance DPX 400 (or 600) spectrometer (Bruker, Berlin, Germany) for identification of the isolated compounds. These were obtained at operating frequencies of 400 MHz ($^1$H) and 100 (or 150) MHz ($^{13}$C) with $CD_3OD$, $(CD_3)_2SO$, $(CD_3)_2CO$, or $D_2O$ and tetramethylsilane (TMS) was used as an internal standard. DPPH was obtained from Sigma Chemical Co. (St. Louis, MO, USA). TPC and TFC were determined using the Folin-Ciocalteu phenol reagent and aluminum chloride colorimetric from Sigma-Aldrich Co. All solvents used in the analysis were HPLC grade obtained from J.T. Baker (Phillipsburg, NJ, USA).

### 4.2. Extraction and Isolation

Twenty COP plants (Table 1) were supplied by the "Kugya Farm" (Chuncheon, Korea). There were 11 variants of *Chrysanthemum morifolium*, six variants of *Chrysanthemum indicum*, one *Aster sphathulifolius* Maxim, one *Chrysanthemum zawadskii* var. *latilobum*, and one *Dendranthema makinoi*. The fresh COP plants were dried at 45 °C in a drying oven and then stored at room temperature. The dried COP plants were then ground to less than 0.5-mm pieces by using a grinder (JL-500, Joy-life, Seoul, Korea). Different plant parts of the dried COP (20 g) were extracted twice with methylene chloride (200 mL) at room temperature, using a shaker for 24 h. Then, the residue was mixed for 24 h with methanol (200 mL × 2). The filtrate was concentrated until dry under reduced pressure on a rotary evaporator at 40 °C. The CA, LE, AP and LA from leaves of Entry 13 methanol extract were isolated by column chromatography. Briefly, Entry 13 (1.0 g) was dissolved in methanol and loaded onto a Sephadex LH-20 column and partitioned to obtain seven fractions. CA and LE were directly obtained from fraction 1 and fraction 4, respectively. Fraction 5 was subsequently separated by a silica-gel column eluted by a solvent mixture of methylene chloride and methanol (from 20:0 to 1:1, *v/v*) to obtain AP. LA was purified from fraction 7 on a Sephadex LH-20 column eluted with 70% methanol.

### 4.3. UPLC Analysis

The samples were analyzed on a Thermo UPLC system (Accela 1250, Thermo). UPLC separation was accomplished on an octadecylsilane (ODS) Hypersil column (2.1 mm × 250 mm, 1.9 μm, Thermo). The mobile phase, consisting of ACN and 0.1% formic acid in water, was used at a flow rate of 0.44 mL/min. The gradient elution program was: 5%–15% ACN (B) (0–1 min), 15% B (1–4 min),

15%–25% B (4–5 min), 25%–40% B (5–6 min), 40%–55% B (6–8 min), 55%–100% B (8–9 min), 100% B (9–11 min), 100%–5% B (11–12 min), and 5% B (12–14 min). The injection volume was 5 μL, and the detection wavelength was 254 nm. The column temperature was maintained at 45 °C using a temperature controller.

### 4 4 Method Validation of Quantitative Analysis

Following the specifications of the International Conference on Harmonization guidelines, the analytical method was validated by determining the linearity, LOD, and LOQ, precision, repeatability, and accuracy for each analyte.

### 4.5. Linearity

Linearity was examined using four flavonoid solutions. The four flavonoids were dissolved in methanol to prepare the standard solutions. Five different concentration of standard solutions dissolved in methanol were prepared at a range of 1, 10, 25, 50 and 100 μg/mL by the serial dilution method. The linearity of the calibration curves was determined by plotting the mean peak area (*y*-axis) versus concentration (*x*-axis) for each analyte in this range.

### 4.6. Limit of Detection and Limit of Quantification

After injecting an aliquot (10 μL) of the serial dilutions of 5 individual standard solutions, the LODs and LOQs under the selected UPLC method were determined at signal to noise (S/N) ratios of 3 and 10, respectively.

### 4.7. Precision and Accuracy

Intra- and inter-day variability of the COP extracts were measured to validate the precision. Intra-day variability was determined by analyzing the samples within 24 h. The solutions were injected 3 times, and the RSD value was calculated for the concentration of each analyte in the extract, and was assumed to represent the measure of precision. Each sample was injected three times a day on three consecutive days to assess inter-day variability. Accuracy was evaluated in a recovery test by calculating the mean recovery (%) of the four flavonoids from a spiked extract solution versus an unspiked extract sample.

### 4.8. DPPH Assay

The stable free radical was used to determine the free radical-scavenging activity of the extracts [28]. Briefly, a 0.32-mM solution of DPPH in methanol was prepared, and then 180 μL of this solution was mixed with 30 μL of each sample (crude extract) at concentrations of 1 mg/mL in methanol. After 15 min incubation in the dark, the decrease in the absorbance of the solution was measured at 570 nm on a microplate reader (EL800 Universal Microplate reader, Bio-Tek Instruments, Inc., Winooski, VT, USA). DPPH inhibitory activity was expressed as the percentage inhibition (%) of DPPH in the above assay system, and was calculated as $(1 - B/A) \times 100$, where A and B are the activities of DPPH without and with the test material, respectively.

### 4.9. Determination of Total Polyphenol Content

The TPC content of different parts of COP plants were determined by using the Folin-Ciocalteu reagent, according to the association of official analytical chemists (AOAC) Folin-Ciocalteu method [29]. An aliquot (25 μL) of samples (1 mg/mL) or standard solution of gallic acid (0.05, 0.1, 0.25, 0.5 and 1 mg/mL) was added to a 1.5 mL test tube, containing 75 μL of Folin-Ciocalteu reagent. After 5 min, 200 μL of 7% $Na_2CO_3$ solution was added to the mixture. After 5 min, 700 μL of distilled water was added, and placed in the dark at room temperature for 60 min. The absorbance of all samples was measured at 750 nm using an EL 800 Universal Microplate Reader (Bio-Tek Instruments, Inc.). The TPC of the COP extracts was expressed as g GAE/100 g dry weight. The reported data are the combined results of the 3 replications.

*4.10. Determination of Total Flavonoid Content*

TFC of different parts of COP plants was determined by using the aluminum chloride colorimetric method [30]. An aliquot (100 μL) of samples (1 mg/mL) or standard solution of catechin (0.05, 0.1, 0.25, 0.5, and 1 mg/mL) was added to an 1.5 mL test tube, containing 400 μL of $D_2O$, with 30 μL of 5% $NaNO_2$ being added to this mixture. After 5 min, 30 μL of 10% $AlCl_3$ solution was added to the mixture. The mixture was allowed to stand at room temperature for 6 min. 200 μL of 1 M NaOH was added, and the total volume was made up to 1 mL with $D_2O$. The absorbance of all samples was measured at 510 nm using an EL 800 Universal Microplate Reader (Bio-Tek Instruments, Inc.). The TFC of the COP extracts was expressed as g CE/100 g dry weight. The reported data are the combined results of three replications.

## 5. Conclusions

Four flavonoids which were isolated from COP, have been reported to exert many pharmacological activities. A simple, accurate and rapid UPLC coupled to DAD method has been developed to quantify these four flavonoids in COP. The method was successfully validated. To the best of our knowledge, it is the first time that a UPLC gradient method has been applied to the simultaneous determination of four flavonoids in the COP. The developed method offers advantages over those previously reported using conventional liquid chromatography by showing a faster chromatographic total run time (14 min) and better chromatographic performance: Separation was performed on a ODS Hypersil (2.1 mm × 250 mm, 1.9 μm) column by using a mobile phase of ACN and water with 0.1% formic acid (*v/v*), at a flow rate 0.44 mL/min. The method was validated in terms of selectivity, linearity, accuracy, precision and recovery. Good linearity was observed over the investigated concentration range (1–100 μg/mL), with correlation coefficient values greater than 0.99. The intra- and inter-day precisions over the concentration range were between 1.46%–2.23% (RSD), and the accuracy was between 95.49%–106.23%. In addition, LA, LE, CA and AP in different parts of COP were analyzed between 0.64–1.47 g/100 g, 0.66–0.89 g/100 g, 0.32–0.52 g/100 g and 0.16–0.18 g/100 g, respectively. Our results suggested that since a high number of samples are needed for quality control analysis, the very fast UPLC separation combined with an adequate efficiency described in this work allows the application of the method for routine and QA/QC analysis.

**Acknowledgments:** This research was supported by Basic Science Research Program through the National Research Foundation of Korea (NRF) funded by the Ministry of Education, Science and Technology (NRF-2012R1A1A2008842) and by Priority Research Centers Program through the National Research Foundation of Korea (NRF) funded by the Ministry of Education, Science and Technology (NRF-2009-0094071) and Hallym University Research Fund (HRF-201609-007).

**Author Contributions:** S.S.L., J.H.P. and H.S.H. designed the experiments. J.H.P. prepared the extract sample and isolated its compounds. H.S.H. conducted the TPC, TFC and DPPH assays. H.S.H. wrote the first draft, and S.S.L. revised the manuscript. All the authors read and approved the final manuscript and all authors' names added in manuscript.

**Conflicts of Interest:** The authors declare no conflict of interest.

## References

1. Marx, J.L. Oxygen free radicals linked to many diseases. *Science* **1987**, *235*, 529–531. [CrossRef] [PubMed]
2. Zheng, R.L.; Lesko, S.A.; Tso, P.O. DNA damage induced in mammalian cells by active oxygen species. *Sci. Sin. B* **1988**, *31*, 676–686. [PubMed]
3. Sarafian, T.A.; Bredesen, D.E. Invited commentary is apoptosis mediated by reactive oxygen species? *Free Radic. Res.* **1994**, *21*, 1–8. [CrossRef] [PubMed]
4. Holst, B.; Williamson, G. Nutrients and phytochemicals: From bioavailability to bioefficacy beyond antioxidants. *Curr. Opin. Biotechnol.* **2008**, *19*, 73–82. [CrossRef] [PubMed]

5. Donnez, D.; Jeandet, P.; Clement, C.; Courot, E. Bioproduction of resveratrol and stilbene derivatives by plant cells and microorganisms. *Trends Biotechnol.* **2009**, *27*, 706–713. [CrossRef] [PubMed]

6. Chen, J.H.; Ho, C.T. Antioxidant activities of caffeic acid and its related hydroxycinnamic acid compounds. *J. Agric. Food Chem.* **1997**, *45*, 2374–2378. [CrossRef]

7. Shabihi, F. In Natural antioxidants: Chemistry, health effects, and applications. *AOCS* **1997**, *48*, 97–150.

8. Akihisa, T., Akihisa, K.; Oinuma, H. Triterpene alcohols from the flowers of Compositae and their anti-inflammatory effects. *Phytochemistry* **1996**, *43*, 1255–1260. [CrossRef]

9. Marino, M.; Bersani, C.; Comi, G. Impedance measurements to study the antimicrobial activity of essential oils from *Lamiaceae* and Compositae. *Int. J. Food Microbiol.* **2001**, *67*, 187–195. [CrossRef]

10. Ukiya, M.; Akihisa, T.; Tokuda, H.; Suzuki, H.; Mukainaka, T.; Ichiishi, E.; Yasukawa, K.; Kasahara, Y.; Nishino, H. Constituents of Compositae plants: III. anti-tumor promoting effects and cytotoxic activity against human cancer cell lines of triterpene diols and triols from edible chrysanthemum flowers. *Cancer Lett.* **2000**, *177*, 7–12. [CrossRef]

11. Zhang, Q.; Li, J.; Wang, C.; Sun, W.; Zhang, Z.; Cheng, W. A gradient HPLC method for the quality control of chlorogenic acid, linarin and luteolin in Flos Chrysanthemi Indici suppository. *J. Pharm. Biomed. Anal.* **2007**, *43*, 753–757. [CrossRef] [PubMed]

12. Velioglu, Y.S.; Mazza, G.; Gao, L.; Oomah, B.D. Antioxidant activity and total phenolics in selected fruits, vegetables, and grain products. *J. Agric. Food Chem.* **1998**, *46*, 4113–4117. [CrossRef]

13. Obón, C.; Rivera, D. Plant pigments and their manipulation. *Econ. Bot.* **2006**, *60*, 92–96. [CrossRef]

14. Shabir, G.A. Validation of high-performance liquid chromatography methods for pharmaceutical analysis understanding the differences and similarities between validation requirements of the US Food and Drug Administration, the US Pharmacopeia and the International Conference on Harmonization. *J. Chromatogr. A* **2003**, *987*, 57–66. [PubMed]

15. Lorenzo, C.D.; Santos, A.D.; Colombo, F.; Moro, E.; Dell'Agli, M.; Restani, P. Development and validation of HPLC method to measure active amines in plant food supplements containing *Citrus aurantium* L. *Food Control* **2014**, *46*, 136–142. [CrossRef]

16. Zhao, M.; Araujo, M.M.; Dal, S.; Sigrist, S.; Bergantzle, M.; Ramanitrahasimbola, D.; Andrianjara, C.; Marchioni, E. Development and validation of a selective and effective pressurized liquid extraction followed by liquid chromatography–mass spectrometry method for the determination of fructosazine analogues in the ammonia treated extract of *Eugenia jambolana* Lamarck seeds. *J. Chromatogr. A* **2016**, *1473*, 66–75. [PubMed]

17. Sena Aquino, A.C.; Azevedo, M.S.; Ribeiro, D.H.B.; Oliveira Costa, A.C.; Amante, E.R. Validation of HPLC and CE methods for determination of organic acids in sour cassava starch wastewater. *Food Chem.* **2015**, *172*, 725–730. [CrossRef] [PubMed]

18. Gudlavalleti, S.K.; Leas, D.A.; Reddy, J.R. UPLC Determination of process impurity hydrazine in *Neisseria meningitidis* A/C/Y/W-135-DT Conjugate Vaccine Formulated in Isotonic Aqueous 1x PBS. *J. Pharm. Pharm. Sci.* **2015**, *1*, 12–16.

19. Zheng, B.; Chen, D.; Yang, X.X.; Igo, L.P.; Li, Z.; Ye, X.; Xiang, Z. Development and validation of an UPLC-PDA method for the determination of corilagin in rat plasma and its application to pharmacokinetic study. *J. Chromatogr. B* **2016**, *1031*, 76–79. [CrossRef] [PubMed]

20. International Conference on Harmonisation of Technical Requirements for Registration of Pharmaceuticals for Human Use, Validation of Analytical Procedures Text. Methodology, 2007. Available online: http://www.ich.org/LOB/media/MEDIA417.pdf (accessed on 27 October 1994).

21. Nugroho, A.; Lim, S.C.; Choi, J.W.; Park, H.J. Identification and quantification of the sedative and anticonvulsant flavone glycoside from *Chrysanthemum boreale*. *Arch. Pharm. Res.* **2013**, *36*, 51–60. [CrossRef] [PubMed]

22. Lai, J.P.; Lim, Y.H.; Shen, H.M.; Ong, C.N. Identification and characterization of major flavonoids and caffeoylquinic acids in three Compositae plants by LC/DAD-APCI/MS. *J. Chromatogr. B* **2007**, *848*, 215–225. [CrossRef] [PubMed]

23. Ju, J.C.; Shin, J.H.; Lee, S.J.; Cho, H.S.; Sung, N.J. Antioxidative activity of hot water extracts from medicinal plants. *J. Korean Soc. Food Sci. Nutr.* **2006**, *35*, 7–14. [CrossRef]

24. Zhang, H.Y.; Xu, X.F.; Huang, Y.; Wamg, D.Q.; Deng, Q.H. Characteristic fingerprint analysis and determination of main components on *Andrographis paniculata* extract by UPLC. *Eur. PMC* **2014**, *37*, 1055–1058.

25. Li, H.; Dong, H.; Su, J.; Yang, B. Development and application of a UPLC method for studying influence of phenological stage on chemical composition of *Scutellariae* Radix. *Arch. Pharm. Res.* **2014**, 1–9. [CrossRef] [PubMed]

26. Wang, Y.; Li, S.; Han, D.; Meng, K.; Wang, M.; Zhao, C. Simultaneous determination of rutin, luteolin, quercetin, and betulinic acid in the extract of *Disporopsis pernyi* (Hua) Diels by UPLC. *J. Anal. Methods Chem.* **2015**, *2015*, 5. [CrossRef] [PubMed]

27. Miranda, T.A.; Silva, P.H.; Pianetti, G.A.; Cesar, I.C. Simultaneous quantitation of chloroquine and primaquine by UPLC-DAD and comparison with a HPLC-DAD method. *Malar. J.* **2015**, *14*, 29. [CrossRef] [PubMed]

28. Molyneux, P. The use of the stable free radical diphenylpicrylhydrazyl (DPPH) for estimating antioxidant activity. *SJST* **2004**, *26*, 211–219.

29. Kumazawa, S.; Taniguchi, M.; Suzuki, Y.; Shimura, M.; Kwon, M.S.; Nakayam, T. Antioxidant activity of polyphenols in carob pods. *J. Agric. Food Chem.* **2002**, *50*, 373–377. [CrossRef] [PubMed]

30. Chang, C.C.; Yang, M.H.; Wen, H.M.; Chern, J.C. Estimation of total flavonoid content in propolis by two complementary colorimetric methods. *JFDA* **2002**, *10*, 178–182.

**Sample Availability:** Samples of linarin, luteolin, chlorogenic acid and apigenin are available from the authors.

*molecules*

MDPI

*Article*

# Inhibitors of the Detoxifying Enzyme of the Phytoalexin Brassinin Based on Quinoline and Isoquinoline Scaffolds

**M. Soledade C. Pedras * , Abbas Abdoli and Vijay K. Sarma-Mamillapalle**

Department of Chemistry, University of Saskatchewan, 110 Science Place, Saskatoon, SK S7N 5C9, Canada;
aba087@mail.usask.ca (A.A.); mas792@mail.usask.ca (V.K.S.-M.)
* Correspondence: s.pedras@usask.ca; Tel.: +1-306-966-4772

Received: 17 July 2017; Accepted: 8 August 2017; Published: 14 August 2017

**Abstract:** The detoxification of the phytoalexin brassinin to indole-3-carboxaldehyde and *S*-methyl dithiocarbamate is catalyzed by brassinin oxidase (BOLm), an inducible fungal enzyme produced by the plant pathogen *Leptosphaeria maculans*. Twenty-six substituted quinolines and isoquinolines are synthesized and evaluated for antifungal activity against *L. maculans* and inhibition of BOLm. Eleven compounds that inhibit BOLm activity are reported, of which 3-ethyl-6-phenylquinoline displays the highest inhibitory effect. In general, substituted 3-phenylquinolines show significantly higher inhibitory activities than the corresponding 2-phenylquinolines. Overall, these results indicate that the quinoline scaffold is a good lead to design paldoxins (phytoalexin detoxification inhibitors) that inhibit the detoxification of brassinin by *L. maculans*.

**Keywords:** antifungal; brassinin oxidase; camalexin; crucifer; *Leptosphaeria maculans*; paldoxin; phenylquinoline; phytoalexin detoxification

## 1. Introduction

The conversion of the phytoalexin brassinin (**1**) to indole-3-carboxaldehyde (**2**) is catalyzed by brassinin oxidase (BOLm), the only enzyme currently known to mediate the transformation of a dithiocarbamate to an aldehyde [1,2]. This transformation is a detoxification process that eliminates the toxophore *S*-methyl dithiocarbamate of brassinin to afford non-toxic products (Scheme 1).

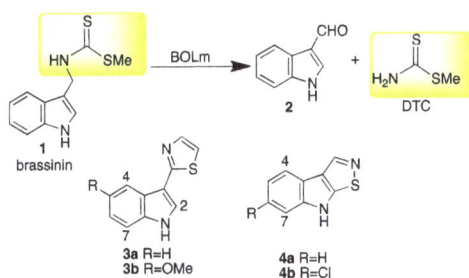

**Scheme 1.** Detoxification of brassinin (**1**) by BOLm (*Leptosphaeria maculans*) and inhibitors camalexin (**3a**), 5-methoxycamalexin (**3b**), brassilexin (**4a**) and 6-chlorobrassilexin (**4b**).

BOLm is an inducible phytoalexin detoxifying enzyme produced by a fungal pathogen of crucifer crops (*Leptosphaeria maculans* (Desm.) Ces. et de Not., asexual stage *Phoma lingam* (Tode ex Fr.) Desm.) that causes major epidemics worldwide. Brassinin (**1**) is a cruciferous phytoalexin produced by plants of the family Brassicaceae (common name crucifer), which include globally cultivated crops belonging to the *Brassica* genus [3]. Brassica crops are of enormous importance worldwide as sources of oil, food, feed, and fuel. Brassinin (**1**) is an important phytoalexin because it functions as antimicrobial plant defense and as biosynthetic precursor of several phytoalexins; depletion of brassinin (**1**) through detoxification is a pathogen's strategy to weaken the defense system of brassicas [3]. In principle, inhibition of such a detoxification transformation could allow brassinin (**1**) to build up in plant cells and stop pathogen growth.

As part of a research program to devise sustainable methods to protect plants against fungal infections, we are particularly interested in the development of paldoxins, i.e., phytoalexin detoxification inhibitors [4]. Paldoxins of BOLm [5,6] are being considered as potential crop protectants having a specific mechanism of action, the inhibition of brassinin detoxification by *L. maculans* [7]. The attraction of this approach lies in the possibility of exploiting paldoxins as selective fungal enzyme inhibitors. It is anticipated that such selective inhibitors will display lower toxicity levels to the encompassing ecosystem and thus are less likely to have a negative environmental impact than conventional fungicides.

Particularly because BOLm has not been expressed in heterologous systems and only relatively small quantities can be obtained using classic chromatographic techniques, in depth structural studies have not been carried out. Consequently, since the tertiary structure of BOLm remains unknown and no relevant model systems have been reported, the design of inhibitors of BOLm is an ongoing challenge. Preliminary screening of a library of more than 80 synthetic brassinin analogues and isosteres, designed by replacement of the dithiocarbamate group of **1** with carbamate, dithiocarbonate, urea, thiourea, sulfamide, sulfonamide, dithiocarbazate, amide, and ester functionalities, plus replacement of the indolyl moiety with naphthalenyl and phenyl, did not identify BOLm inhibitors [8]. Unexpectedly, among several natural products, the phytoalexins camalexin (**3a**) [1] and brassilexin (**4a**) [5] were found to inhibit BOLm. Upon optimization of both lead structures, inhibitors of BOLm more potent than the parent compounds were obtained, 5-methoxycamalexin (**3b**) and 6-chlorobrassilexin (**4b**) became the best competitive inhibitors of BOLm [7]. However, both **3b** and **4b** displayed stronger antifungal activity, a characteristic less desirable in potential paldoxins due to potential toxicity to the plant and surrounding living organisms. Hence, it is of interest to develop new scaffolds containing different heterocyclic systems to establish structural correlations among BOLm inhibitors and their antifungal activities against *L. maculans*. Herein, the inhibition of BOLm and the antifungal activities of a new series of compounds having quinoline and isoquinoline skeletons are reported.

## 2. Results and Discussion

### 2.1. Design and Synthesis of Potential Inhibitors of BOLm

The most significant inhibitors of BOLm discovered to date are derived from indolyl containing scaffolds, namely 5-methoxycamalexin (**3b**) and 6-chlorobrassilexin (**4b**) [7]. Quinolines are monoazanaphthalenes that can be formally considered structural hybrids of indolyl and naphthalenyl skeletons. For this reason, quinolines, and their structural isomers isoquinolines, are heterocyclic scaffolds of interest to us as potential inhibitors of BOLm. A few quinoline derivatives currently used as commercial agricultural fungicides include quinoxyfen and tebufloquin, which are known to affect signal transduction pathways and the respiratory system, respectively [9], and tubulin polymerization [10,11].

quinoxyfen                    tebufloquin

Considering the above framework, the development of quinolines with a new mechanism of action that targets specific fungal pathogens and offers selectivity is of great interest for crop protection. Figure 1 shows the substituted quinolines and isoquinolines designed and synthesized for this purpose.

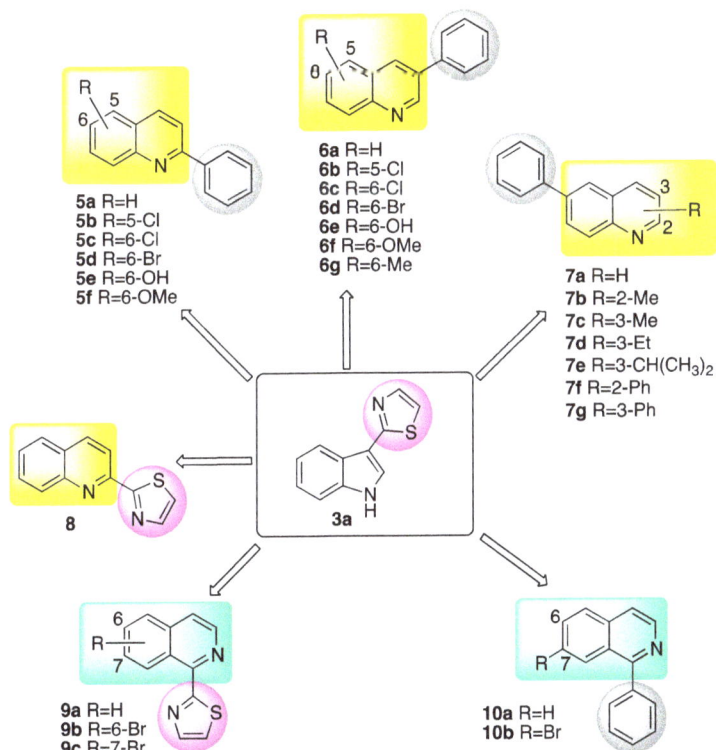

**Figure 1.** Quinolines **5a–8** and isoquinolines **9a–10b** evaluated for inhibition of BOLm.

Syntheses of brassinin (**1**) [8], camalexin (**3a**) [12], and isoquinolines **8** [13] and **9a** [13] were carried out as previously reported. Chemical syntheses of new quinolines **6e**, **7c–7e**, **7g**, improved syntheses of known quinolines **5a–5f**, **6a–6d**, **6f**, **7a**, **7b** and **7f** and syntheses of new isoquinolines **9b**, **9c** and **10b** were carried out as described below.

### 2.1.1. Quinolines

Quinolines can be synthesized using a number of classical methods including the Friedländer, Skraup, Combes, and Doebner–von Miller syntheses from anilines and carbonyl compounds, as well as metal-catalyzed dehydrogenative cyclization and other "greener" methods [14–17]. In the current work, quinolines **5a–5f**, **6a–6f** and **7a–7g** were synthesized using the Friedländer method by condensation of the corresponding 2-aminobenzaldehydes, prepared from substituted 2-nitrobenzaldehydes [18], with aldehydes or ketones. In short, 2-nitrobenzaldehydes **11a–11g** were reduced with iron powder to the corresponding amines by refluxing in EtOH/HCl for 40 min [19]. Aldehydes **12b–12d** and **12f–12h** or ketones **12a**, **12e** and **12i** and KOH (1.2 eq.) were added to the reaction mixture that was kept at RT for 30 min, followed by heating under reflux for 3 h, to afford the corresponding quinolines **5a–5f**, **6a–6f** and **7a–7g** in reasonable yields (Table 1).

**Table 1.** Synthesis of substituted quinolines 5a–5f, 6a–6f and 7a–7g.

| Starting Material | $R_1$ (C#) | 12 | $R_2$ | $R_3$ | Product | % Yield (Reported Yield) |
|---|---|---|---|---|---|---|
| 11a | H | 12a | H | Ph | 5a [a] | 83 (99) |
| 11b | 6-Cl | 12a | H | Ph | 5b [b] | 84 |
| 11c | 5-Cl | 12a | H | Ph | 5c [b] | 63 |
| 11d | 5-Br | 12a | H | Ph | 5d [b] | 64 |
| 11e | 5-OH | 12a | H | Ph | 5e [b] | 42 |
| 11f | 5-OMe | 12a | H | Ph | 5f [b] | 60 |
| 11a | H | 12b | Ph | H | 6a [a] | 63 (87) |
| 11b | 6-Cl | 12b | Ph | H | 6b [b] | 80 |
| 11c | 5-Cl | 12b | Ph | H | 6c [b] | 64 |
| 11d | 5-Br | 12b | Ph | H | 6d [b] | 36 |
| 11e | 5-OH | 12b | Ph | H | 6e [c] | 39 |
| 11f | 5-OMe | 12b | Ph | H | 6f [b] | 53 |
| 11g | 5-Ph | 12c | H | H | 7a [b] | 62 |
| 11g | 5-Ph | 12d | Me | H | 7b [b] | 52 |
| 11g | 5-Ph | 12e | H | Me | 7c [c] | 52 |
| 11g | 5-Ph | 12f | Et | H | 7d [c] | 82 |
| 11g | 5-Ph | 12g | i-Pro | H | 7e [c] | 69 |
| 11g | 5-Ph | 12h | Ph | H | 7f [b] | quantitative |
| 11g | 5-Ph | 12i | H | Ph | 7g [c] | 62 |

[a] Previously synthesized using this method [19]; [b] Previously synthesized by different methods; [c] New compounds.

## 2.1.2. Isoquinolines

The best known classical approaches to isoquinoline scaffolds are the Bischler–Napieralski synthesis, in which a β-arylethylamide is converted into a 3,4-dihydroisoquinoline derivative, the Pictet–Spengler reaction that involves an acid-catalyzed intramolecular cyclization, and the Pomeranz–Fritsch reaction that uses benzylamino acetals [17]. In addition, a large variety of methods that require less harsh conditions are available. In this work, substituted isoquinolines were synthesized from their corresponding bromoisoquinolines **14a–14c**, which were prepared by a modification of the Pomeranz–Fritsch reaction [20,21]. Benzaldehydes **13a–13c** were treated with aminoacetaldehyde dimethylacetal (Dean–Stark conditions) followed by cyclization in the presence of $H_2SO_4/P_2O_5$ to yield isoquinolines **14a–14c**. Thiazole substituted isoquinolines **9a–9c** were prepared from isoquinolines **14a–14c** by treatment with ethyl chloroformate followed by 2-trimethylsilylthiazole and deprotection using *o*-chloranil (Scheme 2).

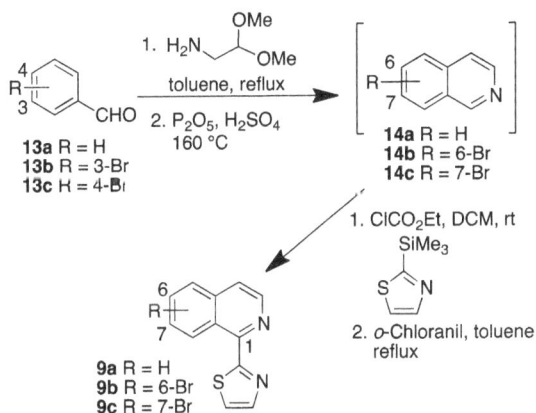

Scheme 2. Synthesis of isoquinolines **9a–9c**.

Syntheses of 1-phenylisoquinolines **10a** and **10b** were carried out by modifying a reported procedure [22]; 1-chloroisoquinolines **15a** and **15b** were treated with iodine followed by PhMgCl and Fe powder in THF (Scheme 3). Unlike in the reported procedure, none of the reactions proceeded in the absence of iodine.

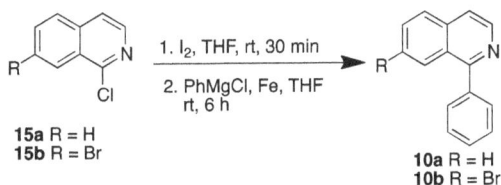

15a R = H
15b R = Br

10a R = H
10b R = Br

Scheme 3. Synthesis of isoquinolines **10a** and **10b**.

## 2.2. Antifungal Activity

The phytoalexins brassinin (**1**) and camalexin (**3a**) were used as reference compounds to compare the antifungal activities of synthetic compounds with those of known phytoalexins. The antifungal activity of **1**, **3a**, quinolines **5a–8** and isoquinolines **9a–10b** against *L. maculans* was determined employing the mycelial growth inhibition assay [23] described in Materials and Methods. The mycelial growth of each plate was measured after incubation for five days and the results were statistically analyzed (Table 2, results of six independent experiments conducted in triplicate). In general, quinoline derivatives (0.50 mM) showed weaker antifungal activity than camalexin (**3a**), except for 3-phenylquinoline (**6a**) and 6-methoxy-3-phenylquinoline (**6f**), whereas 5-chloro-3-phenylquinoline (**6b**) displayed the lowest growth inhibitory activity. 3-Phenylquinoline (**6a**) showed stronger antifungal activity than its structural isomer 6-phenylquinoline (**7a**), whereas structural isomers 6-methyl-3-phenylquinoline (**6g**) and 3-methyl-6-phenylquinoline (**7c**) caused similar mycelial growth inhibition. Interestingly 3,6-diphenylquinoline (**7g**) was not growth inhibitory and 1-(2-thiazolyl)isoquinoline (**9a**) was the most inhibitory of all tested compounds.

**Table 2.** Antifungal activity [a] of the phytoalexins brassinin (**1**), camalexin (**3a**), quinolines **5a–8**, and isoquinolines **9a–10b** against *Leptosphaeria maculans*.

| Compound (#) | Inhibition ± SD (%) [a] | | |
|---|---|---|---|
| | 0.50 mM | 0.20 mM | 0.10 mM |
| Brassinin (**1**) | 50 ± 3 [i,g] | 27 ± 0 [l] | 14 ± 3 [j] |
| Camalexin (**3a**) | 100 ± 0 [c] | 41 ± 2 [i,j] | 24 ± 3 [i] |
| 2-Phenylquinoline (**5a**) | 54 ± 2 [h] | 31 ± 4 [k,l] | 16 ± 5 [j] |
| 5-Chloro-2-phenylquinoline (**5b**) | 39 ± 2 [l] | 18 ± 4 [m,n] | 9 ± 2 [k] |
| 6-Chloro-2-phenylquinoline (**5c**) | 26 ± 4 [m] | 18 ± 5 [m,n] | 5 ± 2 [l] |
| 6-Bromo-2-phenylquinoline (**5d**) | 27 ± 4 [m] | 14 ± 4 [n,o] | 0 [m] |
| 6-Hydroxy-2-phenylquinoline (**5e**) | 68 ± 5 [e,f] | 42 ± 4 [h,i,j] | 32 ± 4 [f,g,h] |
| 6-Methoxy-2-phenylquinoline (**5f**) | 36 ± 4 [l] | 20 ± 4 [m] | 8 ± 2 [k] |
| 3-Phenylquinoline (**6a**) | 100 ± 0 [c] | 77 ± 4 [d] | 39 ± 7 [f] |
| 5-Chloro-3-phenylquinoline (**6b**) | 17 ± 3 [n] | 11 ± 2 [o] | 8 ± 4 [k,l] |
| 6-Chloro-3-phenylquinoline (**6c**) | 28 ± 3 [m] | 17 ± 3 [m,n] | 5 ± 2 [l] |
| 6-Bromo-3-phenylquinoline (**6d**) | 45 ± 2 [k] | 34 ± 2 [k] | 25 ± 0 [i] |
| 6-Hydroxy-3-phenylquinoline (**6e**) | 73 ± 5 [e] | 45 ± 4 [h,i] | 30 ± 4 [g,h] |
| 6-Methoxy-3-phenylquinoline (**6f**) | 100 ± 0 [c] | 68 ± 5 [e] | 55 ± 5 [c] |
| 6-Methyl-3-phenylquinoline (**6g**) | 69 ± 3 [e,f] | 61 ± 2 [f] | 47 ± 0 [e] |
| 6-Phenylquinoline (**7a**) | 64 ± 0 [g] | 55 ± 0 [g] | 30 ± 0 [h] |
| 2-Methyl-6-phenylquinoline (**7b**) | 68 ± 2 [f] | 62 ± 3 [f] | 48 ± 3 [d,e] |
| 3-Methyl-6-phenylquinoline (**7c**) | 74 ± 2 [e] | 67 ± 4 [e] | 53 ± 0 [c] |
| 3-Ethyl-6-phenylquinoline (**7d**) | 67 ± 3 [f] | 61 ± 3 [f] | 48 ± 3 [d,e] |
| 3-Isopropyl-6-phenylquinoline (**7e**) | 72 ± 3 [e] | 66 ± 3 [e] | 56 ± 3 [c] |
| 2,6-Diphenylquinoline (**7f**) | 49 ± 4 [i,j,k] | 40 ± 0 [j] | 33 ± 0 [f,g] |
| 3,6-Diphenylquinoline (**7g**) | 0 [o] | 0 [p] | 0 [m] |
| 2-(2-Thiazolyl)quinoline (**8**) | 54 ± 3 [h] | 46 ± 4 [h] | 8 ± 3 [k,l] |
| 1-(2-Thiazolyl)isoquinoline (**9a**) | 100 ± 0 [c] | 100 ± 0 [c] | 52 ± 4 [c,d] |
| 6-Bromo-1-(2-thiazolyl)isoquinoline (**9b**) | 75 ± 6 [d,e] | 54 ± 3 [g] | 15 ± 4 [j] |
| 7-Bromo-1-(2-thiazolyl)isoquinoline (**9c**) | 82 ± 5 [d] | 63 ± 3 [e,f] | 25 ± 6 [i] |
| 1-Phenylisoquinoline (**10a**) | 54 ± 5 [h,i] | 32 ± 4 [k] | 30 ± 3 [h] |
| 7-Bromo-1-phenylisoquinoline (**10b**) | 47 ± 3 [j,k] | 31 ± 4 [k,l] | 12 ± 2 [j] |

[a] The percentage of inhibition was calculated using the formula: % inhibition = 100 − [(growth on amended/growth in control) × 100]; values are averages of six independent experiments conducted in triplicate; [b] for statistical analysis, one-way ANOVA tests were performed followed by Tukey's test with adjusted $\alpha$ set at 0.05; $n = 6$; different letters in the same column (c–p) indicate significant differences ($p < 0.05$).

## 2.3. Inhibition of Brassinin Oxidase Activity

Cell-free protein extracts of mycelia of *L. maculans* containing BOLm activity induced by 3-phenylindole were employed (prepared as described in Section 3.4) [1] to test the potential paldoxin activity of quinolines **5a–8** and isoquinolines **9a–10b**. Compounds were evaluated at concentrations (0.10 and 0.30 mM) close to the concentration of substrate required for half-maximal activity ($K_M = 0.15$ mM), in the presence of brassinin (**1**, 0.10 mM) and phenazine methosulfate (PMS) as the electron acceptor cofactor (BOLm accepts a wide range of cofactors: PMS, small quinones or flavin mononucleotide, FMN) [1]. To compare the inhibitory activity of all new compounds with those previously reported [7], camalexin (**3a**) was used as a standard due to its established BOLm inhibitory activity (ca. 53% at 0.30 mM) and chemical stability [6]. Results of these assays are presented in Table 3.

**Table 3.** Inhibitory effect of camalexin (**3a**) and quinolines **5a–8** and isoquinolines **9a–10b** on BOLm activity.

| Potential Inhibitor (#) [a] | % Inhibition [b] | | Potential Inhibitor (#) [a] | % Inhibition [b] | |
|---|---|---|---|---|---|
| | 0.10 mM | 0.30 mM | | 0.10 mM | 0.30 mM |
| **3a** | $30 \pm 4$ [c] | $53 \pm 4$ [d] | **8** | n.i. | n.i. |
| **5a** R = H | n.i. | n.i. | **7a** R = H | $16 \pm 2$ [d,e] | $36 \pm 2$ [e,f] |
| **5b** R = 5-Cl | n.i. | n.i. | **7b** R = 2-Me | n.i. | n.i. |
| **5c** R = 6-Cl | n.i. | n.i. | **7c** R = 3-Me | $19 \pm 4$ [d] | $33 \pm 6$ [e,f] |
| **5d** R = 6-Br | n.i. | n.i. | **7d** R = 3-Et | $30 \pm 4$ [c] | $64 \pm 5$ [c] |
| **5e** R = 6-OH | $11 \pm 3$ [e] | $22 \pm 1$ [h] | **7e** R = 3-CH(CH$_3$)$_2$ | $11 \pm 3$ [e] | $33 \pm 5$ [e,f] |
| **5f** R = 6-OMe | n.i. | n.i. | **7f** R = 2-Ph | $10 \pm 2$ [e] | $19 \pm 1$ [h] |
| | | | **7g** R = 3-Ph | $12 \pm 1$ [e] | $25 \pm 1$ [g,h] |
| **6a** R = H | $14 \pm 2$ [d,e] | $30 \pm 4$ [f,g] | **9a** R = H | n.i. | n.i. |
| **6b** R = 5-Cl | n.i. | n.i. | **9b** R = 6-Br | n.i. | n.i. |
| **6c** R = 6-Cl | n.i. | n.i. | **9c** R = 7-Br | n.i. | n.i. |
| **6d** R = 6-Br | n.i. | n.i. | | | |
| **6e** R = 6-OH | $20 \pm 4$ [d] | $34 \pm 2$ [e,f] | | | |
| **6f** R = 6-OMe | $19 \pm 1$ [d] | $40 \pm 3$ [e] | | | |
| **6g** R = 6-Me | $10 \pm 3$ [e] | $32 \pm 3$ [f] | **10a** R = H | n.i. | n.i. |
| | | | **10b** R = 7-Br | n.i. | n.i. |

[a] Brassinin (**1**) was used as substrate (0.10 mM) and PMS (0.10 mM) as cofactor in all experiments. BOLm activity was measured under standard conditions (described in Section 3.4) in the presence of potential inhibitors (0.10 and 0.30 mM). The specific activity of BOLm ($24 \pm 1$ nmol/mg/min) was comparable in all assays; [b] Inhibition is expressed as a percentage of the BOLm activity (100%) of cell-free protein extracts containing brassinin (**1**) and inhibitor; results are means and standard deviations of experiments conducted in triplicate; n.i. = no inhibition. For statistical analysis, one-way ANOVA tests were performed followed by Tukey's test; $n = 6$; different superscript letters within the same column (c–h) indicate significant differences ($p < 0.05$).

It is remarkable that of the 26 compounds tested, 11 inhibited BOLm activity, but none of the halogenated quinolines or isoquinolines affected the activity of BOLm (Table 3). Except for 2-methyl-6-phenyl-quinoline (**7b**), all substituted 6-phenylquinolines inhibited BOLm activity. 3-Ethyl-6-phenylquinoline (**7d**) showed the highest inhibitory effect (ca. 64% at 0.30 mM), somewhat stronger than that of camalexin (**3a**), followed by 6-methoxy-3-phenylquinoline (**6f**) (ca. 40% at 0.30 mM), whereas 6-hydroxy-3-phenylquinoline (**6e**), 6-phenylquinoline (**7a**), 3-methyl-6-phenylquinoline (**7c**), and 3-isopropyl-6-phenylquinoline (**7e**) showed comparable inhibitory effects. Substituted 3-phenylquinolines showed significantly higher inhibitory activities than the corresponding 2-phenylquinolines. The number of structures evaluated is still insufficient to make general conclusions regarding the structure–activity correlations of quinolines and isoquinolines, compounds with substituents at C-4, C-7, and C-8 ought to be considered. Nonetheless, it is apparent and relevant to note that halogenated compounds did not show inhibitory activity against BOLm, contrary to results previously obtained with camalexin (**3a**) and brassilexin (**4a**) derivatives in which halogen-substituted compounds (e.g., **4b**) were somewhat more inhibitory of BOLm activity [7].

## 3. Materials and Methods

### 3.1. General

Chemicals were purchased from Alfa Aesar, Ward Hill, MA or Sigma-Aldrich Canada Ltd., Oakville, ON, Canada; solvents were HPLC grade and used as such. Flash column chromatography (FCC) was carried out using silica gel grade 60, mesh size 230–400 Å or WP C18 prepscale bulk packing 275 Å (J.T. Baker, NJ, USA). Organic extracts were dried over $Na_2SO_4$ and concentrated using a rotary evaporator.

NMR spectra were recorded on Bruker 500 Avance spectrometers (Bruker Corporation, Billerica, MA, USA), for $^1H$, 500.3 MHz and for $^{13}C$, 125.8 MHz; chemical shifts ($\delta$) are reported in parts per million (ppm) relative to TMS; spectra were calibrated using the solvent peaks; spin coupling constants ($J$) are reported to the nearest 0.5 Hz. FTIR data were recorded on a Bio-Rad FTS-40 spectrometer(Bio-Rad Laboratories, Inc., Hercules, CA, USA) and spectra were measured by the diffuse reflectance method on samples dispersed in KBr. HREI-MS were obtained on a VG 70 SE mass spectrometer(Waters Technologies Corporation, Milford, MA, USA) employing a solids probe or on a Jeol AccuToF GCv 4G mass spectrometer (Jeol USA, Peabody, MA, USA) [field desorption (FD)] by direct insertion.

HPLC analysis was carried out with Agilent high performance liquid chromatographs equipped with quaternary pump, automatic injector, and photodiode array detector (DAD, wavelength range 190–600 nm), degasser, and a column Eclipse XDB-C18 (Agilent Technologies, Santa Clara, CA, USA) (5 μm particle size silica, 4.6 mm i.d. × 150 mm), having an in-line filter, using methods: A, mobile phase 80% $H_2O$—20% $CH_3CN$ for 45.0 min, linear gradient, at a flow rate of 0.40 mL/min; B, mobile phase 50% $H_2O$—50% MeOH to 100% MeOH for 25.0 min, linear gradient, at a flow rate 0.75 mL/min.

### 3.2. Synthesis

Syntheses of brassinin (**1**) [8], camalexin (**3a**) [12], and isoquinolines **8** [13] and **9a** [13] were carried out as previously reported. Satisfactory spectroscopic data identical to those previously reported were obtained for all known compounds.

#### 3.2.1. Phenylquinolines **5a–7g**

General Procedure

Iron powder (0.4 eq.) was added to a solution of *o*-nitrobenzaldehyde (**11a–11g**, 1.0 eq.) in EtOH (2.0 mL) followed by HCl (0.05 eq.) [19]. The reaction mixture was refluxed for 40 min followed by cooling to room temperature. The corresponding aldehyde or ketone (1.2 eq.) was added to the reaction mixture followed by KOH (1.2 eq.). The reaction mixture was stirred at RT for 30 min and refluxed for an additional 3 h. The reaction mixture was filtered, the filtrate was neutralized (HCl), extracted with EtOAc, and the combined extracts were dried and concentrated to dryness. The residue was subjected to FCC (EtOAc–hexane, 3:7) to afford substituted quinolines **5a–7g**.

*2-Phenylquinoline* (**5a**): Compound **5a** was prepared from *o*-nitrobenzaldehyde **11a** (151 mg, 1.00 mmol) according to the above procedure in 83% yield (171 mg, 0.83 mmol), obtained as a white powder, m.p. 78–79 °C. Satisfactory spectroscopic data identical to those previously reported [24].

*5-Chloro-2-phenylquinoline* (**5b**): Compound **5b** was prepared from *o*-nitrobenzaldehyde **11b** (200 mg, 1.07 mmol) according to the above procedure in 84% yield (215 mg, 0.90 mmol), obtained as a white powder, m.p. 90–91 °C. Although referenced in the SciFinder database as previously prepared (accessed on 17 April 2017), structure **5b** appears to be a new compound, since it was not published in the cited reference [25].

$^1$H-NMR (500 MHz, CDCl$_3$): δ 8.63 (d, *J* = 9 Hz, 1H), 8.19 (m, 2H), 8.11 (d, *J* = 8 Hz, 1H), 7.99 (d, *J* = 9 Hz, 1H), 7.71–7.59 (m, 2H), 7.56 (m, 2H), 7.50 (dd, *J* = 7, 7 Hz, 1H). $^{13}$C-NMR (125 MHz, CDCl$_3$): δ 158.1, 149.2, 139.2, 133.8, 131.4, 129.9, 129.5, 129.2, 129.1, 127.8, 126.5, 125.5, 119.9. HPLC $t_R$ = 24.4 ± 0.2 min (method B). UV (HPLC, CH$_3$OH-H$_2$O) λ$_{max}$ (nm): 208, 262. FTIR (KBr) ν$_{max}$ cm$^{-1}$: 3058, 1611, 1593, 1580, 1547, 1486, 1461, 1396, 1316, 1278, 1201, 1025, 960, 814, 775, 692, 671. HRMS-EI *m/z* (%): calc. for C$_{15}$H$_{10}$NCl: 239.0502, measured 239.0507 (100), 204.08 (33).

*6-Chloro-2-phenylquinoline* (**5c**): Compound **5c** was prepared from *o*-nitrobenzaldehyde **11c** (200 mg, 1.07 mmol) according to the above procedure in 63% yield (161 mg, 0.670 mmol), obtained as a white powder, m.p. 109–110 °C. Satisfactory spectroscopic data identical to those previously reported [26].

*6-Bromo-2-phenylquinoline* (**5d**): Compound **5d** was prepared from *o*-nitrobenzaldehyde **11d** (59 mg, 0.25 mmol) according to the above procedure in 64% yield (47 mg, 0.16 mmol), obtained as an off-white powder, m.p. 123–125 °C. Satisfactory spectroscopic data identical to those previously reported [27].

*6-Hydroxy-2-phenylquinoline* (**5e**): Compound **5e** was prepared from *o*-nitrobenzaldehyde **11e** (200 mg, 1.19 mmol) according to the above procedure in 42% yield (112 mg, 0.51 mmol), obtained as a brown powder, m.p. 220–223 °C. Satisfactory spectroscopic data identical to those previously reported [28].

*6-Methoxy-2-phenylquinoline* (**5f**): Compound **5f** was prepared from *o*-nitrobenzaldehyde **11f** (150 mg, 0.82 mmol) according to the above procedure in 60% yield (115 mg, 0.49 mmol), obtained as a white powder, m.p. 135–137 °C. Satisfactory spectroscopic data identical to those previously reported [24].

*3-Phenylquinoline* (**6a**): Compound **6a** was prepared from *o*-nitrobenzaldehyde **11a** (200 mg, 1.32 mmol) according to the above procedure in 63% yield (172 mg, 0.84 mmol), obtained as a light yellow powder, m.p. 45–46 °C. Satisfactory spectroscopic data identical to those previously reported [19].

*5-Chloro-3-phenylquinoline* (**6b**): Compound **6b** was prepared from *o*-nitrobenzaldehyde **11b** (200 mg, 1.07 mmol) according to the above procedure in 80% yield (207 mg, 0.87 mmol), obtained as an off-white powder, m.p. 115–116 °C. Compound **6b** was previously synthesized using a different procedure but no spectroscopic data was reported [29].

$^1$H-NMR (500 MHz, CDCl$_3$): δ 9.18 (d, *J* = 2 Hz, 1H), 8.22 (d, *J* = 2 Hz, 1H), 8.08 (d, *J* = 9 Hz, 1H), 7.88 (d, *J* = 2 Hz, 1H), 7.71 (m, 2H), 7.66 (dd, *J* = 9, 2 Hz, 1H), 7.55 (m, 2H), 7.47 (dd, *J* = 7.5, 7.5 Hz, 1H). $^{13}$C-NMR (125 MHz, CDCl$_3$): δ 150.8, 148.1, 137.7, 135.0, 131.7, 130.2, 129.5, 129.2, 128.7, 128.6, 127.8, 127.3, 126.4. HPLC $t_R$ = 22.9 ± 0.2 min (method B). UV (HPLC, CH$_3$OH-H$_2$O) λ$_{max}$ (nm): 208, 238, 257. FTIR (KBr) ν$_{max}$ cm$^{-1}$: 3053, 1487, 1456, 1182, 977, 900, 810, 758, 742, 694. HRMS-EI *m/z* (%): calc. for C$_{15}$H$_{10}$NCl: 239.0502, measured 239.0507 (100), 204.08 (15).

*6-Chloro-3-phenylquinoline* (**6c**): Compound **6c** was prepared from *o*-nitrobenzaldehyde **11c** (200 mg, 1.07 mmol) according to the above procedure in 64% yield (163 mg, 0.68 mmol), obtained as a white powder, m.p. 105–107 °C. Satisfactory spectroscopic data identical to those previously reported [30].

*6-Bromo-2-phenylquinoline* (**6d**): Compound **6d** was prepared from *o*-nitrobenzaldehyde **11d** (200 mg, 0.869 mmol) according to the above procedure in 40% yield (100 mg, 0.35 mmol), obtained as a light yellow powder, m.p. 116–118 °C. Satisfactory spectroscopic data identical to those previously reported [31].

*6-Hydroxy-3-phenylquinoline* (**6e**): Compound **6e** was prepared from *o*-nitrobenzaldehyde **11e** (200 mg, 1.19 mmol) according to the above procedure in 39% yield (102 mg, 0.460 mmol), obtained as a light brown powder, m.p. 223–225 °C.

$^1$H-NMR (500 MHz, DMSO-$d_6$): δ 10.09 (s, 1H), 8.99 (d, *J* = 2 Hz, 1H), 8.41 (s, 1H), 7.89 (d, *J* = 9 Hz, 1H), 7.84 (d, *J* = 7.5 Hz, 2H), 7.53 (dd, *J* = 7.5, 7.5 Hz, 2H), 7.44 (dd, *J* = 7.5, 7.5 Hz, 1H), 7.31 (dd, *J* = 2, 9 Hz,

1H), 7.23 (d, $J$ = 2 Hz, 1H). $^{13}$C-NMR (125 MHz, DMSO-$d_6$): δ 155.9, 146.1, 142.2, 137.4, 132.8, 131.1, 130.1, 129.2, 128.0, 127.1, 122.0, 108.7. HPLC $t_R$ = 14.0 ± 0.2 min (method B). UV (HPLC, CH3OH-H2O) $\lambda_{max}$ (nm): 206, 255, 342. FTIR (KBr) $\nu_{max}$ cm$^{-1}$: 2952, 1620, 1496, 1457, 1376, 1246, 1217, 1166, 950, 898, 763, 701. HRMS-EI $m/z$ (%): calc. for $C_{15}H_{11}NO$: 221.0841, measured 221.0838 (100).

*6-Methoxy-3-phenylquinoline* (**6f**): Compound **6f** was prepared from *o*-nitrobenzaldehyde **11f** (178 mg, 0.98 mmol) according to the above procedure in 53% yield (123 mg, 0.52 mmol), obtained as a yellow powder, m.p. 127–129 °C. Satisfactory spectroscopic data identical to those previously reported [32].

*6-Phenylquinoline* (**7a**): Compound **7a** was prepared from *o*-nitrobenzaldehyde **11g** (20 mg, 0.09 mmol) according to the above procedure in 62% yield (11 mg, 0.053 mmol), obtained as a light brown powder, m.p. 105–106 °C. Satisfactory spectroscopic data identical to those previously reported [33].

*2-Methyl-6-phenylquinoline* (**7b**): Compound **7b** was prepared from *o*-nitrobenzaldehyde **11g** (50 mg, 0.25 mmol) according to the above procedure in 52% yield (25 mg, 0.11 mmol), obtained as a yellow powder, m.p. 89–91 °C. Satisfactory spectroscopic data identical to those previously reported [34].

*3-Methyl-6-phenylquinoline* (**7c**): Compound **7c** was prepared from *o*-nitrobenzaldehyde **11g** (20 mg, 0.090 mmol) and propanal (50 μL) according to the above procedure in 52% yield (10 mg, 0.050 mmol), obtained as a light yellow oil.

$^1$H-NMR (500 MHz, CDCl$_3$): δ 8.78 (d, $J$ = 1.5 Hz, 1H), 8.14 (d, $J$ = 8.5 Hz, 1H), 7.97 (s, 1H), 7.93 (s, 1H), 7.91 (d, $J$ = 2 Hz, 1H), 7.73–7.71 (m, 2H), 7.51 (dd, $J$ = 7.5, 8.0 Hz, 1H), 7.41 (dd, $J$ = 8.5, 7.5 Hz, 1H), 2.55 (s, 3H). $^{13}$C-NMR (125 MHz, CDCl$_3$): δ 152.6, 146.1, 140.7, 139.5, 135.1, 131.1, 129.8, 129.1, 128.5, 128.4, 127.9, 127.7, 125.1, 19.0. HPLC $t_R$ = 23.3 ± 0.2 min (method A). UV (HPLC, CH$_3$CN-H$_2$O) $\lambda_{max}$ (nm): 210, 250. FTIR (KBr) $\nu_{max}$ cm$^{-1}$: 2974, 1600, 1486, 1345, 1174, 900, 835, 759, 697, 570, 479. HRMS-FD $m/z$ (%): calc. for $C_{16}H_{13}N$: 219.1048, measured 219.1051 (100).

*3-Ethyl-6-phenylquinoline* (**7d**): Compound **7d** was prepared from *o*-nitrobenzaldehyde **11g** (60 mg, 0.27 mmol) according to the above procedure in 82% yield (50 mg, 0.21 mmol), obtained as a yellow oil.

$^1$H-NMR (500 MHz, CDCl$_3$): δ 8.80 (s, 1H), 8.15 (d, $J$ = 8.5 Hz, 1H), 7.98 (d, $J$ = 9.5 Hz, 2H), 7.93 (d, $J$ = 8.5 Hz, 1H), 7.73 (d, $J$ = 7.5 Hz, 2H), 7.51 (dd, $J$ = 7, 7.5 Hz, 2H), 7.41 (dd, $J$ = 7.5, 7 Hz, 1H), 2.88 (q, $J$ = 8, 7.5, 7.5 Hz, 2H), 1.39 (t, $J$ = 7.5, 7.5 Hz, 3H).$^{13}$C-NMR (125 MHz, CDCl$_3$): δ 151.9, 146.2, 140.7, 139.6, 137.3, 134.0, 129.6, 129.2, 128.6, 127.9, 127.7, 125.3, 26.5, 15.4. HPLC $t_R$ = 27.2 ± 0.2 min (method A). UV (HPLC, CH$_3$CN-H$_2$O) $\lambda_{max}$ (nm): 210, 252. FTIR (KBr) $\nu_{max}$ cm$^{-1}$: 2965, 1599, 1487, 1348, 1179, 908, 837, 757, 697, 627, 570. HRMS-FD $m/z$ (%): calc. for $C_{17}H_{15}N$: 233.1205, measured 233.1205 (100).

*3-Isopropyl-6-phenyl-quinoline* (**7e**): Compound **7e** was prepared from *o*-nitrobenzaldehyde **11g** (60 mg, 0.27 mmol according to the above procedure in 69% yield (45 mg, 0.18 mmol), obtained as a yellow oil.

$^1$H-NMR (500 MHz, CDCl$_3$): δ 8.85 (s, 1H), 8.17 (d, $J$ = 8.5 Hz, 1H), 7.99 (d, $J$ = 9.5 Hz, 2H), 7.93 (d, $J$ = 10 Hz, 1H), 7.74–7.72 (m, 1H), 7.51 (dd, $J$ = 7.5, 8 Hz, 2H), 7.41 (dd, $J$ = 7.5, 7.5 Hz, 1H), 3.19–3.13 (m, 1H), 1.41 (s, 3H), 1.40 (s, 3H).$^{13}$C-NMR (125 MHz, CDCl$_3$): δ 151.3, 146.4, 141.7, 140.7, 139.4, 132.3, 129.6, 129.1, 128.6, 128.5, 127.8, 127.6, 125.4, 32.1, 23.8. HPLC $t_R$ = 30.1 ± 0.2 min (method A). UV (HPLC, CH$_3$CN-H$_2$O) $\lambda_{max}$ (nm): 210, 252. FTIR (KBr) $\nu_{max}$ cm$^{-1}$: 2960, 1599, 1486, 1351, 1178, 1077, 975, 909, 837, 757, 697, 473. HRMS-FD $m/z$ (%): calc. for $C_{18}H_{17}N$: 247.1361, measured 247.1372 (100).

*3,6-Diphenylquinoline* (**7f**): Compound **7f** was prepared from *o*-nitrobenzaldehyde **11g** (40 mg, 0.18 mmol) according to the above procedure in 100% yield (50 mg, 0.18 mmol), obtained as a light yellow powder, m.p. 131–133 °C.

$^1$H-NMR (500 MHz, CDCl$_3$): δ 9.20 (s, 1H), 8.34 (s, 1H), 8.23 (d, $J$ = 9 Hz, 1H), 8.10 (s, 1H), 8.00 (d, $J$ = 9 Hz, 1H), 7.76–7.74 (m, 4H), 7.57–7.51 (m, 4H), 7.48–7.42 (m, 2H).$^{13}$C-NMR (125 MHz, CDCl$_3$): δ 150.0, 146.9, 140.4, 139.9, 138.0, 134.4, 133.6, 129.8, 129.4, 129.3, 129.2, 128.4, 128.3, 128.0, 127.6, 127.5, 125.8. HPLC $t_R$ = 30.1 ± 0.2 min (method A). UV (HPLC, CH$_3$CN-H$_2$O) $\lambda_{max}$ (nm): 210, 280. FTIR

(KBr) $v_{max}$ cm$^{-1}$: 3052, 1487, 1348, 1077, 910, 834, 758, 699, 571, 531, 497, 470. HRMS-FD *m/z* (%): calc. for $C_{21}H_{15}N$: 281.1205, measured 281.1218 (100).

*2,6-Diphenylquinoline* (**7g**): Compound **7g** was prepared from *o*-nitrobenzaldehyde **11g** (20 mg, 0.090 mmol) according to the above procedure in 62% yield (15 mg, 0.053 mmol), obtained as a light brown powder, m.p. 198–200 °C. Satisfactory spectroscopic data identical to those previously reported [35].

### 3.2.2. Synthesis Isoquinolines **9b**, **9c** and **10b**

Isoquinolines **9b** and **9c**

Aminoacetaldehyde dimethylacetal (3.0 eq.) was added to a solution of bromobenzaldehyde **13b** or **13c** (1.0 eq.) in toluene (30 mL). Each reaction mixture was refluxed (Dean–Stark apparatus) at 120 °C. After consumption of the starting material, each reaction mixture was concentrated to dryness, then dissolved in conc. $H_2SO_4$ (2 mL) and added to a cold solution of $P_2O_5$ in conc. $H_2SO_4$ (0.5 mL). Each reaction mixture was heated at 160 °C for 30 min, allowed to cool to RT, neutralized with NaOH (10 M), extracted with EtOAc, and concentrated to dryness. Each residue was subjected to FCC to afford 6-bromoisoquinoline (**14b**, 30 mg, 0.14 mmol, 14%) and 7-bromoisoquinoline (**14c**, 99 mg, 0.47 mmol, 22%) [20,21]. Ethylchloroformate (1.0 eq.) was added to a solution of isoquinoline **14b** or **14c** (1.0 eq.) in DCM at 0 °C and stirred at the same temperature for 30 min, followed by addition of 2-trimethylsilylthiazole (1.0 eq.). Each reaction mixture was stirred at RT for 3 h, concentrated to dryness, and each residue was subjected to FCC. Each product was dissolved in benzene (5 mL), *o*-chloranil (1.0 eq.) was added, and each reaction mixture was refluxed for 5 h. Each reaction mixture was diluted with 5% NaOH (10 mL), extracted with DCM, and concentrated to dryness. Each reaction mixture residue was subjected to FCC to afford the products **9b** and **9c**.

*6-Bromo-1-(2-thiazolyl)isoquinoline* (**9b**): 6-Bromoisoquinoline (**14b**, 30 mg, 0.14 mmol) was synthesized starting from 4-bromobenzaldehyde (**13c**, 200 mg, 1.08 mmol) in 14% yield. Compound **9b** was synthesized starting from 6-bromoisoquinoline (**14b**, 100 mg, 0.48 mmol) in 15% yield over two steps (21 mg, 0.07 mmol), obtained as an orange powder, m.p. 103–105 °C.

$^1$H-NMR (500 MHz, CDCl$_3$): δ 9.77 (d, *J* = 9 Hz, 1H), 8.58 (d, *J* = 5.5 Hz, 1H), 8.06 (m, 2H), 7.79 (dd, *J* = 2, 9 Hz, 1H), 7.61 (d, *J* = 5.5 Hz, 1H), 7.53 (d, *J* = 3 Hz, 1H). $^{13}$C-NMR (125 MHz, CDCl$_3$): δ 170.6, 149.9, 144.4, 143.0, 138.7, 132.3, 130.1, 129.1, 125.7, 124.0, 122.5, 121.2. HPLC $t_R$ = 22.4 ± 0.2 min (method B). UV (HPLC, CH$_3$OH-H$_2$O) λ$_{max}$ (nm): 240, 345. FTIR (KBr) $v_{max}$ cm$^{-1}$: 2924, 2852, 1782, 1608, 1546, 1426, 1246, 1183, 947, 880, 810. HRMS-EI *m/z* (%): calc. for $C_{12}H_7N_2SBr$: 291.9493, measured 291.9494 (100), 289.95 (95), 247.88 (53), 211.03 (21), 99.96 (39).

*7-Bromo-1-(2-thiazolyl)isoquinoline* (**9c**): 7-Bromoisoquinoline (**14c**, 99 mg, 0.47 mmol) was synthesized starting from compound **13b** in 22% yield [20]. Compound **9c** was synthesized starting from 7-bromoisoquinoline (**14c**, 100 mg, 0.48 mmol) in 24% yield over two steps (35 mg, 0.12 mmol), obtained as an off-white powder m.p. 110–112 °C.

$^1$H-NMR (500 MHz, CDCl$_3$): δ 10.12 (s, 1H), 8.58 (d, *J* = 5.5 Hz, 1H), 8.08 (d, *J* = 3 Hz, 1H), 7.80 (dd, *J* = 9, 2 Hz, 1H), 7.73 (d, *J* = 9 Hz, 1H), 7.67 (d, *J* = 5.5 Hz, 1H), 7.53 (d, *J* = 3 Hz, 1H). $^{13}$C-NMR (500 MHz, CDCl$_3$): δ 170.3, 148.6, 144.5, 142.3, 136.0, 134.2, 130.5, 128.6, 126.3, 123.1, 122.5, 122.1. HPLC $t_R$ = 21.0 ± 0.2 min (method B). UV (HPLC, CH$_3$OH-H$_2$O) λ$_{max}$ (nm): 240, 302, 350. FTIR (KBr) $v_{max}$ cm$^{-1}$: 3089, 1781, 1573, 1544, 1491, 1425, 1247, 1103, 947, 850, 813, 732. HRMS-EI *m/z* (%): calc. for $C_{12}H_7N_2SBr$: 291.9493, measured 291.9483 (100), 289.95 (99), 247.94 (34), 211.03 (32), 99.96 (24).

*1-Phenylisoquinoline* (**10a**): Iodine (18 mg, 0.073 mmol) was added to a solution of compound **15a** (12 mg, 0.073 mmol) in THF (2 mL) and stirred at RT. After 30 min, Fe powder (16 mg, 0.29 mmol) followed by PhMgCl (2 M in THF, 43 μL, 0.087 mmol) were added and stirred at RT. After 6 h, the reaction mixture was diluted with ice-cold water (20 mL) and extracted with EtOAc. The combined organic extracts were dried and concentrated to dryness to afford **10a** in 67% yield (10 mg, 0.050 mmol) as an off-white powder, m.p. 91–92 °C. Satisfactory spectroscopic data identical to those previously reported [36].

*7-Bromo-1-phenylisoquinoline* (**10b**): Iodine (18 mg, 0.073 mmol) was added to a solution of compound **15b** (40 mg, 0.16 mmol) in THF (2 mL) and stirred at RT for 30 min followed by addition of Fe powder (16 mg, 0.294 mmol) and PhMgCl (2 M in THF, 43 μL, 0.087 mmol). The reaction mixture was stirred at RT for 6 h followed by dilution with ice-cold water, extraction with EtOAc, and concentration to dryness to afford **10b** in 47% yield (21 mg, 0.07 mmol) as yellow gum [36].

$^1$H-NMR (500 MHz, CDCl$_3$): δ 8.64 (d, $J$ = 6 Hz, 1H), 8.07 (d, $J$ = 2 Hz, 1H), 7.99 (d, $J$ = 9 Hz, 1H), 7.68 (dd, $J$ = 2, 7.5 Hz, 2H), 7.62 (dd, $J$ = 2, 9 Hz, 1H), 7.58–7.53 (m, 4H). $^{13}$C-NMR (500 MHz, CDCl$_3$): δ 161.2, 143.6, 139.3, 138.2, 130.9, 130.1, 129.7, 129.3, 129.1, 128.7, 125.3, 125.1, 119.0. HPLC $t_R$ = 23.4 ± 0.2 min (method B). UV (HPLC, CH$_3$OH-H$_2$O) λ$_{max}$ (nm): 220, 258, 328. FTIR (KBr) ν$_{max}$ cm$^{-1}$: 3061, 3033, 2923, 1480, 1429, 1261, 1091, 1042, 1008, 903, 800, 728, 697. HRMS-EI *m/z* (%): calc. for C$_{15}$H$_{10}$NBr: 284.9976, measured 284.9969 (10), 237.08 (26), 184.04 (41), 91.05 (100), 82.07 (86).

*3.3. Antifungal Activity*

The antifungal activity of brassinin (**1**), camalexin (**3a**), quinolines **5a–8**, and isoquinolines **9a–10b** against *L. maculans* was evaluated using a mycelial radial growth assay carried out in PDA [23]. Cultures of *L. maculans* were grown on potato dextrose agar (PDA) plates (10 cm diameter) at 23 ± 1 °C under constant light (fluorescent 32W T8 48" daylight) for seven days. DMSO solutions of each compound (0.50, 0.20 and 0.10 mM) in PDA (1% DMSO concentration in PDA) were added to each well (six-well plates, 1.5 mL per well) and inoculated with mycelia (4 mm diameter plugs cut from the edges of 7-day-old solid cultures, placed upside down on the center of each well). Control cultures containing 1% DMSO in PDA were prepared similarly. Plates were incubated under constant light and the diameter of mycelial mat (mm) in each well was measured after five days of incubation. Each assay was conducted in triplicate and repeated twice. The percentage of growth inhibition was calculated as reported in Table 2.

*3.4. Fungal Cultures, Preparation of Cell-Free Protein Extracts, and BOLm Activity*

Liquid cultures of *L. maculans* (isolate UAHM 9410) were handled as described previously. In brief, fungal spores were subcultured on V8 agar under continuous light at 23 ± 1 °C; after 15 days, fungal spores were collected aseptically and stored at −20 °C. Liquid cultures were initiated by inoculating minimal media (100 mL) with fungal spores at 10$^7$/mL in Erlenmeyer flasks, followed by incubation on a shaker under constant light at 23 °C for 48 h; 3-phenylindole (0.050 mM in DMSO) was added to each flask to induce BOLm, and cultures were incubated for an additional 24 h. Cultures were gravity filtered to separate mycelia from culture broth, the mycelia was washed with distilled water, wrung, and stored at −20 °C.

Frozen mycelia (1.2 g) from *L. maculans* (obtained as reported above) were ground in ice-cold extraction buffer (5 mL, mortar and pestle) for 5 min at 4 °C. The extraction buffer consisted of diethanolamine (DEA, 25 mM, pH ~8.3), 10% (*v/v*) glycerol, D,L-dithiothreitol (DTT, 1 mM), and 1/200 (*v/v*) protease inhibitor cocktail (P-8215, Sigma-Aldrich Canada). The homogenate was centrifuged at 4 °C for 30 min at 50,000 *g*. The resulting supernatant was dialyzed three times (twice with 300 mL of dialyzing buffer for 3 h each time and then using 400 mL buffer for 12 h, using dialyzing cassettes in buffer pH 8.3, diethanolamine, 25 mM, 5% glycerol, *v/v*, 10% triton X-100 in deionized water). Dialyzed cell-free protein extracts were used for determination of the specific activity of BOLm and for testing inhibitory activity of compounds.

Protein concentrations were determined as described by Bradford using the Coomassie Brilliant Blue method with BSA as a standard (optical density (OD) was measured at 595 nm).

Determination of BOLm Activity

The reaction mixture contained diethanolamine (20 mM, pH ~8.3), DTT (0.10 mM), 0.1% (*v/v*) triton X-100, brassinin (**1**, 0.10 mM), PMS (0.10 mM), and cell-free protein extract (50.0 µL) in a total volume of 1.0 mL. The reaction was carried out at 24 °C for 20 min. The product was extracted with EtOAc (4 mL) and the extract was concentrated to dryness; the residue was dissolved in $CH_3CN$ (200 µL) and analyzed by HPLC-DAD. The amount of indole-3-carboxaldehyde (**2**) in the reaction assay was determined using a calibration curve built with pure indole-3-carboxaldehyde ($\lambda_{max}$ 220 nm).

## 4. Conclusions

Overall, our results indicate that at higher concentration (0.30 mM) 3-ethyl-6-phenylquinoline (**7d**) inhibited BOLm activity to a larger extent than camalexin (**3a**) and displayed lower antifungal activity against *L. maculans*, both very desirable characteristics. Furthermore, no direct correlation was found between BOLm inhibitory activity of quinolines and their antifungal activity against *L. maculans*. Nevertheless, considering the antifungal activity against *L. maculans*, compounds with a diversity of scaffolds need to be synthesized and assayed before the desirable paldoxin activity of different scaffolds can be predicted. To this end, availability of the crystal structure of BOLm would be of great assistance.

The design of new compounds with new mechanisms of action to protect crops against fungi is expensive and time consuming, providing rather unpredictable outcomes. Nonetheless, the discovery of novel agrochemicals useful for sustainable crop treatments is crucial. It is expected that in the future, broad range fungicides will be replaced with selective crop treatments and that paldoxins will be a rational approach to treat specific fungal diseases such as that caused by *L. maculans*. Further work to uncover paldoxins that can replace current fungicides and prevent crop infestations with *L. maculans* is necessary.

**Supplementary Materials:** The following are available online. $^1$H- and $^{13}$C-NMR spectra of all new compounds: **5b**, **6b**, **6e**, **7c**, **7d**, **7e**, **7g**, **9b**, **9c** and **10b**.

**Acknowledgments:** Financial support for the authors' work was obtained from the Natural Sciences and Engineering Research Council of Canada (Discovery Grant to M.S.C.P.), the Canada Research Chairs program, Canada Foundation for Innovation, the Saskatchewan Government, and the University of Saskatchewan (graduate assistantship to A.A. and V.K.S.-M.). We acknowledge the technical assistance of K. Brown (NMR) and K. Thoms (MS) from the Department of Chemistry and Saskatchewan Structural Sciences Centre.

**Author Contributions:** M.S.C.P., A.A. and V.K.S.-M. conceived and designed the experiments; A.A. and V.K.S.-M. performed the experiments; M.S.C.P., A.A. and V.K.S.-M. analyzed the data; MSCP, A.A. and V.K.S.-M. wrote the paper.

**Conflicts of Interest:** The authors declare no conflict of interest.

## References

1. Pedras, M.S.C.; Minic, Z.; Jha, M. Brassinin oxidase, a fungal detoxifying enzyme to overcome a plant defense—Purification, characterization and inhibition. *FEBS J.* **2008**, *275*, 3691–3705. [CrossRef] [PubMed]
2. Pedras, M.S.C.; Minic, Z.; Sarma-Mamillapalle, V.K. Brassinin oxidase mediated transformation of the phytoalexin brassinin: Structure of the elusive co-product, deuterium isotope effect and stereoselectivity. *Bioorg. Med. Chem.* **2011**, *19*, 1390–1399. [CrossRef] [PubMed]
3. Pedras, M.S.C.; Yaya, E.E.; Glawischnig, E. The phytoalexins from cultivated and wild crucifers: Chemistry and biology. *Nat. Prod. Rep.* **2011**, *28*, 1381–1405. [CrossRef] [PubMed]
4. Pedras, M.S.C. Protecting plants against fungal diseases. *Can. Chem. News* **2005**, *57*, 16–17.
5. Pedras, M.S.C.; Minic, Z.; Sarma-Mamillapalle, V.K.; Suchy, M. Discovery of inhibitors of brassinin oxidase based on the scaffolds of the phytoalexins brassilexin and wasalexin. *Bioorg. Med. Chem.* **2010**, *18*, 2456–2463. [CrossRef] [PubMed]

6.   Pedras, M.S.C.; Minic, Z.; Sarma-Mamillapalle, V.K. Synthetic inhibitors of the fungal detoxifying enzyme brassinin oxidase based on the phytoalexin camalexin scaffold. *J. Agric. Food Chem.* **2009**, *57*, 2429–2435. [CrossRef] [PubMed]

7.   Pedras, M.S.C.; Abdoli, A. Pathogen inactivation of cruciferous phytoalexins: Detoxification reactions, enzymes and inhibitors. *RSC Adv.* **2017**, *38*, 23633–23646. [CrossRef]

8.   Pedras, M.S.C.; Jha, M. Toward the control of *Leptosphaeria maculans*: Design, syntheses, biological activity, and metabolism of potential detoxification inhibitors of the crucifer phytoalexin brassinin. *Bioorg. Med. Chem.* **2006**, *14*, 4958–4979. [CrossRef] [PubMed]

9.   Commercial Agriculture Fungicides are Classified According to Their Target Sites by the International Fungicide Resistance Action Committee (FRAC), cf. Available online: http://www.frac.info/docs/default-source/publications/frac-code-list/frac-code-list-2016.pdf?sfvrsn=2 (accessed on 22 April 2017).

10.  Lamberth, C.; Walter, H.; Kessabi, F.M.; Quaranta, L.; Beaudegnies, R.; Trah, S.; Jeanguenat, A.; Cederbaum, F. The significance of organosulfur compounds in crop protection: Current examples from fungicide research. *Phosphorus Sulfur* **2015**, *190*, 1225–1235. [CrossRef]

11.  Lamberth, C.; Kessabi, F.M.; Beaudegnies, R.; Quaranta, L.; Trah, S.; Berthon, G.; Cederbaum, F.; Knauf-Beiter, G.; Grasso, V.; Bieri, S.; et al. Synthesis and fungicidal activity of quinolin-6-yloxyacetamides, a novel class of tubulin polymerization inhibitors. *Bioorg. Med. Chem.* **2014**, *22*, 3922–3930. [CrossRef] [PubMed]

12.  Ayer, W.A.; Peter, A.C.; Ma, Y.; Miao, S. Synthesis of camalexin and related phytoalexins. *Tetrahedron* **1992**, *48*, 2919–2924. [CrossRef]

13.  Dondoni, A.; Dall'Occo, T.; Galliani, G.; Mastellari, A.; Medici, A. Addition of 2-silylazoles to heteroaryl cations. Synthesis of unsymmetrical azadiaryls. *Tetrahedron Lett.* **1984**, *25*, 3637–3640. [CrossRef]

14.  Batista, V.F.; Pinto, D.C.G.A.; Silva, A.M.S. Synthesis of quinolines: A green perspective. *ACS Sustain. Chem. Eng.* **2016**, *4*, 4064–4078. [CrossRef]

15.  Chelucci, G.; Porcheddu, A. Synthesis of quinolines via a metal-catalyzed dehydrogenative *N*-heterocyclization. *Chem. Rec.* **2017**, *17*, 200–216. [CrossRef] [PubMed]

16.  Prajapati, S.M.; Patel, K.D.; Vekariya, R.H.; Panchal, S.N.; Patel, H.D. Recent advances in the synthesis of quinolines: A review. *RSC Adv.* **2014**, *4*, 24463–24476. [CrossRef]

17.  Cieslik, W.; Serda, M.; Kurczyk, A.; Musiol, R. Microwave assisted synthesis of monoazanaphthalene scaffolds. *Curr. Org. Chem.* **2013**, *17*, 491–503. [CrossRef]

18.  Li, W.Y.; Xiong, X.Q.; Zhao, D.M.; Shi, Y.F.; Yang, Z.H.; Yu, C.; Fan, P.W.; Cheng, M.S.; Shen, J.K. Quinoline-3-carboxamide derivatives as potential cholesteryl ester transfer protein inhibitors. *Molecules* **2012**, *17*, 5497–5507. [CrossRef] [PubMed]

19.  Li, A.H.; Ahmed, E.; Chen, X.; Cox, M.; Crew, A.P.; Dong, H.Q.; Jin, M.; Ma, L.; Panicker, B.; Siu, K.W.; et al. A highly effective one-pot synthesis of quinolines from *o*-nitroarylcarbaldehydes. *Org. Biomol. Chem.* **2007**, *5*, 61–64. [CrossRef] [PubMed]

20.  Jiang, R.; Duckett, D.; Chen, W.; Habel, J.; Ling, Y.Y.; LoGrasso, P.; Kamenecka, T.M. 3,5-Disubstituted quinolines as novel c-Jun N-terminal kinase inhibitors. *Bioorg. Med. Chem. Lett.* **2007**, *17*, 6378–6382. [CrossRef] [PubMed]

21.  Czakó, B.; Kürti, L.; Mammoto, A.; Ingber, D.E.; Corey, E.J. Discovery of potent and practical antiangiogenic agents inspired by cortistatin A. *J. Am. Chem. Soc.* **2009**, *131*, 9014–9019. [CrossRef] [PubMed]

22.  Korn, T.J.; Schade, M.A.; Cheemala, M.N.; Wirth, S.; Guevara, S.A.; Cahiez, G.; Knochel, P. Cobalt-catalyzed cross-coupling reactions of heterocyclic chlorides with arylmagnesium halides and of polyfunctionalized arylcopper reagents with aryl bromides, chlorides, fluorides and tosylates. *Synthesis* **2006**, *21*, 3547–3574. [CrossRef]

23.  Pedras, M.S.C.; Abdoli, A. Metabolism of the phytoalexins camalexins, their bioisosteres and analogues in the plant pathogenic fungus *Alternaria brassicicola*. *Bioorg. Med. Chem.* **2013**, *21*, 4541–4549. [CrossRef] [PubMed]

24.  Movassaghi, M.; Hill, M.D. Synthesis of substituted pyridine derivatives via the ruthenium-catalyzed cycloisomerization of 3-azadienynes. *J. Am. Chem. Soc.* **2006**, *128*, 4592–4593. [CrossRef] [PubMed]

25.  Ji, X.; Huang, H.; Li, Y.; Chen, H.; Jiang, H. Palladium-catalyzed sequential formation of C-C bonds: Efficient assembly of 2-substituted and 2,3-disubstituted quinolines. *Angew. Chem. Int. Ed. Engl.* **2012**, *51*, 7292–7296. [CrossRef] [PubMed]

26. Shi, D.; Rong, L.; Shi, C.; Zhuang, Q.; Wang, X.; Tu, S.; Hu, H. Low-valent titanium reagent-promoted intramolecular reductive coupling reactions of ketomalononitriles: A facile synthesis of benzo[4,5]indene, acridine and quinoline derivatives. *Synthesis* **2005**, *5*, 717–724. [CrossRef]

27. Huo, Z.; Gridnev, I.D.; Yamamoto, Y. A method for the synthesis of substituted quinolines via electrophilic cyclization of 1-azido-2-(2-propynyl)benzene. *J. Org. Chem.* **2010**, *75*, 1266–1270. [CrossRef] [PubMed]

28. Meléndez Gómez, C.M.; Kouznetsov, V.V.; Sortino, M.A.; Álvarez, S.L.; Zacchino, S.A. In vitro antifungal activity of polyfunctionalized 2-(hetero)arylquinolines prepared through imino Diels-Alder reactions. *Bioorg. Med. Chem.* **2008**, *16*, 7908–7920. [CrossRef] [PubMed]

29. Colomb, J.; Billard, T. Palladium-catalyzed desulfitative arylation of 3-haloquinolines with arylsulfinates. *Tetrahedron Lett.* **2013**, *54*, 1471–1474. [CrossRef]

30. Wang, Y.; Xin, X.; Liang, Y.; Lin, Y.; Zhang, R.; Dong, D. A facile and efficient one-pot synthesis of substituted quinolines from α-arylamino ketones under vilsmeier conditions. *Eur. J. Org. Chem.* **2009**, *24*, 4165–4169. [CrossRef]

31. Saunthwal, R.K.; Patel, M.; Verma, A.K. Regioselective synthesis of C-3-functionalized quinolines via Hetero-Diels-Alder cycloaddition of azadienes with terminal alkynes. *J. Org. Chem.* **2016**, *81*, 6563–6572. [CrossRef] [PubMed]

32. Monrad, R.N.; Madsen, R. Ruthenium-catalysed synthesis of 2- and 3-substituted quinolines from anilines and 1,3-diols. *Org. Biomol. Chem.* **2011**, *9*, 610–615. [CrossRef] [PubMed]

33. Shrestha, B.; Thapa, S.; Gurung, S.K.; Pike, R.A.S.; Giri, R. General copper-catalyzed coupling of alkyl-, aryl-, and alkynylaluminum reagents with organohalides. *J. Org. Chem.* **2016**, *81*, 787–802. [CrossRef] [PubMed]

34. Mongin, F.; Mojovic, L.; Guillamet, B.; Trécourt, F.; Quéguiner, G. Cross-coupling reactions of phenylmagnesium halides with fluoroazines and fluorodiazines. *J. Org. Chem.* **2002**, *67*, 8991–8994. [CrossRef] [PubMed]

35. Li, C.; Li, J.; An, Y.; Peng, J.; Wu, W.; Jiang, H. Palladium-catalyzed allylic C-H oxidative annulation for assembly of functionalized 2-substituted quinoline derivatives. *J. Org. Chem.* **2016**, *81*, 12189–12196. [CrossRef] [PubMed]

36. Larivée, A.; Mousseau, J.J.; Charette, A.B. Palladium-catalyzed direct C–H arylation of *N*-iminopyridinium ylides: Application to the synthesis of (±)-anabasine. *J. Am. Chem. Soc.* **2008**, *130*, 52–54. [CrossRef] [PubMed]

**Sample Availability:** Not available.

*molecules*

MDPI

Article

# 4-Hydroxy-7-methyl-3-phenylcoumarin Suppresses Aflatoxin Biosynthesis via Downregulation of *aflK* Expressing Versicolorin B Synthase in *Aspergillus flavus*

Young-Sun Moon [1,†], Leesun Kim [1,†], Hyang Sook Chun [2,‡] and Sung-Eun Lee [1,*,‡]

1   School of Applied Biosciences, Kyungpook National University, Daegu 41566, Korea;
    space92@knu.ac.kr (Y.-S.M.); twosuns.kim@gmail.com (L.K.)
2   Advanced Food Safety Research Group, BK21 Plus, School of Food Science and Technology,
    Chung-Ang University, Anseong 17546, Korea; hschun@cau.ac.kr
*   Correspondence: selpest@knu.ac.kr
†   Authors equally contributed to this paper as first authors.
‡   Authors equally contributed to this paper as corresponding authors.

Academic Editor: Philippe Jeandet
Received: 17 December 2016; Accepted: 27 April 2017; Published: 29 April 2017

**Abstract:** Naturally occurring coumarins possess antibacterial and antifungal properties. In this study, these natural and synthetic coumarins were used to evaluate their antifungal activities against *Aspergillus flavus*, which produces aflatoxins. In addition to control antifungal activities, antiaflatoxigenic properties were also determined using a high-performance liquid chromatography in conjunction with fluorescence detection. In this study, 38 compounds tested and 4-hydroxy-7-methyl-3-phenyl coumarin showed potent antifungal and antiaflatoxigenic activities against *A. flavus*. Inhibitory mode of antiaflatoxigenic action by 4-hydroxy-7-methyl-3-phenyl coumarin was based on the downregulation of *aflD*, *aflK*, *aflQ*, and *aflR* in aflatoxin biosynthesis. In the cases of coumarins, antifungal and aflatoxigenic activities are highly related to the lack of diene moieties in the structures. In structurally related compounds, 2,3-dihydrobenzofuran exhibited antifungal and antiaflatoxigenic activities against *A. flavus*. The inhibitory mode of antiaflatoxigenic action by 2,3-dihydrobenzofuran was based on the inhibition of the transcription factor (*aflS*) in the aflatoxin biosynthesis pathway. These potent inhibitions of 2,3-dihydrobenzofuran and 4-hydroxy-7-methyl-3-phenyl coumarin on the *Aspergillus* growth and production of aflatoxins contribute to the development of new controlling agents to mitigate aflatoxin contamination.

**Keywords:** 4-hydroxy-7-methyl-3-phenyl coumarin; 2,3-dihydrobenzofuran; aflatoxin production; *Aspergillus flavus*; reverse transcription polymerase chain reaction

## 1. Introduction

Aflatoxins including AFB1, AFB2, AFG1, and AFG2 are mycotoxins produced by *Aspergillus flavus* and *A. parasiticus*, with potent carcinogenic activity, especially on human liver [1]. Aflatoxins can be accumulated in humans and livestock through diet of aflatoxin-contaminated foods and feed [2,3]. Outbreaks of aflatoxicosis are notably dependent on the crop species and seasonal changes of a given region [4,5]. Additionally, they are also related to poor agricultural practices [6]. Therefore, alternative agricultural practices may be needed to develop mitigating aflatoxin contamination in crops.

Chemical control of fungal growth and aflatoxin production has been successfully documented using propionic acid in unshelled peanuts on the laboratory scale [7]. In the crop field, usage of fungicides is critical to control fungal growth and mycotoxins with good efficacy [8]. Recently, phytopathogens develop resistance to various fungicides [9,10]. With this reason, alternatives for controlling *Aspergillus* infection and aflatoxin contamination are highly needed, and natural products could be considered as candidate compounds.

Coumarins are naturally occurring compounds produced after cyclization of cinnamic acid via formation of phenylpropanoids through the shikimic acid pathway [11]. Recently, plant-specific coumarins such as umbelliferone and scopoletin have been produced in *E. coli* due to their various applications after enzyme-engineered conversion with or without inexpensive precursors, 4-coumaric acid and ferulic acid [12]. Coumarins possess antibacterial activity against *Ralstonia solanacearum* [13], antimicrobial activity against *Staphylococcus aureus* [14], and antifungal activities against clinically important fungal pathogens [15]. Coumarins with antioxidant activities inhibit aflatoxin formation because aflatoxin formation occurs when fungal species are subject to oxidative stress [16–18]. In addition to this finding, a structure–activity relationship (SAR) study of 24 coumarin derivatives showed that *O*-substitutions seem to be essential for antifungal activity against *A. flavus* and *A. fumigatus* [19]. However, the authors did not study the relationship between the structure of coumarins and the antiaflatoxigenic activity generated by *A. flavus*. In the shikimic pathway, indole is generated and its derivative has shown antifungal activity [20].

In our previous studies, we have found that methylenedioxy-containing natural and synthetic compounds possessed antifungal and antiaflatoxigenic properties against *A. flavus* [21]. In this study, 1-(2-methylpiperidin-1-yl)-3-phenylprop-2-en-1-one showed potent antifungal and antiaflatoxigenic activities against *A. flavus* among the tested 22 compounds, and its mode of inhibitory action on aflatoxin production was caused by inhibition on the expression of some genes involved in aflatoxin biosynthesis such as *aflD*, *aflK*, *aflQ*, *aflR*, and *aflS* [21]. Other reports have shown that natural products are good candidates as preservatives to suppress aflatoxin contamination in cereals and feedstuffs [22–24].

In the present study, 26 coumarins were assessed to determine their antifungal activities against *A. flavus* and the inhibitory effects on aflatoxin production. The mode of inhibitory action on the aflatoxin production was disclosed using real-time PCR. Further studies for antifungal and antiaflatoxigenic activities were undertaken using structurally closed compounds including 2,3-dihydrobenzofuran, indole, 1-methyl indole, 2-methyl indole, 3-methyl indole, and 2-phenyl indole to coumarins for understating relationships between the structure of tested compounds and antifungal and antiaflatoxigenic activities. These antifungal and antiaflatoxigenic substances can be used for controlling *A. flavus* and reducing aflatoxin contamination in agricultural fields before harvest given their ability to decrease aflatoxin production.

## 2. Results and Discussion

The inhibitory effects of the 32 tested compounds on *A. flavus* growth and aflatoxin production were measured and the results are expressed in Table 1. A currently used fungicide thiabendzole (Figure 1) was used as a positive control and all data were calculated on the basis of the inhibition rate (%) in comparison to the solvent-treated controls [21]. Among the tested coumarins, five compounds showed antifungal activities against *A. flavus*. Among them, 4-hydroxy-7-methoxy-3-phenylcoumarin (**1**) and 4-hydroxy-6,7-dimethylcoumarin (**2**) exhibited about 50% inhibition on the fungal growth at the concentration of 100 μg/mL. However, this inhibitory effect of (**2**) disappeared after exposure to 10-fold diluted concentration (Table 1). 6,7-Dimethoxycoumarin (**3**) also possessed inhibitory effects on the fungal growth at concentrations of 1000 μg/mL. However, this inhibition was no longer evident following treatment with a 10-fold lower concentration of the compound. At the concentration of 10 μg/mL, there was no inhibitory effect by Compound **3** (Table 1).

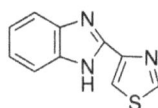

**Figure 1.** Molecular structure of thiabendazole used as a positive control in this study.

**Table 1.** Mycelial growth of *Aspergillus flavus* treated with various coumarins.

| Compounds | Treated Concentration (µg/mL) | Mycelial Growth (mg) |
|---|---|---|
| Thiabendazole (Positive control) | 10 | $1.2 \pm 2.1$ (1.0%) |
| | 5 | $6.3 \pm 10.1$ (5.0%) |
| | 1 | $96.3 \pm 2.4$ (77.0%) |
| 4-Hydroxy-6,7-dimethylcoumarin | 1000 | $0.0 \pm 0.00$ (0%) |
| | 100 | $22.5 \pm 3.7$ (18.0%) |
| | 10 | $73.9 \pm 17.4$ (59.1%) |
| 4-Hydroxy-7-methoxy-3-phenylcoumarin | 1000 | $0.0 \pm 0.00$ (0%) |
| | 100 | $41.1 \pm 27.1$ (32.8%) |
| | 10 | $61.9 \pm 14.1$ (49.5%) |
| 6,7-Dimethoxycoumarin | 1000 | $0.0 \pm 0.00$ (0%) |
| | 100 | $63.0 \pm 44.3$ (50.4%) |
| | 10 | $93.7 \pm 8.3$ (74.9%) |
| 2,3-dihydrobenzofuran | 1000 | $0.0 \pm 0.00$ (0%) |
| | 100 | $152.3.0 \pm 45.1$ (124%) |
| 4-(Bromomethyl)-6,7-dimethoxycoumarin | 1000 | $0.0 \pm 0.00$ (0%) |
| | 100 | $46.4 \pm 3.7$ (37.1%) |
| | 10 | $120.9 \pm 18.5$ (96.7%) |

Mycelial growth for the negative control was $124.0 \pm 23.0$ mg obtained from three experiments.

4-(Bromomethyl)-6,7-dimethoxycoumarin (**4**) and 2,3-dihydrobenzofuran (**5**) showed potent antifungal activities at concentrations of 1000 µg/mL. At the concentration of 100 µg/mL, Compound **4** reduced 63% of fungal growth and Compound **5** completely lost its inhibitory effect. The inhibitory effect of **4** on the *A. flavus* growth was not found at the concentration of 10 µg/mL (Table 1).

The inhibition of aflatoxin production by coumarins was remarkable (Table 2). At a concentration of 10 µg/mL, **1** and **2** showed almost complete inhibition of aflatoxin production. This inhibition was no longer evident following treatment with a 10-fold lower concentration of the compound (Table 2). Compound **1** exhibited potent inhibitory effects on $AFB_1$ and $AFB_2$ production until treatment with the compound at a concentration of 1 µg/mL. This compound significantly enhanced production of $AFG_1$ after the treatment of 100 µg/mL (Table 2). Compound **2** exhibited potent inhibitory effects on $AFB_1$ and $AFB_2$ production until treatment with the compound at a concentration of 10 µg/mL. At a concentration of 1 µg/mL, the antiaflatoxigenic activity of **5** was observed to be 40% inhibition of $AFB_1$ production (Table 2).

The fungicidal and bactericidal activities of coumarin and coumaric acid have been tested against *A. flavus* and *o*-coumaric acid inhibited aflatoxin production, but no correlation with fungal growth was found [25]. In that report, the authors found the complete inhibition of coumarin on fungal growth against *A. flavus* at the level of 10 mmol/L, equivalent to about 1460 µg/mL [25]. This is similar to our result, where most coumarins possessed potent antifungal activities at a concentration of 1000 µg/mL (Table 1).

Coumarins showed similar inhibitory patterns on aflatoxin production, enhancing the production of $AFG_1$ (Table 2). It is likely that coumarins inhibit the production of $AFB_1$, $AFB_2$, and $AFG_2$, but promote that of $AFG_1$. Various coumarins generally use similar target enzymes involved in the aflatoxin biosynthesis pathway to inhibit aflatoxin production; however, the pathway for production of $AFG_1$ escapes inhibition.

Table 2. Inhibitory effects of coumarins on aflatoxin production in *Aspergillus flavus*.

| Compounds | Treated Conc. (µg/mL) | Aflatoxin Production (ng/mL) | | | |
|---|---|---|---|---|---|
| | | Aflatoxin B1 | Aflatoxin B2 | Aflatoxin G1 | Aflatoxin G2 |
| Control | - | 1928.9 ± 403.4 [a] | 37.2 ± 6.3 [a] | 184.3 ± 66.5 [a] | 29.5 ± 5.3 [a] |
| Thiabendazole | 5 | ND [*,b] | ND [b] | 64.4 ± 2.6 [b] | ND [b] |
| | 1 | – | – | 239.2 ± 65.7 [a] | – |
| 3-Acetyl-6-bromocoumarin | 10 | ND [b] | ND [b] | 92.6 ± 25.7 [b] | ND [b] |
| | 1 | – | – | – | – |
| 4-Hydroxy-6,7-dimethyl-coumarin | 10 | ND [b] | ND [b] | ND [b] | ND [b] |
| | 1 | – | ND [b] | – | ND [b] |
| 4-Hydroxy-7-methoxy-3-phenyl-coumarin | 100 | ND [b] | ND [b] | ND [b] | ND [b] |
| | 10 | 158.8 ± 25.6 [c] | ND [b] | 192.6 ± 32.4 [a] | ND [b] |
| | 1 | 1025.4 ± 329.9 [d] | 19.3 ± 3.4 [c] | 137.8 ± 39.0 [a] | 5.3 ± 1.0 [c] |
| Dihydrocoumarin | 1000 | ND [b] | ND [b] | ND [b] | ND [b] |
| | 10 | 1397.5 ± 675.9 [a] | 28.8 ± 7.8 [b] | 99.6 ± 62.0 [b] | ND [b] |
| | 1 | 2517.9 ± 199.8 [a] | 49.7 ± 36.5 [a] | 164.6 ± 120.8 [a] | 5.1 ± 1.7 [c] |
| 2,3-Dihydrobenzofuran | 1000 | ND [b] | ND [b] | ND [b] | ND [b] |
| | 100 | ND [b] | ND [b] | ND [b] | ND [b] |
| | 1 | 1140.0 ± 342.1 [c] | 21.8 ± 5.1 [c] | 123.3 ± 33.7 [c] | 5.8 ± 2.6 [c] |
| 4-(Bromomethyl)-6,7-dimehtoxycoumarin | 1000 | ND [b] | ND [b] | ND [b] | ND [b] |
| | 100 | ND [b] | ND [b] | ND [b] | ND [b] |
| | 10 | – | – | – | ND [b] |

* ND: Not detectable; −: means more than 150% aflatoxin production in comparison to that of the control. Statistical analysis performed and different letters in the same column indicate significantly different from the control group ($p < 0.05$).

Other report using three natural furanocoumarins such as xanthotoxin, bergapten, and psoralen exhibited potent antiaflatoxigenic activities at the 5 mM concentration, but not for antifungal activities due to only 20% inhibition on fungal growth [26]. Holmes et al. [27] reviewed diverse biomolecules for their inhibitory effects on aflatoxin biosynthesis. Coumarins containing bergapten, *p*-coumaric acid, psoralene, and xanthotoxin possessed strong antiaflatoxigenic activities with $IC_{50}$ values below 0.1 mM. Among the known biomolecules, α-ionone ($IC_{50}$ value, 0.4 µM) was the strongest compound to suppress aflatoxin production [28].

In a recent report, authors demonstrated that AFG2 production in *A. flavus* was enhanced after exposure to piperonal, a methylenedioxy-containing compound [24]. In the same report, methyleugenol, a monoterpene, suppressed AFB1 and AFB2 generation, while AFG1 production increased in *A. flavus*. Taken together, chemicals can change the AFB biosynthesis pathway. In contrast, Compound **5** exhibited potent antifungal and antiaflatoxigenic activity in comparison to the positive control, thiabendazole. At concentrations of 1 µg/mL, thiabendazole showed more than 80% inhibition, and Compound **5** showed about 40% inhibition of AFB1 production. These findings are notable, as Compound **5** is a natural product that has potential for use as a major compound in the synthesis of new antiaflatoxigenic compounds.

RT-PCR results showed that Compound **1** downregulated *aflR*, *aflD*, *aflK*, and *aflQ*, thereby inhibiting the expression of several genes involved directly in the biosynthesis of aflatoxins (Figure 2). However, Compound **5** suppressed the expression of *aflS* only, which plays an important role in the transcription of genes involved in the biosynthesis of aflatoxins (Figure 3). The gene *aflD* expresses reductase mediating norsolorinic acid (NOR) to averantin (AVN), while *aflK* expresses versicolorin B (VERB) synthase to form VERB from versiconal (VAL). *aflQ* is responsible for oxidoreductase expression, which mediates the formation of aflatoxins.

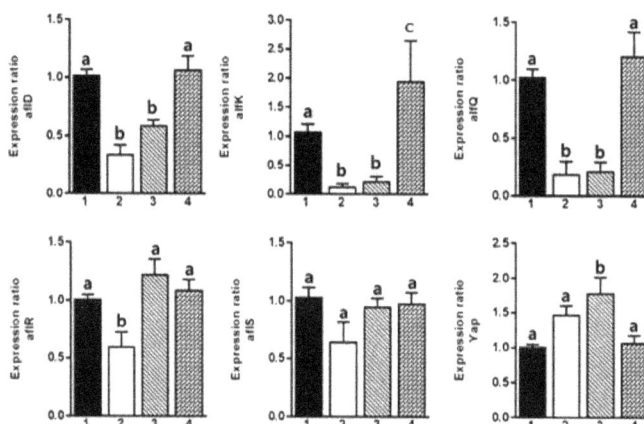

**Figure 2.** RT-PCR results of aflatoxin biosynthesis using six genes (*aflD, aflK, aflQ, aflR, aflS,* and *yap*) regulated by 4-hydroxy-7-methyl-3-phenyl coumarin (**1**). 1: control; 2: 1000 µg/mL of **1**; 3: 100 µg/mL of **1**; 4: 10 µg/mL of **1**). Different letters indicate statistically significant differences between experimental groups analyzed by a Student's *t*-test ($p < 0.05$).

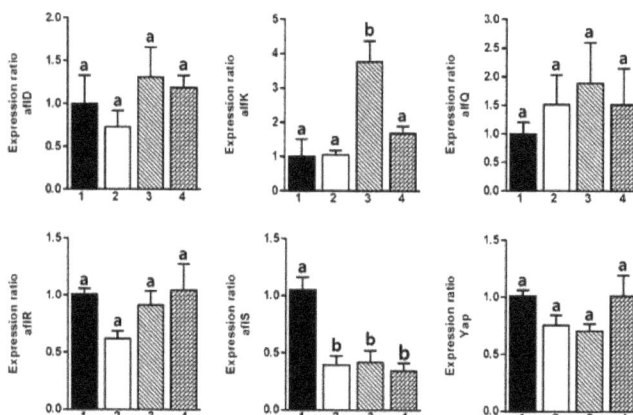

**Figure 3.** RT-PCR results of aflatoxin biosynthesis using six genes (*aflD, aflK, aflQ, aflR, aflS,* and *yap*) regulated by 2,3-dihydrobenzofuran (**5**). 1: control; 2: 1000 µg/mL of **5**; 3: 100 µg/mL of **5**; 4: 10 µg/mL of **5**. Different letters indicate statistically significant differences between experimental groups analyzed by a Student's *t*-test ($p < 0.05$).

The expression of *aflO* and *aflQ* is positively correlated to the production of AFB1 in *A. flavus* [29]. This finding is similar to the results of **1**, which suppressed *aflQ* expression. Recent reports support that the expression pattern of *aflS* gene is related to aflatoxin B1 in *A. flavus* [30,31]. However, the expression of *aflS* varied with aflatoxin-producing ability [32]. Therefore, this finding is related to the non-changeable expression pattern on *aflS* by (**1**), even it possessed inhibitory effects on expression of some genes involved in aflatoxin biosynthesis.

Conclusively, Compounds **1** and **5** among 32 tested compounds exhibited their potent inhibitory effects on *A. flavus* growth and aflatoxin production. The inhibitory effect of Compounds **1** and **5** on fungal growth was observed at a concentration of 100 µg/mL. Compound **1** decreased aflatoxin production at the concentration of 100 µg/mL via downregulation of *aflD, aflK,* and *aflQ* genes, while

Compound **5** downregulated only the expression of the *aflS* gene. The potent inhibition of Compound **1** was related to downregulation of *aflK* gene responsible for VERB synthase expression to form versicolorin B, a key intermediate in aflatoxin biosynthesis. Taken together, Compound **1** can be developed as an antifungal and antiaflatoxigenic agent to control *A. flavus* and aflatoxin contamination in crop plants and stored products.

## 3. Materials and Methods

### 3.1. Chemicals

The following compounds (26 compounds) were all purchased from Sigma-Aldrich Co. (St. Louise, MO, USA). Tested coumarins were 8-acetyl-7-hydroxycoumarin, 3-acetyl-6-bromocoumarin, 6-bromo-3-cyano-4-methyl-coumarin, 4-(bromomethyl)-6,7-dimethoxycoumarin, 3-cyano-7-hydroxy-4-methylcoumarin, 3-cyano-4,6-dimethyl-coumarin, coumarin, 4,6-dichloro-3-formylcoumarin, 6,7-dimethoxy-4-methylcoumarin, 5,7-dimethoxycoumarin, 7-ethoxy-4-methylcoumarin, 7-ethoxycoumarin, 4-hydroxy-6,7-dimethyl-coumarin, 7-methoxy-4-methylcoumarin, 6-methoxy-4-methylcoumarin, 7-methoxycoumarin, 7-hydroxy-6-methoxycoumarin, 4-hydroxy-7-methoxy-3-phenyl-coumarin, 6-methoxycoumarin, 5,6,7-trimethoxycoumarin, 6,7-dimethoxycoumarin, 6-methoxy-[7,8-(1-methoxy)methylenedioxy]coumarin, 6,7-(1-methoxy)methylenedioxy coumarin, 6,7,8-trimethoxycoumarin, 7-hydroxycoumarin, and dihydrocoumarin. In addition to these coumarin derivatives, some benzene-fused compounds were also tested for the evaluation of fungal and antiaflatoxigenic activities. 2,3-Dihydrobenzofuran, indole, 1-methyl indole, 2-methyl indole, 3-methyl indole, and 2-phenyl indole were also purchased from Sigma-Aldrich Co. (Figure 1).

### 3.2. Microorganisms and Preparation of the Spore Solution

*Aspergillus flavus* ATCC 22546 was purchased from the American Type Culture Collection (ATCC, Manassas, VA, USA) and was grown on malt extract agar (MEA: Difco Laboratories, Sparks, MD, USA). This isolate during the development generated aflatoxin B1 and B2, but not for G1 and G2 [23]. It was grown on MEA medium at 30 °C for 5 days until fungal spores were formed. After spore formation, they were collected from slants by shaking under 0.05% ($v/v$) Tween 80 and finally stored at −70 °C in a 20% glycerol solution ($v/v$).

### 3.3. Aflatoxin Analysis Using an HPLC-FLD

Fungal spore suspension adjusted to $10^6$ population was inoculated to the liquid culture media consisting of potato dextrose broth (25 mL) (PDB, Difco Laboratories). All tested compounds were spiked to the corresponding liquid media with a serial basis, and the culture was incubated at 25 °C for 5 days under shaking conditions. All experiments were triplicates for each concentration of the tested compound.

Following liquid medium cultivation for 5 days, the fungal growth was measured using filter paper to weigh the mycelial and sclerotial residues with overnight dryness in a dry oven. Separately, the mycelia from each treatment were subjected to the extraction procedure using an ultrasonic cleaner, and analyses of aflatoxin B (AFB) and G (AFG) was undertaken using an HPLC-FLD [20]. The average of the three replicates were calculated with standard deviation, and the data were compared with a control using one-way ANOVA at a $p < 0.05$ significance level [33].

### 3.4. Real-Time qPCR (RT-qPCR) after Isolation of Total RNA

RT-qPCR was employed to understand the mode of the inhibitory effect on fungal growth and aflatoxin production. Fungal mycelia in liquid media were carefully collected and total RNA was extracted using the QIAzol Lysis reagent supplied by QIAGEN Inc. (Dusseldorf, Germany) after grinding to a fine powder under an appropriate amount of liquid nitrogen. Total RNAs extracted from the treated fungi were quantified by a μDrop™ Plate (Thermo Fisher Scientific Inc., Waltham,

*Molecules* **2017**, *22*, 712

MA, USA), and the extracted RNAs were qualitatively checked using 1% agarose gel with ethidium bromide. Complementary DNA (cDNA) for extracted RNAs (2 μg) was synthesized using Maxima First Strand cDNA Synthesis Kit (Thermo Fisher Scientific Inc., Waltham, MA, USA).

RT-qPCR was undertaken by a Rotor-Gene SYBR Green PCR Kit (QIAGEN Inc.) with an proper amount of cDNA (100 ng). Primers for genes such as *yap*, *aflR*, *aflS*, *aflK*, *aflD*, *aflQ*, and *18S rRNA* were synthesized by Genotech (Daejeon, Korea), and they were used to understand the relationship between aflatoxin biosynthesis and the active compound [20]. Forty cycles of thermal cycling parameters were performed for amplification as follows: denaturation at 95 °C for 30 s, annealing at 60 °C for 20 s, and elongation at 72 °C for 30 s, followed by an additional step at 95 °C for 5 min. RT-qPCR was done triplicates for each treatment. Significant differences in gene expression were calculated using double delta Ct methods [34]. Data were standardized with *18S rRNA*, and gene expressions between the treatment and controls were compared using Prism 6 software (GraphPad, San Diego, CA, USA). Statistically significant differences between experimental groups were analyzed by a Student's *t*-test ($p < 0.05$).

**Acknowledgments:** This research was supported by a grant (15162MFDS044) from Ministry of Food and Drug Safety in 2015.

**Author Contributions:** Hyang Sook Chun and Sung-Eun Lee conceived and designed all experiments; Young-Sun Moon performed the experiments; Hyang Sook Chun and Sung-Eun Lee analyzed the results; Leesun Kim and Sung-Eun Lee wrote the paper.

**Conflicts of Interest:** The authors declare no conflict of interest.

## References

1. Marin, S.; Ramos, A.J.; Cano-Sancho, G.; Sanchis, V. Mycotoxins: Occurrence, toxicology, and exposure assessment. *Food Chem. Toxicol.* **2013**, *60*, 218–237. [CrossRef] [PubMed]
2. Stroka, J.; Anklam, E. Analysis of aflatoxins in various food and feed matrices-results of international validation studies. *Mycotoxin Res.* **2000**, *16* (Suppl. S2), 224–226. [CrossRef] [PubMed]
3. Chen, A.J.; Jiao, X.; Hu, Y.; Lu, X.; Gao, W. Mycobiota and mycotoxins in traditional medicinal seeds from China. *Toxins* **2015**, *7*, 3858–3875. [CrossRef] [PubMed]
4. Fountain, J.C.; Scully, B.T.; Ni, X.Z.; Kemerait, R.C.; Lee, R.D.; Chen, Z.Y.; Guo, B. Environmental influences on maize-*Aspergillus flavus* interactions and aflatoxin production. *Front. Microbiol.* **2014**, *5*, 40. [CrossRef] [PubMed]
5. Atehnkeng, J.; Donner, M.; Ojiambo, P.S.; Ikotun, B.; Augusto, J.; Cotty, P.J.; Bandyopadhyay, R. Environmental distribution and genetic diversity of vegetative compatibility groups determine biocontrol strategy to mitigate aflatoxin contamination of maize by *Aspergillus flavus*. *Microb. Biotechnol.* **2015**, *9*, 75–88. [CrossRef] [PubMed]
6. Diedhiou, P.M.; Bandyopadhyay, R.; Atehnkeng, J.; Ojiambo, P.S. *Aspergillus* colonization and aflatoxin contamination of maize and sesame kernels in two agro-ecological zones in Senegal. *J. Phytopathol.* **2011**, *159*, 268–275. [CrossRef]
7. Calori-Domingues, M.A.; Fonseca, H. Laboratory of chemical control of aflatoxin production in unshelled peanuts (*Arachis hypogaea* L.). *Food Addit. Contam.* **1995**, *12*, 347–350. [CrossRef] [PubMed]
8. D'Mello, J.P.F.; Macdonald, A.M.C.; Postel, D.; Dijksma, W.T.P.; Dujardin, A.; Plancinta, C.M. Pesticide use and mycotoxin production in *Fusarium* and *Aspergillus* phytopathogens. *Eur. J. Plant Path.* **1998**, *104*, 741–751.
9. Deising, H.B.; Reimann, S.; Pascholati, S.F. Mechanisms and significance of fungicide resistance. *Braz. J. Microbiol.* **2008**, *39*, 286–295. [CrossRef] [PubMed]
10. Kanafani, Z.A.; Perfect, J.R. Resistance to antifungal agents: Mechanisms and clinical impact. *Clin. Infect. Dis.* **2008**, *46*, 120–128. [CrossRef] [PubMed]
11. Bourgaud, F.; Hehn, A.; Larbat, R.; Doerper, S.; Gontier, E.; Kellner, S.; Matern, U. Biosynthesis of coumarins in plants: A major pathway still to be unraveled for cytochrome P450 enzymes. *Phytochem. Rev.* **2006**, *5*, 293–308. [CrossRef]
12. Lin, Y.; Sun, X.; Yuan, Q.; Yan, Y. Combinatorial biosynthesis of plant-specific coumarins in bacteria. *Metab. Eng.* **2013**, *18*, 69–77. [CrossRef] [PubMed]

13. Chen, J.; Yu, Y.; Li, S.; Ding, W. Resvertrol and coumarin: Novel agricultural antibacterial agent against *Ralstonia solanacearum* in vitro and in vivo. *Molecules* **2016**, *21*, 1501. [CrossRef] [PubMed]

14. Li, Z.P.; Li, J.; Qu, D.; Hou, Z.; Yang, X.H.; Zhang, Z.D.; Wang, Y.K.; Luo, X.X.; Li, M.K. Synthesis and pharmacological evaluations of 4-hydroxycoumarin derivatives as a new class of anti-*Staphylococcus aureus* agent. *J. Pharm. Pharmacol.* **2015**, *67*, 573–582. [CrossRef] [PubMed]

15 Ji, Q.; Ge, Z.; Ge, Z.; Chen, K.; Wu, H.; Liu, X.; Huang, Y.; Yuan, L.; Yang, X.; Liao, F. Synthesis and biological evaluation of novel phosphoramidate derivatives of coumarin as chitin synthase inhibitors and antifungal agents. *Eur. J. Med. Chem.* **2016**, *108*, 166–176. [CrossRef] [PubMed]

16. Zhang, L.; Hua, Z.; Song, Y.; Feng, C. Monoterpenoid indole alkaloids from *Alstonia rupestris* with cytotoxic, antibacterial and antifungal activities. *Fitoterapia* **2014**, *97*, 142–147. [CrossRef] [PubMed]

17. Roze, L.V.; Laivenieks, M.; Hong, S.Y.; Wee, J.; Wong, S.S.; Vanos, B.; Awad, D.; Ehrlich, K.C.; Linz, J.E. Aflatoxin Biosynthesis Is a Novel Source of Reactive Oxygen Species—A Potential Redox Signal to Initiate Resistance to Oxidative Stress? *Toxins* **2015**, *7*, 1411–1430. [CrossRef] [PubMed]

18. Singh, L.K.; Priyanka; Singh, V.; Katiyar, D. Design, synthesis and biological evaluation of some new coumarin derivatives as potential antimicrobial agents. *Med. Chem.* **2015**, *11*, 128–134. [CrossRef] [PubMed]

19. De Araújo, R.S.A.; Guerra, F.Q.S.; de, O.; Lima, E.; de Simone, C.A.; Tavares, J.F.; Scotti, L.; Scotti, M.T.; de Aquino, T.M.; de Moura, R.O.; Mendonça, F.J.B., Jr.; et al. Synthesis, structure–activity relationships (SAR) and in Silico studies of coumarin derivatives with antifungal activity. *Int. J. Mol. Sci.* **2013**, *14*, 1293–1309.

20. Hussein, K.A.; Joo, J.H. Isolation and characterization of rhizomicrobial isolates for phosphate solubilization and indole acetic acid production. *J. Korean Soc. Appl. Biol. Chem.* **2015**, *58*, 847–855. [CrossRef]

21. Moon, Y.S.; Choi, W.S.; Park, E.S.; Bae, I.K.; Choi, S.D.; Paek, O.; Kim, S.H.; Chun, H.S.; Lee, S.E. Antifungal and antiaflatoxigenic methylenedioxy-containing compounds and piperine-like synthetic compounds. *Toxins* **2016**, *8*, 240. [CrossRef] [PubMed]

22. Lee, S.E.; Mahoney, N.E.; Campbell, B.C. Inhibition of aflatoxin B1 biosynthesis by piperlongumine isolated from *Piper longum* L. *J. Microbiol. Biotechnol.* **2002**, *12*, 679–682.

23. Kohiyama, C.Y.; Yamamoto, R.M.M.; Mossini, S.A.; Bando, E.; Bomfim Nda, S.; Nerilo, S.B.; Rocha, G.H.; Grespan, R.; Mikcha, J.M.; Machinski, M., Jr. Antifungal properties and inhibitory effects upon aflatoxin production of *Thymus vulgaris* L. by *Aspergillus flavus* Link. *Food Chem.* **2015**, *173*, 1006–1010. [CrossRef] [PubMed]

24. Park, E.S.; Bae, I.K.; Kim, H.J.; Lee, S.E. Novel regulation of aflatoxin B1 biosynthesis in *Aspergillus flavus* by piperonal. *Nat. Prod. Res.* **2016**, *30*, 1854–1857. [CrossRef] [PubMed]

25. Paster, N.; Juven, B.J.; Harshemesh, H. Antimicrobial activity and inhibition of aflatoxin B1 formation by olive plant tissue constituents. *J. Appl. Bacteriol.* **1988**, *64*, 293–297. [CrossRef] [PubMed]

26. Mabrouk, S.S.; El-Shayeb, N.M.A. Inhibition of aflatoxin production in *Aspergillus flavus* by natural coumarins and chromones. *World J. Microbiol. Biotechnol.* **1992**, *8*, 60–62. [CrossRef] [PubMed]

27. Holmes, R.A.; Boston, R.S.; Payne, G.A. Diverse inhibitors of aflatoxin biosynthesis. *Appl. Bmicrobiol. Biotechnol.* **2008**, *78*, 559–572. [CrossRef] [PubMed]

28. Norton, R.A. Inhibition of aflatoxin B1 synthesis by *Aspergillus flavus*. *Phytopathology* **1997**, *87*, 814–821. [CrossRef] [PubMed]

29. Jamali, M.; Karimipour, M.; Shams-Ghahfarokhi, M.; Amani, A.; Razzaghi-Abyaneh, M. Expression of aflatoxin genes *aflO* (omtB) and *aflQ* (ordA) differentiates levels of aflatoxin production by *Aspergillus flavus* strains from soils of pistachio orchards. *Res. Microbiol.* **2013**, *164*, 293–299. [CrossRef] [PubMed]

30. Schmidt-Heydt, M.; Rüfer, C.E.; Abdel-Hadi, A.; Magan, N.; Geisen, R. The production of aflatoxin B1 or G1 by *Aspergillus* parasticus at various combinations of temperature and water activity is related to the ratio of *aflS* to *aflR* expression. *Mycotoxin Res.* **2010**, *26*, 241–246. [CrossRef] [PubMed]

31. Mo, H.Z.; Zhang, H.; Wu, Q.H.; Hu, L.B. Inhibitory effects of tea extract on aflatoxin production by *Aspergillus flavus*. *Lett. Appl. Microbiol.* **2013**, *56*, 462–466. [PubMed]

32. Scherm, B.; Palomba, M.; Serra, D.; Marcello, A.; Migheli, Q. Detection of transcripts of the aflatoxin genes *aflD*, *aflO*, and *aflP* by reverse transcription-polymerase chain reaction allows differentiation of aflatoxin-producing and non-producing isolates of *Aspergillus flavus* and *Aspergillus parasiticus*. *Int. J. Food Microbiol.* **2005**, *98*, 201–210. [CrossRef] [PubMed]

33. SAS. *SAS User's Guide*, 4th ed.; SAS Institute: Cary, NC, USA, 2001.
34. Rao, X.; Huang, X.; Zhou, Z.; Lin, X. An improvement of the 2ˆ(−delta delta CT) method for quantitative real-time polymerase chain reaction data analysis. *Biostat. Bioinform. Biomath.* **2013**, *3*, 71–85.

**Sample Availability:** Samples of the tested compounds are available from the authors.

![molecules logo] *molecules*

MDPI

*Article*

# Bactericidal Effect of Pterostilbene Alone and in Combination with Gentamicin against Human Pathogenic Bacteria

Wee Xian Lee, Dayang Fredalina Basri * and Ahmad Rohi Ghazali

School of Diagnostic & Applied Health Sciences, Faculty of Health Sciences, Universiti Kebangsaan Malaysia, Kuala Lumpur 50300, Malaysia; lee.wx@mckl.edu.my (W.X.L.); rohi@ukm.edu.my (A.R.G.)
* Correspondence: dayang@ukm.edu.my; Tel.: +60-3-9289-7652

Academic Editor: Philippe Jeandet
Received: 17 February 2017; Accepted: 11 March 2017; Published: 17 March 2017

**Abstract:** The antibacterial activity of pterostilbene in combination with gentamicin against six strains of Gram-positive and Gram-negative bacteria were investigated. The minimum inhibitory concentration and minimum bactericidal concentration of pterostilbene were determined using microdilution technique whereas the synergistic antibacterial activities of pterostilbene in combination with gentamicin were assessed using checkerboard assay and time-kill kinetic study. Results of the present study showed that the combination effects of pterostilbene with gentamicin were synergistic (FIC index < 0.5) against three susceptible bacteria strains: *Staphylococcus aureus ATCC 25923*, *Escherichia coli O157* and *Pseudomonas aeruginosa 15442*. However, the time-kill study showed that the interaction was indifference which did not significantly differ from the gentamicin treatment. Furthermore, time-kill study showed that the growth of the tested bacteria was completely attenuated with 2 to 8 h treatment with $0.5 \times$ MIC of pterostilbene and gentamicin. The identified combinations could be of effective therapeutic value against bacterial infections. These findings have potential implications in delaying the development of bacterial resistance as the antibacterial effect was achieved with the lower concentrations of antibacterial agents.

**Keywords:** pterostilbene; synergistic; time kill assay; antibiotic; bactericidal

## 1. Introduction

Antibacterial therapy has been a keystone of modern medicine practice for the treatment of several pathological diseases [1,2]. Unfortunately, bacteria have developed genetic modifications and mobilized molecular defense mechanisms that are able to protect them against antibiotics [1]. Due to the emergence of multidrug-resistant pathogens, it is now standard clinical practice to use two or more antibacterial drugs with different mechanisms of action in an attempt to expand the antimicrobial spectrum, to prevent the emergence of resistant organisms, to minimize side effects, and to obtain synergistic antimicrobial activity [3,4].

Natural products and their derivatives have been recognized for many years as a significant source of new leads in the development of new pharmaceutical agents [5]. Phytoalexins are low molecular weight antimicrobial compounds that are produced by plants as part of the plant defense system against a wide range of pathogens and herbivores. Over the past few decades, research and commercial interest in stilbene phytoalexins have escalated. Many of them have been subjected to intense investigations in the light of their potential biological activities and possible pharmacological applications [6,7].

*trans*-3,5-Dimethoxy-4-hydroxystilbene (pterostilbene) is a phytoalexin compound found primarily in *Pterocarpus marsupium* heartwood [8,9] and several foods and drinks, including blueberries and grapevines [10,11]. Pterostilbene, a structural analog of resveratrol, has shown higher bioavailability than resveratrol due to the presence of the methoxy groups, making it advantageous as a therapeutic agent. The useful effects of pterostilbene are well documented and multiple studies have suggested that pterostilbene may have numerous preventive and therapeutic properties in a vast range of human diseases, including cardiovascular diseases, diabetes and cancer, which are attributed to its pharmacological effects such as antioxidant [12], anti-inflammatory, and anticarcinogenic properties leading to improved normal cell function and inhibition of malignant cells [13,14]. Like resveratrol, pterostilbene was also shown to possess antifungal activity against various grapevine pathogens [15]. More recent studies reported that pterostilbene was 5 to 10 times more effective than resveratrol in inhibiting the germination of conidia of *Botrytis cinerea* and sporangia of *Plasmopara viticola* [16]. However, to the best of our knowledge, evaluation of the susceptibilities of several clinical bacterial isolates of different species to pterostilbene has not been reported previously. As a result, this study was undertaken to study the antibacterial potency of pterostilbene and its combination with the standard antibiotic gentamicin against a variety of Gram-positive and Gram-negative bacteria.

## 2. Results

### 2.1. Minimum Inhibitory Concentration (MIC) Values of Pterostilbene and Gentamicin against Gram-Positive Bacteria

The antibacterial property of pterostilbene was quantitatively assessed against both Gram-positive and Gram-negative bacteria by determining their minimum inhibitory concentration (MIC) values. In Table 1, it is shown that the MIC value of pterostilbene was 0.025 mg/mL, whereas gentamicin was effective in the range of 1.56–6.25 µg/mL against Gram-positive bacteria.

**Table 1.** Determination of MIC values of pterostilbene and gentamicin against Gram-positive bacteria.

| Concentrations (mg/mL) | *Staphylococcus aureus* ATCC 25923 | | *Bacillus cereus* ATCC 11778 | |
|---|---|---|---|---|
| | Pterostilbene | Gentamicin | Pterostilbene | Gentamicin |
| 0.20000 | − | − | + | − |
| 0.10000 | − | − | + | − |
| 0.05000 | − | − | + | − |
| 0.02500 | − | − | + | − |
| 0.01250 | + | − | + | − |
| 0.00625 | + | − | + | − |
| 0.00313 | + | − | + | + |
| 0.00156 | + | − | + | + |
| 0.00078 | + | + | + | + |

+: presence of bacterial growth; −: absence of bacterial growth; Positive control comprises bacterial suspension and Mueller-Hinton broth; Negative control comprises Mueller-Hinton broth only.

Pterostilbene was effective in inhibiting the growth of all tested Gram-negative bacteria, except *Acinetobacter baumannii* ATCC 19606, with MIC values in the 0.5–0.025 mg/mL range (Table 2). Out of the two tested Gram-positive bacteria, pterostilbene was the only capable of inhibiting *Staphylococcus aureus* ATCC 25923 at MIC value 0.025 mg/mL. On the other hand, the antibiotic gentamicin used as a standard antibiotic drug was more potent than the tested pterostilbene compound with the MIC values of 1.56 µg/mL, 6.25 µg/mL, 3.13 µg/mL, 6.25 µg/mL against *Staphylococcus aureus* ATCC 25923, *Bacillus cereus* ATCC, *Escherichia coli* O157 and *Pseudomonas aeruginosa* ATCC 15442, respectively (Tables 1 and 2). The most susceptible strains towards pterostilbene were *Staphylococcus aureus* ATCC 25923 and *Escherichia coli* O157 (0.200 mg/mL), followed by *Pseudomonas aeruginosa* ATCC 15442 (MIC = 0.025 mg/mL).

**Table 2.** Determination of MIC values of pterostilbene and gentamicin against Gram-negative bacteria.

| Concentrations (mg/mL) | *Escherichia coli* ATCC 35150 | | *Pseudomonas aeruginosa* ATCC 15442 | | *Acinetobacter baumannii* ATCC 19606 | |
|---|---|---|---|---|---|---|
| | Pterostilbene | Gentamicin | Pterostilbene | Gentamicin | Pterostilbene | Gentamicin |
| 0.20000 | − | − | − | − | + | + |
| 0.10000 | − | − | | − | + | + |
| 0.05000 | − | − | − | − | + | + |
| 0.02500 | + | − | − | − | + | + |
| 0.01250 | + | − | + | − | + | + |
| 0.00625 | + | − | + | − | + | + |
| 0.00313 | + | − | + | + | + | + |
| 0.00156 | + | + | + | + | + | + |
| 0.00078 | + | + | + | + | + | + |

+: presence of bacterial growth; −: absence of bacterial growth; Positive control comprised bacterial suspension and Mueller-Hinton broth; Negative control comprised Mueller-Hinton broth only.

## 2.2. The Minimum Bactericidal Concentration (MBC) Pterostilbene against Gram-Positive and Gram-Negative Bacteria

The minimum bactericidal concentration (MBC) assay results are summarized in Table 3; an antimicrobial agent is considered bactericidal if the MBC value is not more than fourfold higher than the MIC value [17]. The MBC value is the least concentration which has a colony count of less than 10 on the agar plate. From Table 3, the MBC value of pterostilbene was similar to its MIC value against *Staphylococcus aureus* ATCC 25923, which was 0.200 mg/mL, indicating bactericidal activity. However, the MBC value of pterostilbene showed a bacteriostatic mode of action towards the other two susceptible bacteria strains, with MBC values more than fourfold higher than their respective MIC values.

**Table 3.** Determination of MBC values of pterostilbene against Gram-positive and Gram-negative bacteria.

| MIC (mg/mL) | Gram-Positive Bacteria | Gram-Negative Bacteria | |
|---|---|---|---|
| | *Staphylococcus aureus* ATCC 25923 | *Escherichia coli* O157 | *P. aeruginosa* ATCC 15442 |
| 0.20000 | − | + | + |
| 0.10000 | − | ND | + |
| 0.05000 | − | ND | + |
| 0.02500 | − | ND | + |
| 0.00125 | ND | ND | ND |

+: growth of bacteria on agar plate; −: no growth of bacteria on agar plate; ND: not done because the microtiter well at the tested concentration showed the presence of bacteria growth as shown in Tables 1 and 2.

## 2.3. FICI Values and Interaction Effects of Pterostilbene and Gentamicin Combinations against Gram-Positive and Gram-Negative Bacteria

The outcome for the antibiotic combination studies between pterostilbene and standard antibiotic gentamicin, determined by calculation of FIC index values, are presented in Table 4. The combination of pterostilbene and gentamicin displayed FIC values of 0.125, 0.3185 and 0.25 against *Staphylococcus aureus* ATCC 25923, *Escherichia coli* O157 and *Pseudomonas aeruginosa* ATCC 15442, respectively, which indicated a synergistic interaction. It was noted that pterostilbene markedly reduced the MIC value of gentamicin by sixteen fold against *Staphylococcus aureus* ATCC 25923, followed by eightfold and fourfold, respectively, against *Pseudomonas aeruginosa* ATCC 15442 and *Escherichia coli* O157. Our results indicate that the interaction of pterostilbene potentiated the activity of gentamicin. In the present study, only interactions with synergistic effects were selected for time-kill analysis as it would provide descriptive (qualitative) information on the pharmacodynamics of antimicrobial agents [18].

**Table 4.** Determination of FICI values and interaction effects of pterostilbene and gentamicin combinations against Gram-positive and Gram-negative bacteria.

| Species | Antibacterial Agents | MIC (mg/mL) | | FIC | |
|---------|---------------------|-------------|-------------|----------------|-----------|
| | | Alone | Combination | FIC Individual | FIC Index |
| *Staphylococcus aureus* ATCC 25923 | Pterostilbene | 0.025 | 0.00157 | 0.0625 | 0.125 (S) |
| | Gentamicin | 0.00157 | 0.0000981 | 0.0625 | |
| *Escherichia coli* O157 | Pterostilbene | 0.2 | 0.0125 | 0.0625 | 0.3185 (S) |
| | Gentamicin | 0.00313 | 0.0008 | 0.256 | |
| *Pseudomonas aeruginosa* ATCC 15442 | Pterostilbene | 0.2 | 0.025 | 0.125 | 0.25 (S) |
| | Gentamicin | 0.00625 | 0.00078 | 0.125 | |

*2.4. Time Kill Kinetic of Pterostilbene and Gentamicin Combinations against Gram-Positive and Gram-Negative Bacteria*

Time-kill establishes the optimum time of exposure of pterostilbene against bacteria. The time-kill kinetic profiles of pterostilbene at $1 \times$ MIC (Figure 1) displayed bactericidal activity towards *Staphylococcus aureus* ATCC 25923, showing a 3 $\log_{10}$ reduction in viable cell count relatively to the initial inoculum at $1 \times$ MIC value after 5.5 h exposure. As expected from the determined MBC value, the time-kill analysis for pterostilbene against *Staphylococcus aureus* ATCC 25923 was consistent with the bactericidal characteristic. Kinetically, $1 \times$ MIC of pterostilbene alone showed a slower rate of killing (5.5 h) compared to gentamicin alone as well as combination treatment (1.25 h).

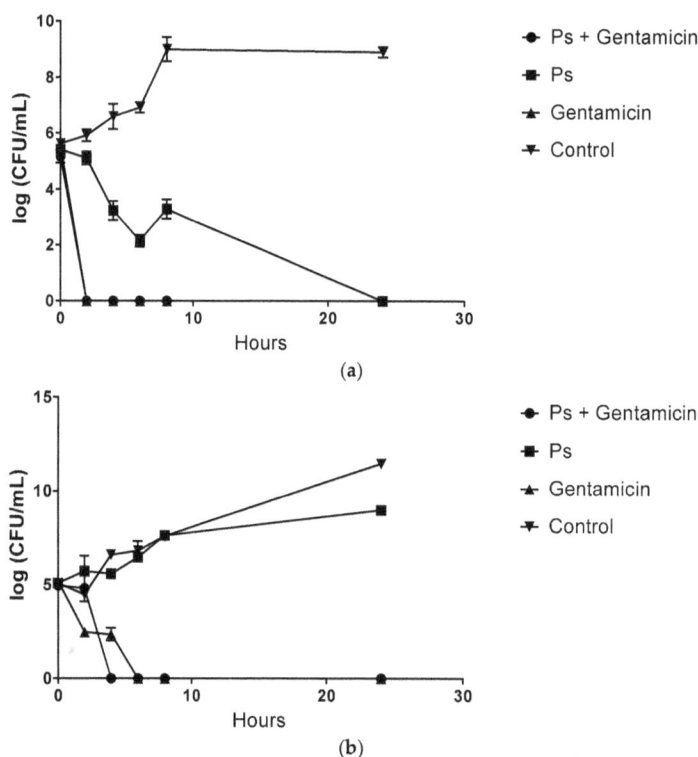

(a)

(b)

**Figure 1.** *Cont.*

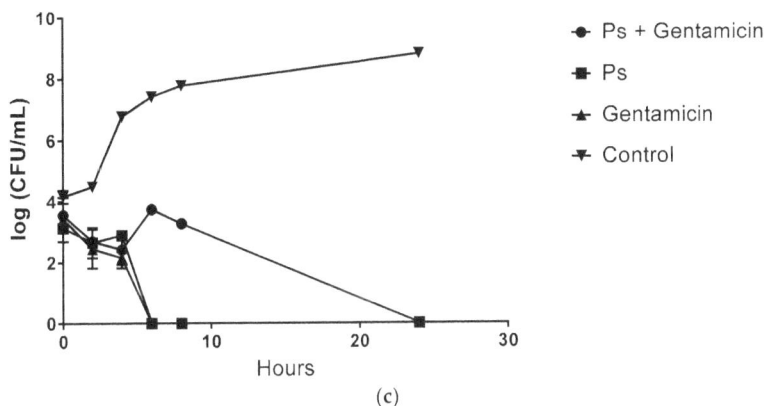

**Figure 1.** Time-kill growth curves of combination of pterostilbene with gentamicin, pterostilbene alone, and gentamicin alone against (**a**) *Staphylococcus aureus* ATCC 25923; (**b**) *Pseudomonas aeruginosa* ATCC 15442 and (**c**) *Escherichia coli* O157. The data is presented as a mean of 3 replicates.

On the other hand, combination treatment at $\frac{1}{2} \times$ MIC of both drugs had a similar killing rate compared to the most active agent gentamicin throughout the 24 h. However, the combination treatment showed indifference interaction with less than 2 $\log_{10}$ decrease in bacterial counts within a specific time period throughout 24 h between the combination and the most active agent. Figure 1b showed that pterostilbene at its MIC value did not demonstrate a reduction in bacterial count over time towards *Pseudomonas aeruginosa* ATCC 15442 relative to the initial inoculum. However, incubating the bacteria for 4 h resulted in an increase of the viable cell count ranging between 5.6 and 8.9 $\log_{10}$ CFU/mL. The time-kill assay curve in Figure 1b also indicates that the combination treatment showed an indifferent interaction between pterostilbene and gentamicin against *Pseudomonas aeruginosa* ATCC 15442. This is justified by the not more than 2 $\log_{10}$ increase in bacterial counts within a specific time period over 24 h between the combination and the most active agent. However, the combinations demonstrated more rapid killing after 3.8 h than gentamicin alone and the bacterial growth was completely attenuated after 4 h, whereas for gentamicin alone at the MIC complete bactericidal activity was exhibited after only 6 h. Our results suggest that pterostilbene combined with antibiotics might be microbiologically beneficial and synergistic. Our findings have potential implications in delaying the development of bacterial resistance as the antibacterial effect was achieved with lower concentrations of both drugs (antibiotics and stilbenes). In Figure 1c, the time-kill kinetics also showed a decrease in the number of surviving *Escherichia coli* O157 over the first 4 h, but the increase in bacterial counts from 2.59 and 3.72 $\log_{10}$ CFU/mL after 6 h indicated that after the initial bacteriostatic effect, regrowth of the organism could occur. Combination treatment failed to show a synergistic effect as indicated in time-kill curves showing an indifferent interaction through 24 h.

*2.5. Scanning Electron Microscope (SEM) Analysis*

To investigate the interaction between pterostilbene and the bacterial cells, we decided to perform SEM analysis of bacterial cultures exposed to pterostilbene compound, in order to obtain additional evidence of the different effect observed in previous tests. As shown in Figure 2a,c, untreated *S. aureus* and *E. coli* exist as colonies of round-shaped and rod-shaped microbes, respectively, in the control culture. They have both smooth and intact cell walls. Morphological observation using SEM analysis of the *S. aureus* treated at $1 \times$ MIC with pterostilbene showed that the quantity of *S. aureus* has greatly decreased and the bacteria became more irregular and unhomogenous in shape. A lot of blebs also appeared on the bacterial surface.

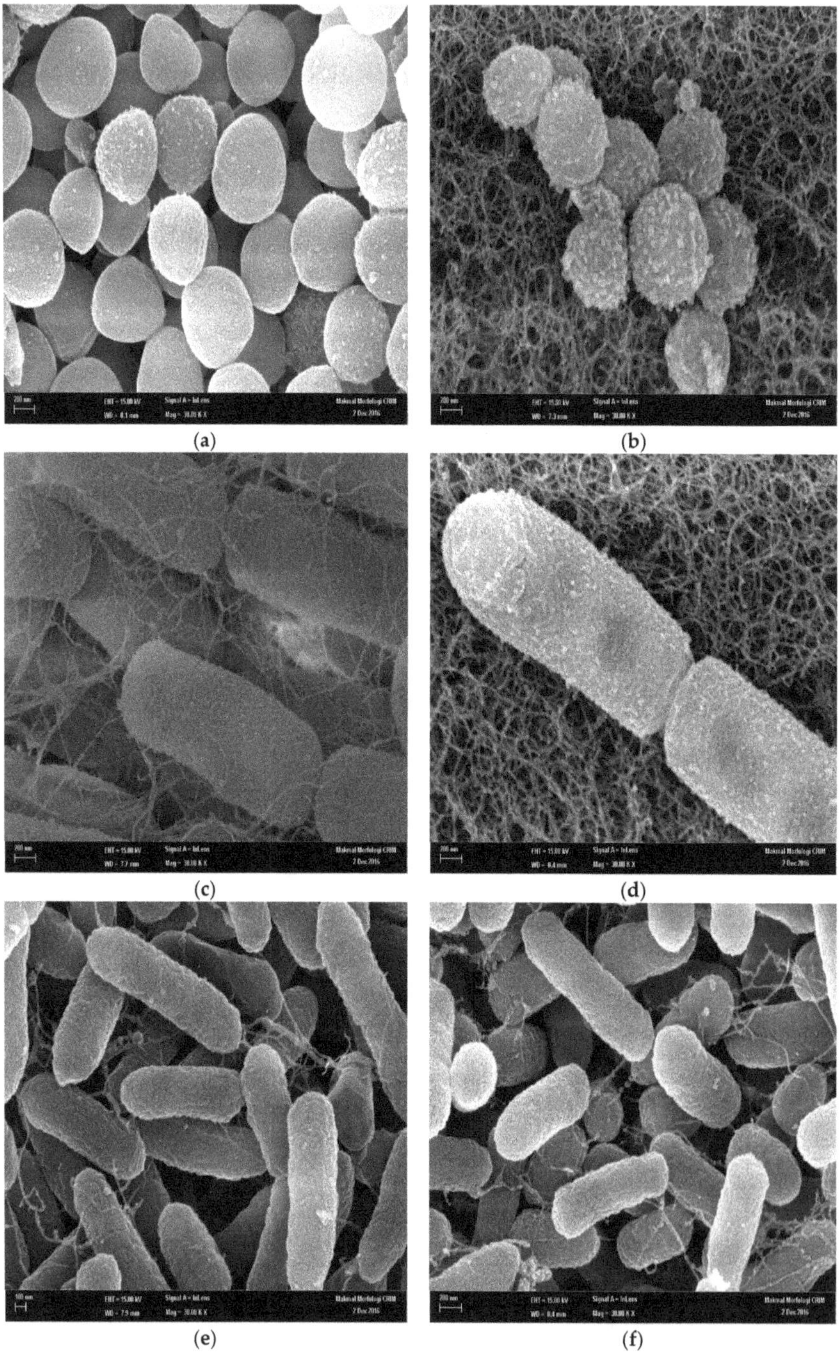

**Figure 2.** (**a–f**): Scanning electron microscope images of *Staphylococcus aureus* ATCC 25923 treated with pterostilbene at $1 \times$ MIC (**a**) and control (**b**); *Escherichia coli* O157 treated with pterostilbene at $1 \times$ MIC (**c**) and control (**d**) and *Pseudomonas aeruginosa* ATCC 15442 treated with pterostilbene at $1 \times$ MIC (**e**) and control (**f**).

On the other hand, most of *E. coli* treated with pterostilbene at 1 × MIC maintained their rod shape but blebbing was notably visible on their surface. As for the *Pseudomonas aeruginosa* ATCC 15442, the mophologies of most of the survival cells remained unchanged, with rod-shapes and smooth surfaces (Figure 2e).

## 3. Discussion

Gram-positive and Gram-negative bacteria can cause serious infectious diseases in mammals, and the emergence of antimicrobial resistance has led to the evaluation of other agents with potential antimicrobial activity. For instance, gentamicin is an effective aminoglycoside when used against most bacterial infections, but its small therapeutic window, potential side effects and emergence of resistant strains limit its application [19]. Stilbenes are a new antimicrobial agent belonging to the polyphenol class which is primarily active against Gram-positive and negative pathogens [20]. Organic chemists have found *trans*-stilbenes, including resveratrol and pterostilbene (Figure 3), to be an intriguing structural class of compounds that may be considered 'privileged structures' because of their diverse array of biological and potential therapeutic properties [21,22]. As indicated by a huge number of published studies, resveratrol and its derivatives have yielded excellent potency and bactericidal activity. On the basis of the many promising reports on the activity of resveratrol and substituted stilbenes, the design of and synthesis of chemically novel analogs of these compounds was undertaken. Recent studies revealed that *E*-phenoxystyrenes and *E*-phenylthiostyrenes together with several substituted *E*-stilbenoid analogs were found to exhibit promising activity against Gram-positive bacteria [21]. An antimicrobial phenolic stilbene, (*E*)-3-hydroxy-5-methoxystilbene was recently isolated and shown to possess inhibitory activity against Gram positive bacteria, including isolates of methicillin-resistant *Staphylococcus aureus* (MRSA), *Mycobacterium bovis* BCG, and a virulent *Bacillus anthracis* (Sterne strain), among others. [23]. Pterostilbene is a resveratrol analogue with enhanced fungitoxic activity with respect to its precursor [24]. It was previously reported that the presence of a methoxy group in position 5′ together with a methoxy group in position 3′ confer high antimicrobial activity [20].

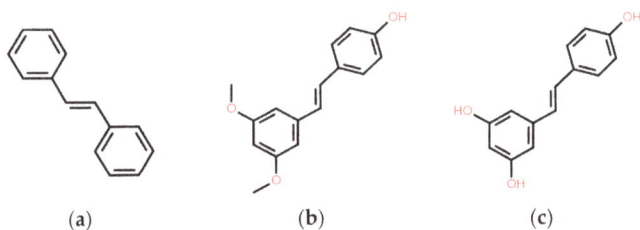

**Figure 3.** Chemical structures of (**a**) stilbene; (**b**) pterostilbene and (**c**) resveratrol.

Preliminary study showed that pterostilbene possessed antibacterial activity against the reference strains of *Staphylococcus aureus* ATCC 25923, *Escherichia coli* O157 and *Pseudomonas aeruginosa* ATCC 15442. In comparison with gentamicin used as broad spectrum therapy against Gram-positive and Gram-negative infections, pterostilbene still showed 16 times lower antimicrobial activity against *Staphylococcus aureus* ATCC 25923 and *Escherichia coli* O157 and four times less inhibitory potency against *Pseudomonas aeruginosa* ATCC 15442. The same phenomenon was seen in the anti-MRSA effect of pterostilbene whereby both MRSA ATCC and clinical strains are less susceptible to pterostilbene compared to standard antibiotics [25]. On the other hand, the killing kinetics of *Pseudomonas aeruginosa* ATCC 15442 demonstrated bacteriostatic activity with less than 3 $\log_{10}$ CFU/mL reductions in colony count at 24 h compared with the starting inoculum. Consistent with the previous study, pterostilbene has been reported to exert bacteriostatic activity against MRSA strains with MBC values exceeding their MIC values [25]. In general terms, most of the phytochemicals or secondary metabolites from the plants are capable of inhibiting or slowing the growth of bacteria rather than killing the pathogen [23]. However, our study has showed that pterostilbene exert bactericidal effect on both *Staphylococcus aureus* ATCC 25923 and *Escherichia coli* O157. These results concur with the conclusions of bactericidal activity for pterostilbene from time-kill analysis, typically achieving 3 $\log_{10}$ CFU/mL kill over a period of 24 h.

The fundamental morphological differences in the cell wall and membrane organization of Gram negative and Gram positive organisms modulate their susceptibilities to phytoalexins in plants. Hence, Gram positive bacteria are often nevertheless susceptible to plant products. More importantly, studies also suggested that the combined effects of the mixture of natural compounds found in plants might be necessary to produce a synergistic antibacterial activity against Gram negative organisms [19]. However, pterostilbene was generally found active against Gram-negative bacteria with activity comparable with that against Gram-positive bacteria. Of course, several successful plant pathogens are nevertheless able to circumvent the toxic effects of these plant metabolites which can be observed from our study in both *Bacillus cereus* ATCC 11778 and *Acinetobacter baumannii* ATCC 19606, demonstrating no susceptibility to the pterostilbene compound. Combine antibiotic therapy may produce synergistic effects in the treatment of bacterial infection [26] as supported by previous study that phytochemical has the ability to delay the emergence of antimicrobial resistance [19]. Previous in vitro studies have reported synergistic combinations of various stilbene compounds, such as ε-viniferin and johorenol A against MRSA, both ATCC and clinical strains [3,4]. Synergistic antibacterial activity of two stilbene analogues (3,4,5-trihydroxystilbene and 3,5-dihydroxy-4-isopropylstilbene) purified from a *Bacillus* sp. N strain associated with entomopathogenic nematode *Rhabditis* (*Oscheius*) in combination with ciprofloxacin was also reported [20]. However, there was no literature concerning the role of pterostilbene in the antibacterial activity particularly the combination therapy.

In our experiments, the in vitro interactive effects of the antibiotics were determined by the broth microdilution checkerboard method as previously described. All combination treatment groups demonstrated synergism were subjected to retesting by time-kill kinetics studies. However, these results somewhat differed from the time-kill kinetics reports in which indifference bactericidal activity of pterostilbene against *Staphylococcus aureus* ATCC 25923 was observed. Comparison of the time-kill plots for the three organisms studied showed that the killing rate was the greatest for *Staphylococcus aureus* ATCC 25923, then *Escherichia coli* O157 and then with no bactericidal activity for *Pseudomonas aeruginosa* ATCC 15442. Likewise, gentamicin consistently showed bactericidal activity against *Staphylococcus aureus* ATCC 25923, then *Escherichia coli* O157 and *Pseudomonas aeruginosa* ATCC 1544. Considering the wide range of combination outcomes produced and the different killing rate from each compound, it is most likely that the antibacterial activity of pterostilbene and gentamicin is not attributable to one specific mechanism, but to several targets in the cell.

Damage to the structural integrity of the cells and considerable morphological alteration were confirmed by SEM analysis. The marked irregular shape of *Staphylococcus aureus* ATCC 25923 exposed to pterostilbene may have been the result of the extensive loss of cell contents, the exit of molecules and ions and cell lysis [22]. This was also supported by our results from the time kill assays demonstrating bactericidal action of pterostilbene against *Staphylococcus aureus*. It might be specific to the cocci bacteria, as no significant changes in cell shape were observed in *Escherichia coli* O157 which has remained its rod-shaped. In addition, the cell membrane surface in both treated *Staphylococcus aureus* ATCC 25923 and *Escherichia coli* O157 were rougher than that of the control, with pores and blebs appearing on the membrane surfaces. This considerable alteration in cell morphology was not observed in *Pseudomonas aeruginosa* ATCC 15442 treated with pterostilbene. This may be explained by the regrowth of the bacterial colonies after 4 h treatment as observed in time-kill analysis.

## 4. Materials and Methods

### 4.1. Preparation of Pterostilbene Compound

Purchased pterostilbene was in an amber bottle and each bottle contained 10 mg extracted pterostilbene (molecular weight: 256.3 g/mole). Pterostilbene (10 mg) was dissolved in 10% dimethyl sulfoxide (DMSO, 1 mL) to achieve the concentration of 10 mg/mL. The 10% DMSO solvent was prepared by adding 100% DMSO stock solution (0.1 mL) in distilled water (0.9 mL). The stock solution was kept in −4 °C till used.

### 4.2. Preparation of Bacterial Inoculum

The bacteria strains used in this study were cultured and maintained in Mueller Hinton agar (MHA) and incubated at 37 °C for 18 h. Isolated colonies were inoculated onto slant agar in a bijou bottle for storage at 37 °C for 24 h. The bacteria suspension was adjusted to a turbidity corresponding to a spectrophotometric absorbance at 0.08 at wavelength 650 nm, which is equivalent to 0.5 McFarland's standard or a bacteria inoculum size of approximately 106 CFU/mL.

### 4.3. Determination of Minimal Inhibitory Concentration (MIC) and Minimum Bactericidal Concentration (MBC)

For the initial determination of the antibacterial activity of pterostilbene and gentamicin, a susceptibility screening study was conducted. The MIC values of pterostilbene and gentamicin against five bacterial strains were determined using the twofold serial microdilution method based on [27]. The tested pterostilbene and gentamicin were pippetted in a 96-well plate containing the sterile Mueller-Hinton broth enriched with 2% NaCl before the bacterial suspension at final inoculum of 106 CFU/mL was added. The range of final concentration of pterostilbene and gentamicin was 0.00078 mg/mL–0.2 mg/mL. A positive control comprised bacteria inoculum in Mueller-Hinton broth whereas the tested compounds in Mueller-Hinton broth were used as negative control to ensure medium sterility. The 96-well plate was then incubated at 37 °C for 24 h. MIC was the lowest concentration of compound and antibiotic showing no turbidity after 24 h, where the turbidity was interpreted as visible growth of bacteria. For consistency, each test was carried out in triplicate in a final volume of 0.1 mL per test well.

The MBC value was later determined by subculturing the well which showed no apparent bacterial growth on a sterile Mueller Hinton agar in an absence of antibacterial agent. The plate was incubated for 18 to 24 h. Bacterial growth on agar was observed and the concentration which has a colony count of less than 10 was considered as the MBC value. The MBC value was defined as the least concentration of antimicrobial agent that can kill >99% of the microorganism population where there is no visible growth on nutrient agar [28].

### 4.4. Determination of Fractional Inhibitory Concentration (FIC)

The combined antibiotic of pterostilbene with gentamicin against five strains of bacteria was evaluated from the FICI values for each combination using microdilution checkerboard technique [29]. The concentration of pterostilbene and gentamicin were prepared in five concentrations, namely 1/16 × MIC, 1/8 × MIC, 1/4 × MIC, 1/2 × MIC and 1 × MIC. Along the x-axis across the checkerboard plate, 50 µL of each pterostilbene was added into each well in the following sequence: 1/16 × MIC, 1/8 × MIC, 1/4 × MIC, 1/2 × MIC and 1 × MIC. As for the y-axis, 50 µL of vancomycin was added into each well in the same sequence as the tested pterostilbene. Inoculum size of approximately 106 CFU/mL was then added into all the wells. The well containing MHB and bacterial suspension served as positive control whereas well comprised only MHB and the tested antimicrobial agents served as negative control. The concentration of individual compound in the combination of pterostilbene and gentamicin which prevented visible bacterial growth was recorded as the MIC of the individual compound in the respective combination. The FICI value was then calculated as follows [29]:

$$\text{FIC}_{index} \left( \sum \text{FIC} \right) = \text{FIC}_A + \text{FIC}_B = A/\text{MIC A} + B/\text{MIC B}$$

where A = MIC of drug A in combination; B = MIC of drug B in combination; $\text{MIC}_A$ = MIC of drug A alone; $\text{MIC}_B$ = MIC of drug B alone Synergistic effect is defined as FICI of $\leq 0.5$; partial synergism as 0.5 > FICI < 1; additivity as FICI = 1; indifference as 1 > FICI < 4; and antagonism as FICI of more than 4.

*4.5. Time-Kill Study*

Time-kill assays were performed by the broth macro-dilution method [29]. The rate of bacterial killing over time was performed only on the antibiotic combinations found to be synergistic by the microdilution checkerboard method. The time-kill kinetic study of the pterostilbene in combination with gentamicin against five strains of bacteria was performed in the microtiter 96-well plates. In each well, the combined agent was added to 0.04 mL Mueller-Hinton broth and 0.05 mL Inoculum suspension with approximately $10^6$ CFU/mL of exponentially growing cells were subsequently added. The growth control wells comprised only bacteria and 0.05 mL Mueller-Hinton broth. The wells were then incubated at 37 °C and viable counts were performed at 0, 2, 4, 6, 8, and 24 h after addition of antibacterial agents. At each hour, 0.01 mL of the sample was removed from the wells to be diluted twofold with normal saline (0.9% NaCl). The sample was then spread on Mueller-Hinton agar plates using cotton swab rod and incubated for 24 h at 37 °C. Colony count of bacteria of between 30 and 300 CFU/mL for each plate was determined to obtain time-mortality curves by plotting the $\log_{10}$ CFU/mL on the x-axis and time (h) on the y-axis. Synergistic interaction was defined as $\log_{10}$ decrease in CFU/mL between the combination and the most active agent at 24 h. Additive or indifference was defined as log10 CFU/mL reduction in colony count at 24 h by the combination compared with the most active single agent. Antagonism was described as $\log_{10}$ increase in CFU/mL after 24 h between the combination and the most active agent [30]. The time-kill curves were recorded as a decrease in CFU/m within a specific time period. Bactericidal and bacteriostatic were, respectively, defined as $\geq 3 \log_{10}$ or $<3 \log_{10}$ reduction in colony count at 24 h compared with the starting inoculum [31].

*4.6. Scanning Electron Microscopic (SEM) Analysis*

In order to investigate the effect of pterostilbene on morphological changes of bacteria, treated and control cells were examined by scanning electron microscope (SEM) after 24 h treatment. The bacteria cells treated with 10% dimethyl sulfoxide (DMSO) were used as control. The cells were collected by centrifugation and washed with distilled water. The cells were fixed with 2% glutaraldehyde in 0.1 M phosphate buffer solution (PBS) and pH 7.4 for 15 min subsequently; the cells were washed three times and fixed with 1% osmium tetroxide in distilled water for 5 min at room temperature. The samples were dehydrated with a series of graded ethanol (70%, 90% and absolute ethanol), respectively for 5 min each, then coated with 42 nm thickness gold and examined on a Philips XL30 ESEM instrument (FEI Company, Hillsboro, OR, USA) at 28–30 kV [32].

## 5. Conclusions

The interaction between pterostilbene and gentamicin was bacteriostatic and enhancement of bactericidal activity of gentamicin against *Staphylococcus aureus* ATCC 25923, *Escherichia coli* O157 and *Pseudomonas aeruginosa* ATCC shown that these combinations have great potential worthy of further study and possible development as an alternative treatment in combating bacterial infections in the future.

**Acknowledgments:** The work was supported by Universiti Kebangsaan Malaysia grant (ERGS/1/2012/SKK03/UKM/02/2 and GUP-2016-036).

**Author Contributions:** Wee Xian Lee: Performed the experiments and drafting the article; Ahmad Rohi Ghazali: Designed the experiments, contributing to context and provided the laboratory facilities for the study; Dayang Fredalina Basri: Cosupervised the project, provided financial support and edited the final manuscript.

**Conflicts of Interest:** The authors declare no conflict of interest. This article is original and contains unpublished material. The corresponding author confirms that all of the other authors have read and approved the manuscript and no ethical issues involved.

## References

1. Sköld, O. Antibiotics: The greatest triumph of scientific medicine. In *Antibiotics and Antibiotic Resistance*; John Wiley & Sons, Inc.: Hoboken, NJ, USA, 2011; pp. 1–3.
2. Paulo, L.; Ferreira, S.; Gallardo, E.; Queiroz, J.; Domingues, F. Antimicrobial activity and effects of resveratrol on human pathogenic bacteria. *World J. Microbiol. Biotechnol.* **2010**, *8*, 1533–1538. [CrossRef]
3. Basri, D.F.; Chan, K.L.; Azmi, A.M.; Jalifah, L. Evaluation of the combined effects of stilbenoid from *Shorea gibbosa* and vancomycin against methicillin-resistant *Staphylococcus aureus* (MRSA). *Pharmaceuticals* **2012**, *5*, 1032–1043. [CrossRef] [PubMed]
4. Basri, D.F.; Lee, W.X.; Shukor, N.I.A.; Jalifah, L. Bacteriostatic antimicrobial combination: Antagonistic interaction between epsilon-viniferin and vancomycin against methicillin-resistant *Staphylococcus aureus*. *BioMed Res. Int.* **2014**, *2014*, 461756. [CrossRef] [PubMed]
5. Lahlou, M. The success of natural products in drug discovery. *Pharmacol. Pharm.* **2013**, *4*, 17–31. [CrossRef]
6. Mahady, G.B. Resveratrol as an antibacterial agent. In *Resveratrol in Health and Disease*; Aggarwal, B., Shishodia, S., Eds.; Taylor & Francis Group: Boca Raton, FL, USA; London, UK; New York, NY, USA, 2006; pp. 465–474.
7. Jeandet, P.; Hébrard, C.; Deville, M.A.; Cordelier, S.; Dorey, S.; Aziz, A.; Crouzet, J. Deciphering the role of phytoalexins in plant-microorganism interactions and human health. *Molecules* **2014**, *19*, 18033–18056. [CrossRef] [PubMed]
8. Roupe, K.A.; Remsberg, C.M.; Ya'nez, J.A.; Davies, N.M. Pharmacometrics of stilbenes: Seguing towards the clinic. *Curr. Clin. Pharmacol.* **2006**, *1*, 81–101. [CrossRef] [PubMed]
9. Lin, H.S.; Yue, B.D.; Ho, P.C. Determination of pterostilbene in rat plasma by a simple HPLC-UV method and its application in pre-clinical pharmacokinetic study. *Biomed. Chromatogr.* **2009**, *12*, 1308–1315. [CrossRef] [PubMed]
10. Adrian, M.; Jeandet, P.; Douillet-Breuil, AC.; Tesson, L.; Bessis, R. Stilbene content of mature *Vitis vinifera* berries in response to UV-C elicitation. *J. Agric. Food Chem.* **2000**, *12*, 6103–6105. [CrossRef]
11. Rimando, A.M.; Kalt, W.; Magee, J.B.; Dewey, J.; Ballington, J.R. Resveratrol, pterostilbene, and piceatannol in vaccinium berries. *J. Agric. Food Chem.* **2004**, *15*, 4713–4719. [CrossRef] [PubMed]
12. Pari, L.; Satheesh, M.A. Effect of pterostilbene on hepatic key enzymes of glucose metabolism in streptozotocin- and nicotinamide-induced diabetic rats. *Life Sci.* **2006**, *79*, 641–645. [CrossRef] [PubMed]
13. Rimando, A.M.; Cuendet, M.; Desmarchelier, C.; Mehta, R.G.; Pezzuto, J.M.; Duke, S.O. Cancer chemopreventive and antioxidant activities of pterostilbene, a naturally occurring analogue of resveratrol. *J. Agric. Food Chem.* **2002**, *12*, 3453–3457. [CrossRef]
14. Remsberg, C.M.; Ya'nez, J.A.; Ohgami, Y.K.; Vega-Villa, R.; Rimando, A.M.; Davies, N.M. Pharmacometrics of pterostilbene: Preclinical pharmacokinetics and metabolism, anticancer, antiinflammatory, antioxidant and analgesic activity. *Phytother. Res.* **2008**, *2*, 169–179. [CrossRef] [PubMed]
15. Langcake, P.; Cornford, C.A.; Pryce, R.J. Identification of pterostilbene as a phytoalexin from *Vitis vinifera* leaves. *Phytochemistry* **1979**, *6*, 1025–1027. [CrossRef]
16. Jeandet, P.; Douillet-Breuil, A.C.; Bessis, R.; Debord, S.; Sbaghi, M.; Adrian, M. Phytoalexins from the *Vitaceae*: Biosynthesis, phytoalexin gene expression in transgenic plants, antifungal activity, and metabolism. *J. Agric. Food Chem.* **2002**, *50*, 2731–2741. [CrossRef] [PubMed]
17. Levison, M.E. Pharmacodynamics of antimicrobial drugs. *Infect. Dis. Clin. N. Am.* **2004**, *18*, 451–465. [CrossRef] [PubMed]
18. Tam, V.H.; Schilling, A.N.; Nikolaou, M. Modelling time-kill studies to discern the pharmacodynamics of meropenem. *J. Antimicrob. Chemother.* **2005**, *55*, 699–706. [CrossRef] [PubMed]

19. Ali, B.H. Agents ameliorating or augmenting experimental gentamicin nephrotoxicity: Some recent research. *Food Chem. Toxicol.* **2003**, *41*, 1447–1452. [CrossRef]

20. Malik, C.; Agnès, K.; Abdelwahad, E.; Philippe, M.; Dominique, V.F.; Marielle, A. Antimicrobial activity of resveratrol analogues. *Molecules* **2014**, *19*, 7679–7688.

21. Roberti, M.; Pizzirani, D.; Recanatini, M.; Simoni, D.; Grimaudo, S.; Cristina, A.D.; Abbadessa, V.; Gebbia, N.; Tolomeo, M.J. Synthesis and biological evaluation of resveratrol and analogues as apoptosis-inducing agents. *J. Med. Chem.* **2003**, *46*, 3546–3554. [CrossRef] [PubMed]

22. Denyer, S.P. Mechanisms of action of biocides. *Int. Biodeterior.* **1990**, *26*, 89–100. [CrossRef]

23. Nazzaro, F.; Fratianni, F.; De Martino, L.; Coppola, R.; De Feo, V. Effects of essential oils on pathogenic bacteria. *Pharmaceuticals* **2013**, *6*, 1451–1474. [CrossRef] [PubMed]

24. Castellarin, S.D.; Bavaresco, L.; Falginella, L.; Gonçalves, M.I.V.Z.; Di Gaspero, G. Phenolics in grape berry and key antioxidants. In *The Biochemistry of the Grape Berry*; Gerós, H., Chaves, M.M., Delrot, S., Eds.; Bentham Science Publishers: Bussum, The Netherlands, 2012; pp. 90–91.

25. Ishak, S.F.; Ghazali, A.R.; Zin, N.M.; Basri, D.F. Pterostilbene enhanced anti-methicillin resistant *Staphylococcus aureus* (MRSA) activity of oxacillin. *Am. J. Infect. Dis.* **2016**, *1*, 1–10. [CrossRef]

26. Kumar, S.N.; Siji, J.V.; Nambisan, B.; Mohandas, C. Activity and synergistic interactions of stilbenes and antibiotic combinations against bacteria in vitro. *World J. Microbiol. Biotechnol.* **2012**, *28*, 3143–3150. [CrossRef] [PubMed]

27. Cockeril, F.R.; Wikler, M.A.; Alder, J.; Dudley, M.N.; Eliopoulos, G.M.; Ferrado, M.J.; Hardy, D.J.; Hecht, D.W. *Methods for Dilution Antimicrobial Susceptibility Tests for Bacteria That Grow Aerobically; Approved Standard*, 9th ed.; Clinical and Laboratory Standards Institute: Wayne, PA, USA, 2012; Volume 2, p. 18.

28. Pankey, G.A.; Sabath, L.D. Clinical relevance of bacteriostatic versus bactericidal mechanisms. *Clin. Infect. Dis.* **2004**, *6*, 864–870. [CrossRef] [PubMed]

29. Basri, D.F.; Khairon, R. Pharmacodynamic interaction of *Quercus infectoria* galls extract in combination with vancomycin against MRSA using microdilution checkerboard and time-kill assay. *Evid. Based Complement. Altern. Med.* **2012**, *2012*, 493156. [CrossRef] [PubMed]

30. Neu, H.C.; Gootz, T.D. Antimicrobial chemotherapy. In *Medical Microbiology*, 4th ed.; Boron, S., Ed.; University of Texas Medical Branch: Galveston, TX, USA, 1996.

31. Lee, G.C.; Burgess, D.S. Polymyxins and doripenem combination against KPC-producing *Klebsiella pneumonia*. *J. Clin. Med. Res.* **2003**, *2*, 97–100.

32. Basri, D.F.; Jaffar, N.; Zin, N.M.; Raj, S. Electron microscope study of gall extract from *Quercus infectoria* in combination with vancomycin against MRSA using post-antibiotic effect determination. *Int. J. Pharm.* **2013**, *2*, 150–156. [CrossRef]

**Sample Availability:** Samples of the compounds are available from the authors.

*molecules*

MDPI

*Review*

# Antiproliferative Effect of Indole Phytoalexins

**Martina Chripkova [1,2], Frantisek Zigo [3] and Jan Mojzis [1,*]**

[1]   Department of Pharmacology, Faculty of Medicine, P.J. Šafárik University, 040 11 Košice, Slovakia; chripkova.martina@gmail.com
[2]   Department of Human and Clinical Pharmacology, University of Veterinary Medicine and Pharmacy, 040 11 Košice, Slovakia
[3]   Department of Animal Breeding, University of Veterinary Medicine and Pharmacy, 040 11 Košice, Slovakia; frantisek.zigo@uvlf.sk
*   Correspondence: jan.mojzis@upjs.sk; Tel.: +421-552-303-485

Academic Editor: Philippe Jeandet
Received: 18 October 2016; Accepted: 22 November 2016; Published: 26 November 2016

**Abstract:** Indole phytoalexins from crucifers have been shown to exhibit significant anti-cancer, chemopreventive, and antiproliferative activity. Phytoalexins are natural low molecular antimicrobial compounds that are synthesized and accumulated in plants after their exposure to pathogenic microorganisms. Most interestingly, crucifers appear to be the only plant family producing sulfur-containing indole phytoalexins. The mechanisms underlying its anti-cancer properties are unknown. Isolation from cruciferous plants does not provide sufficient quantities of indole phytoalexins and, for biological screening, they are usually obtainable through synthesis. Understanding the molecular mechanism of the action of these substances and their structure-activity relationships is quite important in the development of new analogs with a more favorable profile of biological activities. In this review, we present the key features of indole phytoalexins, mainly their antiproliferative ativities.

**Keywords:** indole phytoalexins; crucifers; brassinin; antiproliferative; cancer

---

## 1. Introduction

Substances of plant origin have a significant position among the drugs used for cancer treatment. Information about the healing effects of extracts from various plants can be found in Chinese, Indian, as well as ancient literature [1]. The tradition in their use and claimed successes in anti-tumor therapy are an inspiration for a more detailed study. Modern methods in biochemistry and pharmacology allow their chemical identification and exact testing.

The discovery of the anti-cancer effects of foods rich in vegetables from the *Cruciferae* (*Brassicaceace*) family led to an increased interest in the natural substances contained therein [2]. Their frequent consumption, along with other kinds of vegetables and fruits, can significantly reduce the risk of cancer and prolong a patient's life [3]. Although cruciferous vegetables contain many compounds with anti-cancer properties, they are unique because of the high content of sulfur-containing phytochemicals, the glucosinolates [4]. After metabolic degradation, glucosinolates are converted to a variety of products, from which the isothiocyanates play probably the most important role in cancer chemoprevention [5]. During the last 20 years, multiple mechanisms of their anti-cancer action have been recognized, including modulation of carcinogen metabolism [6–8], induction of apoptosis and cell cycle arrest, inhibition of neovascularization [9,10], inhibition of cancer cell migration [11,12], as well as the blocking of signaling pathways associated with malignant transformation or cell survival [13–15].

Furthermore, a special group of substances in cruciferous plants are the indole phytoalexins. Indole phytoalexins have been reviewed with regard to their isolation, occurrence, synthesis, biosynthesis, biotransformation, and role in plant defenses [16–21]. These substances show a wide range of biological effects, including antiproliferative [22–24], antifungal [25], antiprotozoal [26], chemopreventive [27], and anti-cancer effects [28]. While a variety of possible molecular mechanisms have been proposed to explain this activity, none has been directly validated in vivo.

## 2. Occurrence, Structure, and Biological Activity of Indole Phytoalexins

Phytoalexins are antimicrobial secondary metabolites with a low molecular weight produced *de novo* by plants after exposure to biological (bacteria, fungi, viruses), physical (UV radiation, heat shock, injury), or chemical (heavy metals) stress [20,29]. They include a number of different types of substances that are species-specific, such as terpenoids, alkaloids, flavonoids, and the like. The name of phytoalexins comes from their function of defending the plant organism, as it is derived from the Greek words phyton–plant, alexos–defend. They are not located in the tissues of healthy plants. The synthesis of phytoalexins is triggered by specific substances, so-called elicitors, in plants. These substances initiate the defense response of a plant after it is attacked by a pathogen. Phytoalexins appear in an infected plant several hours or days after such an attack [30–32]. These substances are produced by plants in small amounts, and isolation of them from plants is difficult. The introduction of the chemical synthesis and their analogs provided the appropriate quantities necessary for an evaluation of their environmental function and biological activity [21].

Phytoalexins were first described by Müller and Börger in 1940 as antifungal substances produced by *Solanum tuberosum* after being attacked by the fungus *Phytophora infestans* [33]. The first phytoalexin to be isolated and chemically characterized was (+)-pisatin. Pisatin was obtained from *Pisum sativum* (pea) in a small concentration after the plant was infected by the *Ascochyta pisi* pathogen [34].

The structure of phytoalexins depends on the type of plant that synthesizes them and partially on the elicitor inducing their synthesis. Isoflavones prevail in plants of the *Fabaceae* family, sesquiterpenoids in the *Solanaceae* family, diterpenes in the *Poaceae* family, and for plants of the *Brassicaceae* family it is a characteristic that it is the only family that produces indole phytoalexins containing sulphurous phytoalexins [16,35].

At present, 44 different species of indole phytoalexins are known. Some of them are produced by more than one plant species and may be elicited by a number of pathogens or abiotic factors [21]. What is interesting concerning the indole phytoalexins' structure is the unique connection of the indole nucleus with the side chain or a heterocycle containing nitrogen and sulfur atoms [36]. The side chain is most often in the $-CH_2-$ group in position 3 of the indole nucleus. A heterocycle may be joined by condensation (cyclobrassinin), by a single bond (camalexin) or through a spiroatom (spirobrassinin) (Figure 1). Methyl-1-methoxy-indole-3-carboxylate has a carbonyl group (–C=O) at position 3 of the indole nucleus, discovered as the first of two phytoalexins of cruciferous plants that do not contain sulfur. It was isolated from wasabi, and it is assumed that its presence in this plant results in the crusting resistance caused by the fungus *Phoma lingam*. Brasicanal A, B, and C belong to the indole phytoalexins having an aldehyde group (–CH=O) at position 3 of the indole nucleus [18,37]. The bifunctional structural characteristic of natural brassinin, i.e., the position of both the indole nuclei and the dithiocarbamoyl aminomethyl moiety, is similar to that of the chemopreventive agents, such as indole-3-carbinol and benzyl isothiocyanate [38]. As was mentioned above, isothiocyanates and indoles derived from the hydrolysis of glucosinolates, such as sulforaphane and indole-3-carbinol, have been implicated in a variety of anticarcinogenic mechanisms [4]. With the presence of both components, a positive biological effect can be achieved. An aliphatic analog of brassinin, (±)-4-methylsulfinyl-1-(S-methyldithiocarbamyl)-butane (Sulforamate), has structural similarities to sulforaphane [38].

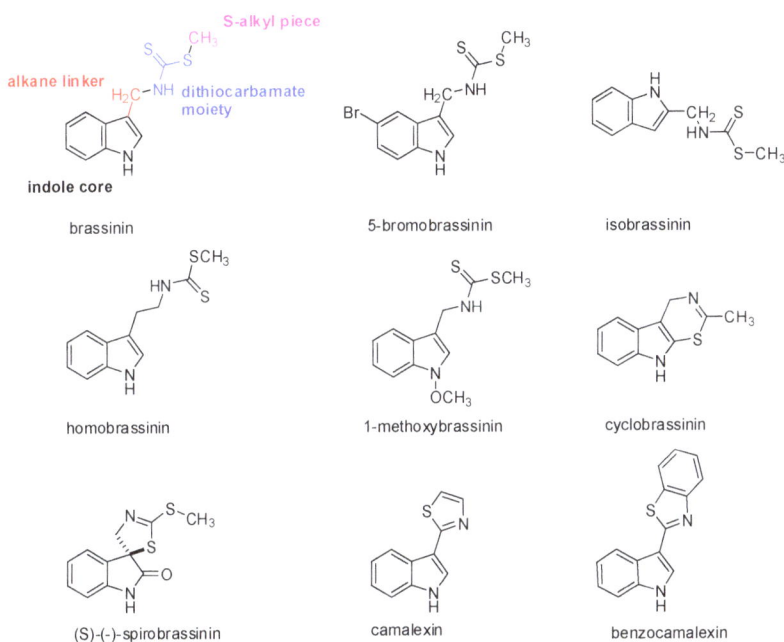

**Figure 1.** Molecular structures of the representative indole phytoalexins and their derivatives.

The biosynthetic pathway of indole phytoalexins begins with the conversion of tryptophan to indolyl-3-acetohydroxamic acid and then opens into a number of metabolic pathways, including indolyl glucosinolates, brassinins, and camalexin [39]. Brassinin is a biosynthetic precursor of several other phytoalexins [35].

Brassinin, cyclobrassinin, and 1-methoxybrassinin (Figure 1) are indole phytoalexins first isolated from Chinese cabbage which had been infected by the *Pseudomonas cichorii* bacterium [36]. Cyclobrassinin is a natural product of oxidative brassinin cyclization [27], while 1-methoxybrassinin is 1-methoxyindol alkaloid with a methoxy group bound to the indole nitrogen atom. Other typical representatives of 1-methoxyindol phytoalexins are 1-methoxybrassitin, also isolated from Chinese cabbage; 1-methoxybrasenin A and 1-methoxybrasenin B of cabbage; (*R*)-(+)-1-methoxyspirobrassinin isolated from kohlrabi; (2*R*,3*R*)-(−)-1-methoxyspirobrassinol methyl ether isolated from the Japanese radish; and sinalbin B from white mustard [17,18]. Later, arvelexin, located in *Thlaspi arvense* (field pennycress), was described [40]. Isalexin, brassicanate, and rutalexin were isolated from *Brassica napus*, ssp. *rapifera* (rutabaga) after abiotic elicitation through UV light or after being infected by the pathogenic fungus *Rhizoctonia solani* [25]. Erucalexin was obtained from *Erucastrum gallicum* (common dogmustard). The presence of caulilexin, caulilexin B, and caulilexin C was discovered in cauliflower [41]. *Thellungiella salsuginea* (saltwater cress) produces wasalexin A and B. In addition to wasalexins, this plant also produces the phytoalexins 1-methoxybrasenin B and rapalexin A [42]. Brussels sprouts were shown to produce a unique thiolcarbamate, brussalexin A. It is the first naturally occurring thiolcarbamate in which the sulfur atom is attached to the 3-methylindolyl moiety [43]. The phytoalexins 4-methoxycyclobrassinin and dehydrocyclobrassinin were isolated for the first time from canola roots infected with the biotroph *Plasmodiophora brassicae* (clubroot disease) [44].

(*S*)-(−)-Spirobrassinin (Figure 1) was isolated in 1987 from Japanese radish [45]. Spirobrassinin resembles other anticarcinogenic substances in its structure, such as pentacyclic oxindole alkaloids found in *Uncaria tomentosa* (cat's claw) from the Andean region and Peru, a plant used in folk medicine as an anti-cancer and anti-inflammatory substance as well as a contraceptive [46]. These alkaloids have an antiproliferative effect on HL-60 and U-937 leukemic cell lines, without inhibiting the growth of progenitor cells, with the highest activity observed for Uncarina F ($IC_{50}$ = 21.7–29 µmol/L) [47].

Camalexin (Figure 1) has a remarkable position among the indole phytoalexins in that it is produced by the *Arabidopsis thaliana* plant (Arabidopsis willow) after being infected by the *Alternaria brassicicola* fungus and *Pseudomonas syringae* bacterium. The presence of this substance has also been demonstrated in the plants *Capsella bursa-pastoris* (shepherd's pocket) and *Camelina sativa* (gold-of-pleasure). The genetic data show that camalexin is synthesized from tryptophan. The reaction is catalyzed by two kinds of P450cytochrome (CYP79B2 and CYP71B15) [48]. Biosynthesis is localized at the site of pathogen infection and takes place in the endoplasmic reticulum [49]. This phytoalexin has a cytostatic effect on the *Trypanosoma cruzi* pathogenic flagellate [26].

What is remarkable in indole phytoalexins is the presence of a dithiocarbamate group (NH-CS-SR) (Figure 1), which is part of some organic fungicides [50]. Dithiocarbamates are distinguished through strong antioxidant and antitumor effects. It has been confirmed that the dithiocarbamate side chain is very crucial for the anti-cancer activity [51]. Reactive metabolites of certain dithiocarbamates (proline-dithiocarbamate, diethyldithiocarbamate) induce the expression of p21KIP1/CIP1 in the p53 dependent pathway, leading to the cessation of the cell cycle in a G1/S HepG2 cell line. Furthermore, they affect the phosphorylation of cyclin E, the cyclin of dependent kinase inhibition 2, and cyclin E degradation in these cells during apoptosis. They also cause a decrease of Bcl-2 anti-apoptotic proteins and an increase in the level of p53 protein. 4(3*H*)-Quinazolinone dithiocarbamate exhibited anti-cancer activity against human myelogenous leukaemia cells [52,53].

Duan et al., [54] reported the synthesis of a series of novel 1,2,3-triazole-dithiocarbamate hybrids and evaluated them for anti-cancer activity against several human tumor cell lines (MGC-803, MCF-7, PC-3, EC-109). Another study reported novel dithiocarbamte derivatives where benzimidazole replaced brassinin in the indole moiety. Their chemotherapeutic activity was evaluated. This docking study revealed that benzimidazoledithiocarbamate derivatives are more selective for anti-cancer activity than antimicrobial activity [55]. Some dithiocarbamates act by modulating the responses of heat shock proteins (HSP—they synthesize in the cell as a response to the influence of any stress) or by inhibiting the activity of the NF-κB transcription factor [19]. In the case of tumor cells, the activation of NF-κB plays a role in the protection against apoptosis induced by e.g., the action of TNF-α, ionizing radiation, or other inducements. The activated signaling pathway of NF-κB inhibits the apoptotic potential of chemotherapeutic agents and thereby contributes to the resistance of cancer cells to these agents. Therefore, it follows that the substance inhibiting the signaling pathway may be used to overcome the drug resistance of tumor cells [56]. NF-κB also regulates the expression of a large number of genes that play an important role in the non-adaptive (innate) immune response [57]. In connection to this fact, the anti-inflammatory effects of arvelexin, which inhibits the activation of NF-κB in macrophages with a subsequent decrease of the expression of pro-inflammatory inducible enzymes (iNOS, COX-2) and cytokines (TNF-a, IL-6 and IL-1b), were demonstrated [58].

## 3. The Antiproliferative Effect of Naturally Occurring Indole Phytoalexins

The ability of indole phytoalexins to inhibit the growth of cells was tested in vitro in a number of cancer cell lines. The mechanism of the antiproliferative effect of these substances is still unclear. Available data indicate that the antiproliferative activity of indole phytoalexins is more a result of modulating the activity of transcription factors regulating the cell cycle, cell differentiation, and apoptosis than a direct interaction with DNA [19,59]. The possible anti-cancer effects of representative indole phytoalexins and their derivatives are summarized in Table 1.

The antiproliferative effect of brassinin, spirobrassinin, and cyclobrassinin was tested in a (B16) mouse melanoma and (L1210) leukemia cancer cell line. Brassinin showed the highest inhibitory effect in that, at a 100 µmol/L concentration, it reduced the cell growth by 35% and at 10 µmol/L concentration by 15% (L1210) and 9% (B16), respectively, after 24 h of incubation. Spirobrassinin was less effective; a reduction of the number of cells occurred only at a concentration of 100 µmol/L (about 13%). Cyclobrassinin did not show antiproliferative activity [59].

Significant chemopreventive activity was also recorded in these substances in a DMBA model (7,12-dimethylbenzathracene) of induced mammary gland carcinogenesis in mice. Brassinin inhibited the formation of pre-neoplastic lesions of the mammary gland by 73%, cyclobrassinin by 90.9%, and spirobrassinin by 76% at 10 µmol/L concentration. The mechanism of the chemopreventive effect is unknown, but probably the induction of phase II detoxification enzymes occurs. Regarding the inhibition of tumor growth, these substances may have a chemopreventive effect in the initiation and promotional phases of carcinogenesis (Table 1) [27,38,60].

Cyclobrassinin, brassilexin, and their synthetic analogs (homocyclobrassinin and 5-methoxybrassilexin) caused growth inhibition of the KB cell line (epidermoid carcinoma), while the highest efficiency was observed in brassilexin ($IC_{50}$ 8 µg/mL). Brassilexin had the same values of $IC_{50}$ for human KB carcinoma and normal monkey kidney cells, which indicates a lack of selectivity for cancer cells [61].

Brassinin and its derivatives are inhibitors of indoleamine 2,3-dioxygenase (IDO), which is a new target in cancer immunotherapy. IDO is an extrahepatic enzyme that catalyzes the initial and rate-limiting step in the degradation of tryptophan along the kynurenine pathway that leads to the biosynthesis of nicotinamide adenine dinucleotide ($NAD^+$) [62–64]. The degradation of tryptophan reduces the immune response to tumor cells. Therefore, blocking IDO could lead to greater efficiency in tumor immunotherapy [65]. The role of the inhibition of IDO in the antitumor mechanism of the bioavailable analog of brassinin (5-bromobrassinin (Figure 1)) was confirmed in vivo, where this compound suppressed growth of B16-F10 melanoma xenografts in C57BL/6 mice but not in athymic NCr-nu/nu and IDO knock-out mice (Table 1) [28].

Gaspari et al. [66] undertook a structure-activity relationship study of brassinin with the goal of obtaining a more potent IDO inhibitor. They divided the brassinin structure into four components: the indole core, the alkane linker, the dithiocarbamate moiety, and the S-alkyl piece (Figure 1). The study showed that replacement of the indole moiety with other aromatic rings retained the activity, whereas an increase in the length of alkyl chain increased the potency of compounds, but upon replacement of the dithiocarbamate side chain with any other group, the compounds exhibited very little or no activity at all.

Brassinin has been reported to induce G1 phase arrest through the increase of p21 and p27 by inhibition of the phosphatidylinositol 3-kinase signaling pathway in colorectal cancer cells [67]. The latest data suggests possible brassinin interference with the PI3K/Akt/mTOR/S6K1 signaling pathway [68]. Regulation of the mTOR (mammalian target of rapamycine) protein kinase plays an important role in the cellular metabolism of proliferation and angiogenesis. It is an attractive therapeutic target because it is a key point at which a number of signaling pathways converge. The activation of the PI3K/Akt/mTOR/S6K1 signaling pathway is closely linked with the development of prostate cancer, its metastasis, and angiogenesis [69]. The ability of brassinin to inhibit this cascade is pre-tagging it as a potential candidate for the treatment and prevention of prostate cancer [68]. Brassinin can inhibit the constitutive and inducible STAT3 (Signal transducer and activator of transcription 3) signaling pathway, thereby attenuating tumor growth (Table 1) [70]. STAT are proteins that regulate gene expression by affecting transcription. They are part of the signal transduction pathway of many growth factors and cytokines and are activated by phosphorylation of tyrosine and serine residues by upstream kinases [71]. Constitutive activation of STAT3 has been reported in many types of malignancies, such as myeloma, head and neck cancer, breast cancer, prostate cancer, and non-small cell lung cancers (NSCLC) [72–77]. There is evidence showing that inhibition of STAT3 leads to cessation of tumor cell growth and apoptosis. Brassinin suppressed STAT3 activation through the modulation of two groups of signaling proteins known to inactivate STAT proteins, the protein inhibitors of activated STAT (PIAS) and the suppressors of cytokine signaling (SOCS). In addition, brassinin enhanced the antitumor effects of paclitaxel, a chemotherapeutic drug used extensively to treat NSCLC (non-small cell lung cancer) patients [70].

Kim et al. [78] analyzed the potential synergistic anti-tumor effects of brassinin combined with capsaicin on prostate cancer PC-3 cells (Table 1). Capsaicin, an alkaloid derived from the chilli pepper, has been shown to promote cell death in a variety of tumor cells [79]. After treatment with brassinin and capsaicin at various concentrations, the synergistic cytotoxic effect of PC-3 cells was observed [78].

In our studies, we examined the antiproliferation effects of 1-methoxybrassinin [24] and homobrassinin (Table 1), (Figure 1) [80]. We found the redistribution of the cell contents into the G2/M phase after just 24 hours of incubation with homobrassinin. The accumulation of cells in the G2/M phase could indicate a possible interaction of this substance with tubulins, which are involved in the construction of the spindle apparatus. This theory is supported by the results of the study by Smith et al. [81], in which they reported that the degradation products of glucosinolates, substances biogenetically related to indole phytoalexins, caused the condensation of α-tubulin and subsequent blocking of the mitotic phase in colorectal cancer cells (Caco-2). The results of cell cycle analysis led us to monitor the expression of selected genes involving the formation of microtubules. We found changes in the expression of tubulin subunits after exposure to homobrassinin in the form of a reduced expression of β5-tubulin and increased expression of α-tubulin. The ability of the studied substances to induce the apoptosis of Caco-2 cells was associated with changes in the balance between pro- and anti-apoptotic representatives of proteins of the Bcl-2 family and caspase-3 activation [80]. 1-methoxybrassinin showed significant antiproliferative effects on the Jurkat (human acute T lymphoblastic leukemia) cell line (IC$_{50}$ 10 µmol/L). Cell cycle analysis showed a reduction in the number of cells in the S and G2/M phase of the cell cycle with an increased fraction of sub-G0/G1 DNA, which is considered a marker of apoptosis. After 72 h of the incubation of Jurkat cells with 1-methoxybrassinin, the amount of this fraction increased to more than 90% [22]. The increase in ROS levels, reduction in the mitochondrial membrane potential levels, and decrease of GSH in the Caco-2 cells after treatment with 1-methoxybrassinin probably also contribute to the triggering of the apoptotic cascade. The potential of this substance to cause GSH depletion in tumor cells could be used to increase their sensitivity to chemotherapeutic drugs [24]. The significant potentiation of vincristine cytotoxicity to U-87 MG (human glioblastoma astrocytoma) cells by brassinin, spirobrassinin, 1-methoxyspirobrassinin, and 1-methoxyspirobrassinol, as well as drug-like characters of these compounds, suggest the possibility of their future role in combination chemotherapy [82].

While several reports showed the antiproliferative effect of cruciferous phytoalexins, a study by Mezencev et al. revealed a contradictory effect. While spirobrassinin and 1-methoxyspirobrassinol methyl ether reduced the growth of MCF-7 (breast cancer cell line, estrogen receptor positive) and Caco-2, brassinin, 1-methoxyspirobrassinol, and 1-methoxyspirobrassinin in contrast stimulated the proliferation of these cells. All tested substances inhibited the growth of the MDA-MB-231 (breast cancer cell line). It can therefore be assumed that MCF-7 growth stimulation may be caused by the partial estrogen-receptor agonism of these indole phytoalexins and their metabolites [82].

Camalexin has demonstrated antiproliferative activity on SKBr3 (human breast carcinoma cell line) with the increased expression of topoisomerase IIα. The inhibition of tumor cell growth induced by camalexin (IC$_{50}$ 2.7 µmol/L) was even more evident when compared to conventional cytostatic agents, such as melphalan (IC$_{50}$ 13.0 µmol/L) and cisplatin (IC$_{50}$ 7.4 µmol/L) [83]. The mechanism of the cytotoxic effect of camalexin on Jurkat cells can be compared to the action of the ATO drug (arsenic trioxide) used for the therapy of relapse and resistant acute promyelocytic leukemia [84,85]. Data from recent years suggest that camalexin causes the accumulation of reactive forms of oxygen in tumor cells, resulting in the formation of oxidative stress, the activation of caspases, and induction of apoptosis. This effect was shown in the metastatic prostate cancer cell line and leukemia cell line. Inhibition of the growth of prostate cancer cells may be associated with a change of expression and activity of the cathepsin lysosomal enzyme in these cells due to camalexin influence. This hypothesis was also confirmed by an experiment with pepstatin A, an inhibitor of cathepsins activity, which blocked the cytotoxic effect of camalexin. Cathepsins are secreted into the cytosol during the initiation of apoptosis. Various incentives—such as oxidative stress, TNF-α, and p53—can increase lysosomal membrane permeabilization, thus triggering the translocation of these enzymes into the cytosol. Affecting the activity of lysosomal proteases, such as cathepsin, represents a great potential—particularly in the treatment of metastatic prostate cancer [86]. The advantage of camalexin, as well as its derivatives is its minimal cytotoxic effect on non-tumor cells (Table 1) [23,85]. Furthermore, in our study we have found structure-activity relationship. The fusion of benzene with thiazole ring of camalexin significantly enhances its cytotoxicity. On the other hand, further modulation of chemical structure (e.g., methylation of benzocamalexin) resulted in decreased antiproliferative activity and neither addition of methoxy-, fluoro-, nor cyano-group increased it (Figure 2) [85].

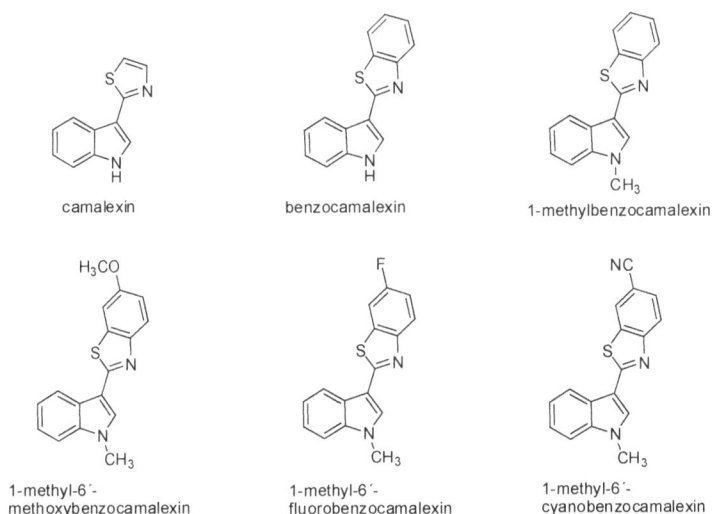

**Figure 2.** Chemical structure of camalexin and its synthetic analogs.

## 4. Antiproliferative Effect of Synthetic Derivatives of Indole Phytoalexins

Indole phytoalexins represent a natural template for the synthesis of several substituted derivatives in order to find more favorable antiproliferative and chemopreventive effects of these substances.

Glyoxylic analogs of natural phytoalexins, such as brassinin, brassitin, and some 1-methoxyindole phytoalexins, were synthesized. When comparing the antiproliferative activity of these derivatives to natural phytoalexins, it was shown that the most effective was an analog of glyoxylic 1-methoxybrassenin B ($IC_{50}$ 3.3–66.1 µmol/L), which reduced the growth of cells in the most acute lymphoblastic leukemia line (CCRF-CEM) from a number of tested cancer cell lines (Jurkat, HeLa, MCF-7, MDA-MB231, A-549, CCRF-CEM) [87]. There are also several other studies on the anti-tumor effects of substances containing the indolyl glyoxylic group. Glyoxylic derivatives have been found as intermediates in the synthesis of anti-tumor indolocarbazole alkaloids. The indolyl glyoxylic group is part of the natural marine product, hyrtiosin B (isolated from the sea sponge *Hyrtios erecta*), which has shown in vitro antiproliferative activity against the KB cell line [88]. Synthetic indolyl glyoxylic amides have been identified as anti-tumor substances destabilizing the microtubules of cells, with indibulin as the most active derivative. This derivative was characterized as demonstrating in vitro activity against tumor SKOV3 (ovarian cancer), U87 (glioblastoma), and ASPC-1 (pancreas adenocarcinoma) cell lines [89].

Various synthetic 2-amino derivatives of spiroindoline phytoalexins have shown remarkable anti-tumor features. Through the introduction of a substituted phenylamino group into position 2 of the indole ring of 1-methoxyspirobrassinol methyl ether derivatives with a better anti-tumor effect than natural phytoalexins themselves were obtained. Some even achieved a better antiproliferative effect on cancer cell lines, such as cisplatin, etoposide, and doxorubicin [90]. Similarly, the 2-amino derivative of 1-methoxyspirobrassinol, *trans*-1-Boc-2-deoxy-2-(1-piperidinyl) spirobrassinol created an anti-tumor effect against small cell lung carcinoma, renal, ovarian, prostate carcinoma, and colorectal carcinoma and induced glutathione depletion in MCF-7 breast cancer cells [91]. The findings of these studies are useful for the suggestion of more amino analogs. By substituting the bis (2-chloroethyl) amino alkyl group into the above-mentioned amino derivatives of 1-methoxyspirobrassinol, the following compounds arose: *cis*- and *trans*-1-methoxy-2-deoxy-2 [*N,N*-bis(2-chloroethyl)amino] spirobrassinol. It follows that these synthetic analogs acquired the feature of alkylating substances to destabilize dsDNA and the capability of inducing the depletion of glutathione in tumor cells. Both compounds demonstrated in vitro antiproliferative activity against tumor cells of the ovarian adenocarcinoma and leukemia cell lines. Compared to the antitumor alkylating agent melphalan, the *cis*-amino derivative had a more notable antiproliferative effect on the ovarian adenocarcinoma cell line. Jurkat-M tumor cells (melphalan-resistant) showed less resistance against this amino compound [92].

In order to improve the anti-cancer activity of the natural phytoalexin cyclobrassinin, its new analogs with NR1R2 group instead of SCH3 were synthesized and evaluated. Several new analogs demonstrated higher antiproliferative potency than natural phytoalexin on at least one evaluated cancer cell line (Jurkat, MCF-7, MDA-MB-23, HeLa, CCRF-CEM, and A-549). The substance *N*-[1-(tert-Butoxycarbonyl)indol-3-yl]methyl-*N'*-phenylthiourea, which was found to be the most potent among all tested compounds on the MCF-7 cells, displays a potency very close to that of doxorubicin on these cells. Replacement of the 2-methylthio moiety of cyclobrassinin with an Ar-NH group resulted in a considerable increase in potency relative to the parental compound [93].

As part of the continuous development of the synthesis of potential antitumor derivatives of indole phytoalexins, the nucleoside analogs of 1-methoxybrassenin B, 1-(α-D-ribofuranosyl) brassenin B, and 1-(β-D-ribofuranosyl) brassenin B were suggested as well. The testing of the antiproliferative activity of these analogs on the Jurkat, CEM, CEM-VCR, MCF-7, and HeLa cancer cell lines revealed the significant activity of natural 1-methoxybrasenin B. The antiproliferative effect of individual nucleoside analogs were likely to decline with the loss of lipophilic features [94]. In general, indole nucleosides represent a rare type of natural products with interesting biological properties. Among them, the nucleoside rebeccamycin antibiotic and its analogs were identified as antineoplastic drugs. Through the glycosylation of natural indololcarbazole and its subsequent modifications, a potential anti-cancer drug J-107088 (edotecarin) was created showing an effect on MKN-45 cell (gastric cancer) implanted in mice. This substance belongs to the group of topoisomerase inhibitors [95–97].

Based on knowledge of the biological activity of indole phytoalexins, their isomers were obtained and tested as prospective anti-cancer substances. Regioisomer (isobrassinin (Figure 1)) showed the interesting antiproliferative effects of brassinin in cervical carcinoma, breast carcinoma, and epidermoid carcinoma cell lines. It inhibited from 70.7% to 89% of cell growth at a concentration of 30 μmol/L [98].

Enantiomeric forms of a 1-methoxyspirobrassinin and 2*R*,3*R*-(−)-1-methoxyspirobrassinol methyl ether were obtained through the spirocyclization method. The enantiomers of these indole phytoalexins were compared within the Jurkat, MCF-7, and HeLa tumor cell growth inhibition. The results of the study showed that a significant difference in antiproliferative activity among enantiomers occurred only with 1-methoxyspirobrassinol methyl ether, and only on the Jurkat cells. The concentration of 100 μmol/L of the 2*R*,3*R*-(−) form reduced the growth of these cells to 36.9% compared to the control, while the 2*S*,3*S*-(+) enantiomer at the same concentration slightly affected cell survival (79.8%). Other isomers showed a slight antiproliferative effect on all tested cell lines [99].

**Table 1.** Summarizing possible anti-cancer properties of representative indole phytoalexins and their derivatives.

| Indole Phytoalexin | Possible Anti-Cancer Properties | Reference |
|---|---|---|
| Brassinin | Reduces the cell growth of mouse melanoma (B16) and leukemic cancer cell line (L1210) | [59] |
| | Exhibits cancer chemopreventive activity: inhibits the formation of preneoplastic mammary lesions in culture | [27] |
| | Induces phase II enzymes that metabolically inactivate chemical carcinogens | |
| | Enhances the effectiveness of tumor immunotherapy by blocking indoleamine 2,3-dioxygenase (IDO), the enzyme that drives immune escape in cancer | [65] |
| | Induces G1 phase arrest through increase of p21 and p27 by inhibition of the PI3K signaling pathway in human colon cancer cells (HT-29) | [67] |
| | Induces apoptosis in human prostate cancer cells (PC-3) through the suppression of PI3K/Akt/mTOR/S6K1 signaling cascades | [69] |
| | Inhibits STAT3 signaling through modulation of PIAS-3 and SOCS-3, thereby reducing tumor cell growthEnhances the antitumor effects of paclitaxel in human lung cancer xenograft in nude mice | [70] |
| | In combination with capsaicin, enhances apoptotic and anti-metastatic effects in human prostate cancer cells (PC-3) | [78] |
| | Potentiates vincristine cytotoxicity to U-87 MG (human glioblastoma astrocytoma) | [82] |
| Isobrassinin | Antiproliferative effect on cervical carcinoma (HeLa), breast carcinoma (MCF-7), and epidermoid carcinoma (A431) cell lines | [98] |
| 5-Bromobrassinin | Suppresses growth of B16-F10 melanoma xenografts in C57BL/6 mice by inhibiting IDO enzyme | [28] |
| Homobrassinin | Induces mitotic phase arrest via inhibition of microtubule formation (dysregulation of α-tubulin, α1-tubulin, and β5-tubulin expression) in colorectal cancer cells (Caco-2) | [80] |
| | Induction of apoptosis in Caco-2 is associated with the loss of mitochondrial membrane potential, caspase-3 activation as well as intracellular reactive oxygen species (ROS) production. | |
| 1-Methoxybrassinin | Exhibits antiproliferative effects on the human acute T lymphoblastic leukemia cell line (Jurkat) IC$_{50}$ 10 μmol/L | [22] |
| | Induces apoptosis in Caco-2 cells, which is associated with the: <br> - upregulation of pro-apoptotic genes expression (Bax) <br> - downregulation of anti-apoptotic genes expression (Bcl-2) <br> - activation of caspase-3,-7 <br> - cleaveage of Poly (ADP-ribose) polymerase (*PARP*) <br> - decrease intracellular GSH content | [24] |
| Cyclobrassinin | Exhibits antiproliferative effects on the epidermoid carcinoma cell line (KB) IC$_{50}$ 8 μg/mL | [61] |
| | Exhibits cancer chemopreventive activity: inhibits the formation of preneoplastic mammary lesions in culture | [27] |
| | Induces phase II enzymes that metabolically inactivate chemical carcinogens | |
| Spirobrassinin | Reduces the cell growth of mouse melanoma (B16) and the leukemic cancer cell line (L1210) | [59] |
| | Exhibits cancer chemopreventive activity: inhibits the formation of preneoplastic mammary lesions in culture | [27] |
| | Induces phase II enzymes that metabolically inactivate chemical carcinogens | |
| | Potentiates vincristine cytotoxicity to U-87 MG (human glioblastoma astrocytoma) | [82] |

## 5. Conclusions

Research in the development and synthesis of new derivatives of indole phytoalexins which may have more favorable antiproliferative and chemopreventive properties than natural substances themselves is continuing. In the future, further experiments aimed at the elucidation of the mechanism of these noteworthy phytochemicals are necessary not only under in vitro but also in vivo conditions. The results of these studies have shown that these substances, due to their simple structure and antiproliferative activity, are potentially effective in developing new anti-tumor drugs that originate from nature.

**Acknowledgments:** The authors gratefully acknowledge G. Gönciova for the technical assistance. This work was supported by the research grant VEGA 1/0103/16 and VEGA 1/0322/14.

**Author Contributions:** Martina Chripkova developed the concept and wrote the paper; Frantisek Zigo provided critical revision of the article; Jan Mojzis revised the paper and gave final approval.

**Conflicts of Interest:** The authors declare no conflict of interest.

## References

1.  Wachtel-Galor, S.; Benzie, I.F.F. Herbal Medicine: An Introduction to Its History, Usage, Regulation, Current Trends, and Research Needs. In *Herbal Medicine: Biomolecular and Clinical Aspects*, 2nd ed.; Benzie, I.F.F., Wachtel-Galor, S., Eds.; CRC Press/Taylor & Francis: Boca Raton, FL, USA, 2011.
2.  Verhoeven, D.T.H.; Verhagen, H.; Goldbohm, R.A.; van den Brandt, P.A.; van Poppel, G. A review of mechanisms underlying anticarcinogenicity by brassica vegetables. *Chem. Biol. Interact.* **1997**, *103*, 79–129. [CrossRef]
3.  Herr, I.; Lozanovski, V.; Houben, P.; Schemmer, P.; Buchler, M.W. Sulforaphane and related mustard oils in focus of cancer prevention and therapy. *Wien. Med. Wochenschr.* **2013**, *163*, 80–88. [CrossRef] [PubMed]
4.  Higdon, J.V.; Delage, B.; Williams, D.E.; Dashwood, R.H. Cruciferous vegetables and human cancer risk: Epidemiologic evidence and mechanistic basis. *Pharmacol. Res.* **2007**, *55*, 224–236. [CrossRef] [PubMed]
5.  Johnson, I.T. Glucosinolates: Bioavailability and importance to health. *Int. J. Vitam. Nutr. Res.* **2002**, *72*, 26–31. [CrossRef] [PubMed]
6.  Brooks, J.D.; Paton, V.G.; Vidanes, G. Potent induction of phase 2 enzymes in human prostate cells by sulforaphane. *Cancer Epidemiol. Biomark.* **2001**, *10*, 949–954.
7.  Fimognari, C.; Lenzi, M.; Hrelia, P. Interaction of the isothiocyanate sulforaphane with drug disposition and metabolism: Pharmacological and toxicological implications. *Curr. Drug Metab.* **2008**, *9*, 668–678. [CrossRef] [PubMed]
8.  Fuentes, F.; Paredes-Gonzalez, X.; Kong, A.T. Dietary Glucosinolates Sulforaphane, Phenethyl Isothiocyanate, Indole-3-Carbinol/3,3′-Diindolylmethane: Anti-Oxidative Stress/Inflammation, Nrf2, Epigenetics/Epigenomics and In Vivo Cancer Chemopreventive Efficacy. *Curr. Pharmacol. Rep.* **2015**, *1*, 179–196. [CrossRef] [PubMed]
9.  Boreddy, S.R.; Sahu, R.P.; Srivastava, S.K. Benzyl isothiocyanate suppresses pancreatic tumor angiogenesis and invasion by inhibiting HIF-α/VEGF/Rho-GTPases: Pivotal role of STAT-3. *PLoS ONE* **2011**, *6*, e25799. [CrossRef] [PubMed]
10. Hudson, T.S.; Perkins, S.N.; Hursting, S.D.; Young, H.A.; Kim, Y.S.; Wang, T.C.; Wang, T.T. Inhibition of androgen-responsive LNCaP prostate cancer cell tumor xenograft growth by dietary phenethyl isothiocyanate correlates with decreased angiogenesis and inhibition of cell attachment. *Int. J. Oncol.* **2012**, *40*, 1113–1121. [PubMed]
11. Lai, K.C.; Hsu, S.C.; Kuo, C.L.; Ip, S.W.; Yang, J.S.; Hsu, Y.M.; Huang, H.Y.; Wu, S.H.; Chung, J.G. Phenethyl isothiocyanate inhibited tumor migration and invasion via suppressing multiple signal transduction pathways in human colon cancer HT29 cells. *J. Agric. Food Chem.* **2010**, *58*, 11148–11155. [CrossRef] [PubMed]
12. Lai, K.C.; Lu, C.C.; Tang, Y.J.; Chiang, J.H.; Kuo, D.H.; Chen, F.A.; Chen, I.L.; Yang, J.S. Allyl isothiocyanate inhibits cell metastasis through suppression of the MAPK pathways in epidermal growth factorstimulated HT29 human colorectal adenocarcinoma cells. *Oncol. Rep.* **2014**, *31*, 189–196. [PubMed]

13. Cheung, K.L.; Kong, A.N. Molecular targets of dietary phenethyl isothiocyanate and sulforaphane for cancer chemoprevention. *AAPS J.* **2010**, *12*, 87–97. [CrossRef] [PubMed]
14. Gupta, P.; Wright, S.E.; Kim, S.H.; Srivastava, S.K. Phenethyl isothiocyanate: A comprehensive review of anti-cancer mechanisms. *Biochim. Biophys. Acta* **2014**, *1846*, 405–424. [CrossRef] [PubMed]
15. Qin, C.Z.; Zhang, X.; Wu, L.X.; Wen, C.J.; Hu, L.; Lv, Q.L.; Shen, D.Y.; Zhou, H.H. Advances in molecular signaling mechanisms of beta-phenethyl isothiocyanate antitumor effects. *J. Agric. Food Chem.* **2015**, *63*, 3311–3322. [CrossRef] [PubMed]
16. Gross, D. Phytoalexins of the Brassicaceae. *J. Plant. Dis. Protect.* **1993**, *100*, 433–442.
17. Pedras, M.S.; Okanga, F.I.; Zaharia, I.L.; Khan, A.Q. Phytoalexins from crucifers: Synthesis, biosynthesis, and biotransformation. *Phytochemistry* **2000**, *53*, 161–176. [CrossRef]
18. Pedras, M.S.C.; Jha, M.; Ahiahonu, P.W.K. The synthesis and biosynthesis of phytoalexins produced by cruciferous plants. *Curr. Org. Chem.* **2003**, *7*, 1635–1647. [CrossRef]
19. Mezencev, R.; Mojzis, J.; Pilatova, M.; Kutschy, P. Anti-proliferative and cancer chemopreventive activity of phytoalexins: Focus on indole phytoalexins from crucifers. *Neoplasma* **2003**, *50*, 239–245. [PubMed]
20. Pedras, M.S.C.; Zheng, Q.A.; Sarma-Mamillapalle, V.K. The phytoalexins from Brassicaceae: Structure, biological activity, synthesis and biosynthesis. *Nat. Prod. Commun.* **2007**, *2*, 319–330.
21. Pedras, M.S.; Yaya, E.E. Phytoalexins from Brassicaceae: News from the front. *Phytochemistry* **2010**, *71*, 1191–1197. [CrossRef] [PubMed]
22. Pilatova, M.; Sarissky, M.; Kutschy, P.; Mirossay, A.; Mezencev, R.; Curillova, Z.; Suchy, M.; Monde, K.; Mirossay, L.; Mojzis, J. Cruciferous phytoalexins: Anti-proliferative effects in T-Jurkat leukemic cells. *Leuk. Res.* **2005**, *29*, 415–421. [CrossRef] [PubMed]
23. Pilatova, M.; Ivanova, L.; Kutschy, P.; Varinska, L.; Saxunova, L.; Repovska, M.; Sarissky, M.; Seliga, R.; Mirossay, L.; Mojzis, J. In vitro toxicity of camalexin derivatives in human cancer and non-cancer cells. *Toxicol. In Vitro* **2013**, *27*, 939–944. [CrossRef] [PubMed]
24. Chripkova, M.; Drutovic, D.; Pilatova, M.; Mikes, J.; Budovska, M.; Vaskova, J.; Broggini, M.; Mirossay, L.; Mojzis, J. Brassinin and its derivatives as potential anticancer agents. *Toxicol. In Vitro* **2014**, *28*, 909–915. [CrossRef] [PubMed]
25. Pedras, M.S.; Montaut, S.; Suchy, M. Phytoalexins from the crucifer rutabaga: Structures, syntheses, biosyntheses, and antifungal activity. *J. Org. Chem.* **2004**, *69*, 4471–4476. [CrossRef] [PubMed]
26. Mezencev, R.; Galizzi, M.; Kutschy, P.; Docampo, R. Trypanosoma cruzi: Anti-proliferative effect of indole phytoalexins on intracellular amastigotes in vitro. *Exp. Parasitol.* **2009**, *122*, 66–69. [CrossRef] [PubMed]
27. Mehta, R.G.; Liu, J.; Constantinou, A.; Thomas, C.F.; Hawthorne, M.; You, M.; Gerhuser, C.; Pezzuto, J.M.; Moon, R.C.; Moriarty, R.M. Cancer chemopreventive activity of brassinin, a phytoalexin from cabbage. *Carcinogenesis* **1995**, *16*, 399–404. [CrossRef] [PubMed]
28. Banerjee, T.; Duhadaway, J.B.; Gaspari, P.; Sutanto-Ward, E.; Munn, D.H.; Mellor, A.L.; Malachowski, W.P.; Prendergast, G.C.; Muller, A.J. A key in vivo antitumor mechanism of action of natural product-based brassinins is inhibition of indoleamine 2,3-dioxygenase. *Oncogene* **2008**, *27*, 2851–2857. [CrossRef] [PubMed]
29. Bailey, J.A.; Mansfield, J.W. *Phytoalexins*; Blackie: Glasgow, UK, 1982.
30. Dixon, R.A.; Lamb, C.J. Molecular Communication in Interactions between Plants and Microbial Pathogens. *Annu. Rev. Plant Physiol. Plant Mol. Biol.* **1990**, *41*, 339–367. [CrossRef]
31. Daniel, M.; Purkayastha, R.P. *Handbook of Phytoalexin Metabolism and Action*; CRC Press: Boca Raton, FL, USA, 1994.
32. Heil, M.; Bostock, R.M. Induced systemic resistance (ISR) against pathogens in the context of induced plant defences. *Ann. Bot.* **2002**, *89*, 503–512. [CrossRef] [PubMed]
33. Müller, K.O.; Börger, H. Experimentelle Untersuchungen Über die Phytophthorainfestans—Resistenz der Kartoffel. *Arb. Biol. Reichsanst. Land Forstwirtsch* **1940**, *97*, 189–231.
34. Cruickshank, I.A.; Perrin, D.R. Isolation of a phytoalexin from *Pisum sativum* L. *Nature* **1960**, *187*, 799–800. [CrossRef] [PubMed]
35. Pedras, M.S.; Ahiahonu, P.W. Metabolism and detoxification of phytoalexins and analogs by phytopathogenic fungi. *Phytochemistry* **2005**, *66*, 391–411. [CrossRef] [PubMed]
36. Takasugi, M.; Katsui, N.; Shirata, A. Isolation of three Novel Sulfur-Containing Phytoalexins from the Chinese-Cabbage *Brassica campestris* L. Ssp *pekinensis* (Cruciferae). *J. Chem. Soc. Chem. Commun.* **1986**, *14*, 1077–1078. [CrossRef]

37. Kutschy, P.; Dzurilla, M.; Takasugi, M.; Sabova, A. Synthesis of some analogs of indole phytoalexins brassinin and methoxybrassenin B and their positional isomers. *Collect. Czech. Chem. C* **1999**, *64*, 348–362. [CrossRef]

38. Mehta, R.G.; Naithani, R.; Huma, L.; Hawthorne, M.; Moriarty, R.M.; McCormick, D.L.; Steele, V.E.; Kopelovich, L. Efficacy of chemopreventive agents in mouse mammary gland organ culture (MMOC) model: A comprehensive review. *Curr. Med. Chem.* **2008**, *15*, 2785–2825. [CrossRef] [PubMed]

39. Pedras, M.S.; Okinyo, D.P. Remarkable incorporation of the first sulfur containing indole derivative: Another piece in the biosynthetic puzzle of crucifer phytoalexins. *Org. Biomol. Chem.* **2008**, *6*, 51–54. [CrossRef] [PubMed]

40. Pedras, M.S.; Chumala, P.B.; Suchy, M. Phytoalexins from *Thlaspi arvense*, a wild crucifer resistant to virulent *Leptosphaeria maculans*: Structures, syntheses and antifungal activity. *Phytochemistry* **2003**, *64*, 949–956. [CrossRef]

41. Pedras, M.S.; Suchy, M.; Ahiahonu, P.W. Unprecedented chemical structure and biomimetic synthesis of erucalexin, a phytoalexin from the wild crucifer *Erucastrum gallicum*. *Org. Biomol. Chem.* **2006**, *4*, 691–701. [CrossRef] [PubMed]

42. Pedras, M.S.; Adio, A.M. Phytoalexins and phytoanticipins from the wild crucifers *Thellungiella halophila* and *Arabidopsis thaliana*: Rapalexin A, wasalexins and camalexin. *Phytochemistry* **2008**, *69*, 889–893. [CrossRef] [PubMed]

43. Pedras, M.S.; Zheng, Q.A.; Sarwar, M.G. Efficient synthesis of brussalexin A, a remarkable phytoalexin from Brussels sprouts. *Org. Biomol. Chem.* **2007**, *5*, 1167–1169. [CrossRef] [PubMed]

44. Pedras, M.S.; Zheng, Q.A.; Strelkov, S. Metabolic changes in roots of the oilseed canola infected with the biotroph *Plasmodiophora brassicae*: Phytoalexins and phytoanticipins. *J. Agric. Food Chem.* **2008**, *56*, 9949–9961. [CrossRef] [PubMed]

45. Takasugi, M.; Monde, K.; Katsui, N.; Shirata, A. Spirobrassinin, a Novel Sulfur-Containing Phytoalexin from the Daikon *Rhaphanus sativus* L. var. *hortensis* (Cruciferae). *Chem. Lett.* **1987**, *8*, 1631–1632. [CrossRef]

46. Keplinger, K.; Laus, G.; Wurm, M.; Dierich, M.P.; Teppner, H. *Uncaria tomentosa* (Willd.) DC.—Ethnomedicinal use and new pharmacological, toxicological and botanical results. *J. Ethnopharmacol.* **1999**, *64*, 23–34. [CrossRef]

47. Falkiewicz, B.; Lukasiak, J. *Uncaria tomentosa* (Willd) DC. and *Uncaria guianensis* (Aublet) Gmell.—A review of published scientific literature. *Case Rep. Clin. Pract. Rev.* **2001**, *2*, 305–316.

48. Glawischnig, E. Camalexin. *Phytochemistry* **2007**, *68*, 401–406. [CrossRef] [PubMed]

49. Glawischnig, E. The role of cytochrome P450 enzymes in the biosynthesis of camalexin. *Biochem. Soc. Trans.* **2006**, *34*, 1206–1208. [CrossRef] [PubMed]

50. Leroux, P. Modes of action of agrochemicals against plant pathogenic organisms. *C. R. Biol.* **2003**, *326*, 9–21. [CrossRef]

51. Navneetha, O.; Saritha Jyostna, T. Therapeutic potentials of brassinin and its analogues—A review. *Adv. J. Pharm. Life Sci. Res.* **2016**, *4*, 76–82.

52. Cao, S.L.; Feng, Y.P.; Jiang, Y.Y.; Liu, S.Y.; Ding, G.Y.; Li, R.T. Synthesis and in vitro antitumor activity of 4(3H)-quinazolinone derivatives with dithiocarbamate side chains. *Bioorg. Med. Chem. Lett.* **2005**, *15*, 1915–1917. [CrossRef] [PubMed]

53. Liu, S.; Liu, F.; Yu, X.; Ding, G.; Xu, P.; Cao, J.; Jiang, Y. The 3D-QSAR analysis of 4(3H)-quinazolinone derivatives with dithiocarbamate side chains on thymidylate synthase. *Bioorg. Med. Chem.* **2006**, *14*, 1425–1430. [CrossRef] [PubMed]

54. Duan, Y.C.; Ma, Y.C.; Zhang, E.; Shi, X.J.; Wang, M.M.; Ye, X.W.; Liu, H.M. Design and synthesis of novel 1,2,3-triazole-dithiocarbamate hybrids as potential anti-cancer agents. *Eur. J. Med. Chem.* **2013**, *62*, 11–19. [CrossRef] [PubMed]

55. Navneetha, O.; Anuradha, B.S.; Sandhya, M.S.N.; Sree Kanth, S.; Vijjulatha, M.; Saritha Jyostna, T. Bioisosteres of brassinin: Synthesis, molecular docking and chemotherapeutic activity. *Indo Am. J. Pharm. Sci.* **2016**, *6*, 4070–4079.

56. Xue, W.; Meylan, E.; Oliver, T.G.; Feldser, D.M.; Winslow, M.M.; Bronson, R.; Jacks, T. Response and resistance to NF-κB inhibitors in mouse models of lung adenocarcinoma. *Cancer Discov.* **2011**, *1*, 236–247. [CrossRef] [PubMed]

57. Zhang, G.; Ghosh, S. Toll-like receptor-mediated NF-κB activation: A phylogenetically conserved paradigm in innate immunity. *J. Clin. Investig.* **2001**, *107*, 13–19. [CrossRef] [PubMed]

58. Shin, J.S.; Noh, Y.S.; Lee, Y.S.; Cho, Y.W.; Baek, N.I.; Choi, M.S.; Jeong, T.S.; Kang, E.; Chung, H.G.; Lee, K.T. Arvelexin from *Brassica rapa* suppresses NF-κB-regulated pro-inflammatory gene expression by inhibiting activation of I κB kinase. *Br. J. Pharmacol.* **2011**, *164*, 145–158. [CrossRef] [PubMed]
59. Sabol, M.; Kutschy, P.; Siegfried, L.; Mirossay, A.; Suchy, M.; Hrbkova, H.; Dzurilla, M.; Maruskova, R.; Starkova, J.; Paulikova, E. Cytotoxic effect of cruciferous phytoalexins against murine L1210 leukemia and B16 melanoma. *Biologia* **2000**, *55*, 701–707.
60. Mehta, R.G.; Liu, J.; Constantinou, A.; Hawthorne, M.; Pezzuto, J.M.; Moon, R.C.; Moriatly, R.M. Structure-activity relationships of brassinin in preventing the development of carcinogen-induced mammary lesions in organ culture. *Anti Cancer Res.* **1994**, *14*, 1209–1213.
61. Tempete, C.; Devys, M.; Barbier, M. Growth inhibitions on human cancer cell cultures with the indole sulphur-containing phytoalexins and their analogues. *Z. Naturf. C J. Biosci.* **1991**, *46*, 706–707.
62. Sono, M.; Roach, M.P.; Coulter, E.D.; Dawson, J.H. Heme-containing oxygenases. *Chem. Rev.* **1996**, *96*, 2841–2887. [CrossRef] [PubMed]
63. Botting, N.P. Chemistry and neurochemistry of the kynurenine pathway of tryptophan metabolism. *Chem. Soc. Rev.* **1995**, *24*, 401. [CrossRef]
64. Sono, M.; Taniguchi, T.; Watanabe, Y.; Hayaishi, O. Indoleamine 2,3-dioxygenase. Equilibrium studies of the tryptophan binding to the ferric, ferrous, and CO-bound enzymes. *J. Biol. Chem.* **1980**, *255*, 1339–1345. [PubMed]
65. Munn, D.H.; Mellor, A.L. Indoleamine 2,3-dioxygenase and tumor-induced tolerance. *J. Clin. Investig.* **2007**, *117*, 1147–1154. [CrossRef] [PubMed]
66. Gaspari, P.; Banerjee, T.; Malachowski, W.P.; Muller, A.J.; Prendergast, G.C.; DuHadaway, J.; Bennett, S.; Donovan, A.M. Structure-activity study of brassinin derivatives as indoleamine 2,3-dioxygenase inhibitors. *J. Med. Chem.* **2006**, *49*, 684–692. [CrossRef] [PubMed]
67. Izutani, Y.; Yogosawa, S.; Sowa, Y.; Sakai, T. Brassinin induces G1 phase arrest through increase of p21 and p27 by inhibition of the phosphatidylinositol 3-kinase signaling pathway in human colon cancer cells. *Int. J. Oncol.* **2012**, *40*, 816–824. [CrossRef] [PubMed]
68. Kim, S.M.; Park, J.H.; Kim, K.D.; Nam, D.; Shim, B.S.; Kim, S.H.; Ahn, K.S.; Choi, S.H.; Ahn, K.S. Brassinin induces apoptosis in PC-3 human prostate cancer cells through the suppression of PI3K/Akt/mTOR/S6K1 signaling cascades. *Phytother. Res.* **2014**, *28*, 423–431. [CrossRef] [PubMed]
69. Gao, N.; Zhang, Z.; Jiang, B.H.; Shi, X. Role of PI3K/AKT/mTOR signaling in the cell cycle progression of human prostate cancer. *Biochem. Biophys. Res. Commun.* **2003**, *310*, 1124–1132. [CrossRef] [PubMed]
70. Lee, J.H.; Kim, C.; Sethi, G.; Ahn, K.S. Brassinin inhibits STAT3 signaling pathway through modulation of PIAS-3 and SOCS-3 expression and sensitizes human lung cancer xenograft in nude mice to paclitaxel. *Oncotarget* **2015**, *6*, 6386–6405. [CrossRef] [PubMed]
71. Ihle, J.N. STATs and MAPKs: Obligate or opportunistic partners in signaling. *BioEssays News Rev. Mol. Cell. Dev. Biol.* **1996**, *18*, 95–98. [CrossRef] [PubMed]
72. Grandis, J.R.; Drenning, S.D.; Chakraborty, A.; Zhou, M.Y.; Zeng, Q.; Pitt, A.S.; Tweardy, D.J. Requirement of Stat3 but not Stat1 activation for epidermal growth factor receptor-mediated cell growth in vitro. *J. Clin. Investig.* **1998**, *102*, 1385–1392. [CrossRef] [PubMed]
73. Catlett-Falcone, R.; Landowski, T.H.; Oshiro, M.M.; Turkson, J.; Levitzki, A.; Savino, R.; Ciliberto, G.; Moscinski, L.; Fernandez-Luna, J.L.; Nunez, G.; et al. Constitutive activation of Stat3 signaling confers resistance to apoptosis in human U266 myeloma cells. *Immunity* **1999**, *10*, 105–115. [CrossRef]
74. Epling-Burnette, P.K.; Liu, J.H.; Catlett-Falcone, R.; Turkson, J.; Oshiro, M.; Kothapalli, R.; Li, Y.; Wang, J.M.; Yang-Yen, H.F.; Karras, J.; et al. Inhibition of STAT3 signaling leads to apoptosis of leukemic large granular lymphocytes and decreased Mcl-1 expression. *J. Clin. Investig.* **2001**, *107*, 351–362. [CrossRef] [PubMed]
75. Buettner, R.; Mora, L.B.; Jove, R. Activated STAT signaling in human tumors provides novel molecular targets for therapeutic intervention. *Clin. Cancer Res.* **2002**, *8*, 945–954. [PubMed]
76. Zimmer, S.; Kahl, P.; Buhl, T.M.; Steiner, S.; Wardelmann, E.; Merkelbach-Bruse, S.; Buettner, R.; Heukamp, L.C. Epidermal growth factor receptor mutations in non-small cell lung cancer influence downstream Akt, MAPK and Stat3 signaling. *J. Cancer Res. Clin. Oncol.* **2009**, *135*, 723–730. [CrossRef] [PubMed]

77.    Looyenga, B.D.; Hutchings, D.; Cherni, I.; Kingsley, C.; Weiss, G.J.; Mackeigan, J.P. STAT3 is activated by JAK2 independent of key oncogenic driver mutations in non-small cell lung carcinoma. *PLoS ONE* **2012**, *7*, e30820. [CrossRef] [PubMed]

78.    Kim, S.M.; Oh, E.Y.; Lee, J.H.; Nam, D.; Lee, S.G.; Lee, J.; Kim, S.H.; Shim, B.S.; Ahn, K.S. Brassinin Combined with Capsaicin Enhances Apoptotic and Anti-metastatic Effects in PC-3 Human Prostate Cancer Cells. *Phytother. Res.* **2015**, *29*, 1828–1836. [CrossRef] [PubMed]

79.    Sanchez, A.M.; Sanchez, M.G.; Malagarie-Cazenave, S.; Olea, N.; Diaz-Laviada, I. Induction of apoptosis in prostate tumor PC-3 cells and inhibition of xenograft prostate tumor growth by the vanilloid capsaicin. *Apoptosis* **2006**, *11*, 89–99. [CrossRef] [PubMed]

80.    Kello, M.; Drutovic, D.; Chripkova, M.; Pilatova, M.; Budovska, M.; Kulikova, L.; Urdzik, P.; Mojzis, J. ROS-dependent anti-proliferative effect of brassinin derivative homobrassinin in human colorectal cancer Caco2 cells. *Molecules* **2014**, *19*, 10877–10897. [CrossRef] [PubMed]

81.    Smith, T.K.; Lund, E.K.; Parker, M.L.; Clarke, R.G.; Johnson, I.T. Allyl-isothiocyanate causes mitotic block, loss of cell adhesion and disrupted cytoskeletal structure in HT29 cells. *Carcinogenesis* **2004**, *25*, 1409–1415. [CrossRef] [PubMed]

82.    Mezencev, R.; Mojžiš, J.; Pilátová, M.; Kutschy, P.; Čurillová, Z. Effects of indole phytoalexins from cruciferous plants on the growth of cancer cells.Implications for cancer chemoprevention and chemotherapy. *Int. J. Canc. Prev.* **2004**, *1*, 105–112.

83.    Moody, C.J.; Roffey, J.R.; Stephens, M.A.; Stratford, I.J. Synthesis and cytotoxic activity of indolyl thiazoles. *Anti Cancer Drugs* **1997**, *8*, 489–499. [CrossRef] [PubMed]

84.    Smith, B.A.; Neal, C.L.; Chetram, M.; Vo, B.; Mezencev, R.; Hinton, C.; Odero-Marah, V.A. The phytoalexin camalexin mediates cytotoxicity towards aggressive prostate cancer cells via reactive oxygen species. *J. Nat. Med.* **2013**, *67*, 607–618. [CrossRef] [PubMed]

85.    Mezencev, R.; Updegrove, T.; Kutschy, P.; Repovska, M.; McDonald, J.F. Camalexin induces apoptosis in T-leukemia Jurkat cells by increased concentration of reactive oxygen species and activation of caspase-8 and caspase-9. *J. Nat. Med.* **2011**, *65*, 488–499. [CrossRef] [PubMed]

86.    Smith, B.; Randle, D.; Mezencev, R.; Thomas, L.; Hinton, C.; Odero-Marah, V. Camalexin-induced apoptosis in prostate cancer cells involves alterations of expression and activity of lysosomal protease cathepsin D. *Molecules* **2014**, *19*, 3988–4005. [CrossRef] [PubMed]

87.    Kutschy, P.; Sykora, A.; Curillova, Z.; Repovska, M.; Pilatova, M.; Mojzis, J.; Mezencev, R.; Pazdera, P.; Hromjakova, T. Glyoxyl Analogs of Indole Phytoalexins: Synthesis and Anti-cancer Activity. *Collect. Czech. Chem. C* **2010**, *75*, 887–903. [CrossRef]

88.    Kobayashi, J.; Murayama, T.; Ishibashi, M.; Kosuge, S.; Takamatsu, M.; Ohizumi, Y.; Kobayashi, H.; Ohta, T.; Nozoe, S.; Sasaki, T. Hyrtiosin-a and Hyrtiosin-B, New Indole Alkaloids from the Okinawan Marine Sponge *Hyrtios erecta. Tetrahedron* **1990**, *46*, 7699–7702. [CrossRef]

89.    Bacher, G.; Nickel, B.; Emig, P.; Vanhoefer, U.; Seeber, S.; Shandra, A.; Klenner, T.; Beckers, T. D-24851, a novel synthetic microtubule inhibitor, exerts curative antitumoral activity in vivo, shows efficacy toward multidrug-resistant tumor cells, and lacks neurotoxicity. *Cancer Res.* **2001**, *61*, 392–399. [PubMed]

90.    Kutschy, P.; Salayova, A.; Curillova, Z.; Kozar, T.; Mezencev, R.; Mojzis, J.; Pilatova, M.; Balentova, E.; Pazdera, P.; Sabol, M.; et al. 2-(Substituted phenyl)amino analogs of 1-methoxyspirobrassinol methyl ether: Synthesis and anti-cancer activity. *Bioorg. Med. Chem.* **2009**, *17*, 3698–3712. [CrossRef] [PubMed]

91.    Mezencev, R.; Kutschy, P.; Salayova, A.; Curillova, Z.; Mojzis, J.; Pilatova, M.; McDonald, J. Anti-cancer properties of 2-piperidyl analogues of the natural indole phytoalexin 1-methoxyspirobrassinol. *Chemotherapy* **2008**, *54*, 372–378. [CrossRef] [PubMed]

92.    Mezencev, R.; Kutschy, P.; Salayova, A.; Updegrove, T.; McDonald, J.F. The design, synthesis and anti-cancer activity of new nitrogen mustard derivatives of natural indole phytoalexin 1-methoxyspirobrassinol. *Neoplasma* **2009**, *56*, 321–330. [CrossRef] [PubMed]

93.    Budovska, M.; Pilatova, M.; Varinska, L.; Mojzis, J.; Mezencev, R. The synthesis and anti-cancer activity of analogs of the indole phytoalexins brassinin, 1-methoxyspirobrassinol methyl ether and cyclobrassinin. *Bioorg. Med. Chem.* **2013**, *21*, 6623–6633. [CrossRef] [PubMed]

94.    Curillova, Z.; Kutschy, P.; Solcaniova, E.; Pilatova, M.; Mojzis, J.; Kovacik, V. Synthesis and anti-proliferative activity of 1-methoxy-, 1-(α-D-ribofuranosyl)- and 1-(β-D-ribofuranosyl)brassenin B. *ARKIVOC* **2008**, *8*, 85–104.

95.  Kojiri, K.; Arakawa, H.; Satoh, F.; Kawamura, K.; Okura, A.; Suda, H.; Okanishi, M. New antitumor substances, BE-12406A and BE-12406B, produced by a streptomycete. I. Taxonomy, fermentation, isolation, physico-chemical and biological properties. *J. Antibiot.* **1991**, *44*, 1054–1060. [CrossRef] [PubMed]

96.  Yoshinari, T.; Ohkubo, M.; Fukasawa, K.; Egashira, S.; Hara, Y.; Matsumoto, M.; Nakai, K.; Arakawa, H.; Morishima, H.; Nishimura, S. Mode of action of a new indolocarbazole anti-cancer agent, J-107088, targeting topoisomerase I. *Cancer Res.* **1999**, *59*, 4271–4275. [PubMed]

97.  Arakawa, H.; Morita, M.; Kodera, T.; Okura, A.; Ohkubo, M.; Morishima, H.; Nishimura, S. In vivo anti-tumor activity of a novel indolocarbazole compound, J-107088, on murine and human tumors transplanted into mice. *Jpn. J. Cancer Res.* **1999**, *90*, 1163–1170. [CrossRef] [PubMed]

98.  Csomos, P.; Zupko, I.; Rethy, B.; Fodor, L.; Falkay, G.; Bernath, G. Isobrassinin and its analogues: Novel types of anti-proliferative agents. *Bioorg. Med. Chem. Lett.* **2006**, *16*, 6273–6276. [CrossRef] [PubMed]

99.  Monde, K.; Taniguchi, T.; Miura, N.; Kutschy, P.; Curillova, Z.; Pilatova, M.; Mojzis, J. Chiral cruciferous phytoalexins: Preparation, absolute configuration, and biological activity. *Bioorg. Med. Chem.* **2005**, *13*, 5206–5212. [CrossRef] [PubMed]

*molecules*

MDPI

Communication

# Cytotoxicity of Labruscol, a New Resveratrol Dimer Produced by Grapevine Cell Suspensions, on Human Skin Melanoma Cancer Cell Line HT-144

Laetitia Nivelle [1], Jane Hubert [2], Eric Courot [3], Nicolas Borie [2], Jean-Hugues Renault [2],
Jean-Marc Nuzillard [2], Dominique Harakat [2], Christophe Clément [3], Laurent Martiny [1],
Dominique Delmas [4], Philippe Jeandet [3,*] and Michel Tarpin [1]

[1] Unité Matrice Extracellulaire et Dynamique Cellulaire, UMR CNRS 7369, SFR Cap-Santé FED 4231,
   UFR des Sciences Exactes et Naturelles, Université de Reims Champagne-Ardenne, BP 1039,
   51687 Reims CEDEX 2, France; nivellelaetitia@gmail.com (L.N.); laurent.martiny@univ-reims.fr (L.M.);
   michel.tarpin@univ-reims.fr (M.T.)
[2] Institut de Chimie Moléculaire de Reims, UMR CNRS 7312, SFR Cap-Santé FED 4231, UFR de Pharmacie,
   Université de Reims Champagne-Ardenne, 51687 Reims CEDEX 2, France; jane.hubert@univ-reims.fr (J.H.);
   nicolas.borie@univ-reims.fr (N.B.); jh.renault@univ-reims.fr (J.-H.R.);
   jean-marc.nuzillard@univ-reims.fr (J.-M.N.); dominique.harakat@univ-reims.fr (D.H.)
[3] Unité de Recherche Vignes et Vins de Champagne EA 4707, SFR Condorcet FR CNRS 3417,
   UFR des Sciences Exactes et Naturelles, Université de Reims Champagne-Ardenne, BP 1039,
   51687 Reims CEDEX 2, France; eric.courot@univ-reims.fr (E.C.); christophe.clement@univ-reims.fr (C.C.)
[4] Centre de Recherche Inserm U866, Université de Bourgogne, 21000 Dijon, France;
   dominique.delmas@u-bourgogne.fr
*   Correspondence: philippe.jeandet@univ-reims.fr; Tel.: +33-3-26913-341; Fax: +33-3-26913-340

Received: 4 October 2017; Accepted: 6 November 2017; Published: 9 November 2017

**Abstract:** A new resveratrol dimer (**1**) called labruscol, has been purified by centrifugal partition chromatography of a crude ethyl acetate stilbene extract obtained from elicited grapevine cell suspensions of *Vitis labrusca* L. cultured in a 14-liter stirred bioreactor. One dimensional (1D) and two dimensional (2D) nuclear magnetic resonance (NMR) analyses including $^1$H, $^{13}$C, heteronuclear single-quantum correlation (HSQC), heteronuclear multiple bond correlation (HMBC), and correlation spectroscopy (COSY) as well as high-resolution electrospray ionisation mass spectrometry (HR-ESI-MS) were used to characterize this compound and to unambiguously identify it as a new stilbene dimer, though its relative stereochemistry remained unsolved. Labruscol was recovered as a pure compound (>93%) in sufficient amounts (41 mg) to allow assessment of its biological activity (cell viability, cell invasion and apoptotic activity) on two different cell lines, including one human skin melanoma cancer cell line HT-144 and a healthy human dermal fibroblast (HDF) line. This compound induced almost 100% of cell viability inhibition in the cancer line at a dose of 100 µM within 72 h of treatment. However, at all tested concentrations and treatment times, resveratrol displayed an inhibition of the cancer line viability higher than that of labruscol in the presence of fetal bovine serum. Both compounds also showed differential activities on healthy and cancer cell lines. Finally, labruscol at a concentration of 1.2 µM was shown to reduce cell invasion by 40%, although no similar activity was observed with resveratrol. The cytotoxic activity of this newly-identified dimer is discussed.

**Keywords:** resveratrol; labruscol; melanoma; fibroblasts; cytotoxic activity; bioreactor; *Vitis labrusca* L.

## 1. Introduction

Stilbenoids, which are naturally-occurring secondary metabolites widely represented in the plant kingdom, can be divided into monomeric and oligomeric compounds [1]. The most studied stilbene monomer is resveratrol and its biological activity as a phytoalexin in plants or its preventing action against human diseases has already led to a considerable number of works. Though the biological properties of resveratrol are well known, those of the oligomeric resveratrol derivatives are less documented. Most of our knowledge on the activity of stilbene oligomers mainly concerns dimeric structures such as ε-viniferin, pallidol and δ-viniferin. Among dimers, ε-viniferin is the most studied and there are works reporting ε-viniferin to act as an antioxidant [2], anti-inflammatory [3], anticancer [4–9], as well as a cardioprotective agent [10]. Limitations in studying the biological properties of resveratrol dimers essentially reside in the difficulty to recover these compounds in large amounts by using conventional plant extraction procedures or chemical synthesis. We have previously described the bioproduction of various phytostilbenes in stirred bioreactors from grapevine cell suspensions [9]. Cultivating grapevine cells in bioreactors in the presence of various defense-inducing compounds, the so-called elicitors [11–15], has indeed been shown to constitute a useful technique for the production of tens to hundreds milligrams of dimeric stilbenes with high purity [9].

Here we report on the characterization of a new stilbene dimer called labruscol (**1**) produced by grapevine cell suspensions of *Vitis labrusca* L. var. Concord as well as the determination of its biological activity (inhibition of cancer cell viability, effect on tumor cell invasion and apoptosis) on the human skin melanoma cancer cell line HT-144, considered to be an aggressive cancer line. Its properties were compared to those of a resveratrol bioproduced from the same cell suspensions.

## 2. Results and Discussion

### 2.1. CPC Purification and Chemical Characterization of Labruscol (**1**)

The crude ethyl acetate extract (1.5 g) obtained from the elicited grapevine cell suspensions of *V. labrusca* cultured in a 14-L stirred bioreactor was fractionated by centrifugal partition chromatography (CPC) in a single run of 160 min using a normal phase gradient elution method, including a biphasic solvent system composed of *n*-heptane, ethyl acetate, methanol and water in the ascending mode (see Section 3.3). CPC is a solid support-free separation technique involving the differential partition of solutes between at least two immiscible liquid phases according to their distribution coefficient [16]. The CPC column used here, with a total capacity of 303 mL and 231 partition cells, offers the possibility to collect fractions in sufficient amounts (all > 20 mg) to allow their chemical characterization and determine their biological activity.

All fractions eluted with the organic mobile phase over the gradient contained various stilbene derivatives that together represented 73% of the crude extract injected mass. Besides, the most hydrophilic compounds of the extract, mainly including residual cyclodextrins (used for stilbene elicitation) and culture medium nutrients, were well-retained inside the column (i.e., in the aqueous stationary phase) over the whole CPC experiment. Metabolite identification was performed in the collected fractions by using a $^{13}$C-nuclear magnetic resonance (NMR)-based dereplication procedure [17,18], revealing *trans*-resveratrol, δ-viniferin, pallidol, ε-viniferin, leachianol F, and leachianol G as the major phytostilbenes bioproduced under the used elicitation conditions [9]. A new resveratrol dimer was also detected in fractions eluted from 115 to 127 min with a purity greater than 93% (41 mg in total). It was not possible to identify its structure during a search in the database, suggesting that the structure of this compound was original. Structure elucidation of this resveratrol dimer, called labruscol (**1**) (Figure 1), was unambiguously achieved by 1D and 2D NMR analyses including $^1$H, $^{13}$C, heteronuclear single-quantum correlation (HSQC), heteronuclear multiple bond correlation (HMBC), and correlation spectroscopy (COSY). The molecular formula of $C_{28}H_{24}O_7$ was confirmed by high-resolution electrospray ionisation mass spectrometry (HR-ESI-MS) analysis, revealing the molecular ions $[M - H]^-$ at *m*/*z* 471 and $[2M - H]^-$ at *m*/*z* 943 in the negative ionization mode.

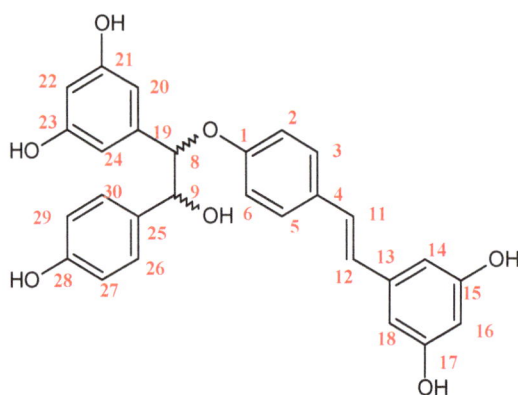

| Carbon n° | ¹³C (ppm) | ¹H (ppm) | J$_{HH}$ Coupling Constants | HMBC Correlations | NOESY Correlations |
|---|---|---|---|---|---|
| 1 | 157.9 | - | - | - | - |
| 2 | 115.9 | 6.90 | 9 Hz (d) | C1, C4, C6 | H9, H8, H3 |
| 3 | 127.0 | 7.34 | 9 Hz (d) | C1, C11, C5 | H2 |
| 4 | 130.2 | - | - | - | - |
| 5 | 127.0 | 7.34 | 9 Hz (d) | C1, C11, C3 | H6 |
| 6 | 115.9 | 6.90 | 9 Hz (d) | C1, C4, C2 | H9, H8, H5 |
| 8 | 85.0 | 5.00 | 7.4 Hz (d) | C1, C19, C25, C20, C24, C9 | H24, H20, H2, H6, H26, H30 |
| 9 | 77.6 | 4.81 | 7.4 Hz (d) | C8, C26, C30, C19 | H24, H20, H2, H6, H26, H30 |
| 11 | 127.7 | 6.94 | 17 Hz trans (d) | C13, C3, C5, C12 | H14 |
| 12 | 126.4 | 6.81 | 17 Hz trans (d) | C14, C18, C11, C4, C13 | H18, H14 |
| 13 | 139.7 | - | - | - | - |
| 14 | 104.4 | 6.45 | 2.0 Hz (d) | C15, C12, C18, C16 | H11, H12, H16 |
| 15 | 158.2 | - | - | - | - |
| 16 | 101.4 | 6.17 | 2.0 Hz (t) | C14, C18, C15, C17 | H14, H18 |
| 17 | 158.2 | - | - | - | - |
| 18 | 104.4 | 6.45 | 2.0 Hz (d) | C14, C12, C17, C16 | H12, H16 |
| 19 | 140.5 | - | - | - | - |
| 20 | 105.8 | 6.13 | 2.0 Hz (d) | C21, C24, C22, C8 | H9, H8, H22 |
| 21 | 157.8 | - | - | - | - |
| 22 | 101.5 | 6.08 | 2 Hz (t) | C20, C24, C21, C23 | H20, H24 |
| 23 | 157.8 | - | - | - | - |
| 24 | 105.8 | 6.13 | 2.0 Hz (d) | C23, C20, C22, C8 | H9, H8, H22 |
| 25 | 131.1 | - | - | - | - |
| 26 | 128.3 | 7.05 | 9 Hz (d) | C27, C28, C9 | H9, H8, H27 |
| 27 | 114.3 | 6.68 | 9 Hz (d) | C28, C25, C26 | H26 |
| 28 | 156.6 | - | - | - | - |
| 29 | 114.3 | 6.68 | 9 Hz (d) | C28, C25, C30 | H30 |
| 30 | 128.3 | 7.05 | 9 Hz (d) | C29, C28, C9 | H9, H8, H29 |

HMBC: heteronuclear multiple bond correlation; NOESY: nuclear Overhauser effect spectroscopy.

**Figure 1.** Chemical structure and nuclear magnetic resonance (NMR) data for labruscol (**1**).

In order to determine the relative stereochemistry of the two stereogenic centers on C8 and C9, a rotating frame Overhauser effect spectroscopy (ROESY) experiment (Figure 1 and Table) together with a conformational optimization of the two putatively relative stereochemistries using a molecular mechanics (MM3*) force field was carried out. Unfortunately, the dihedral angle H8-H9 measured on the two minimized structures corresponding to the 8R9R (or 8S9S) or 8R9S (or 8S9R) is about 170° in the two cases, which is consistent with the coupling constant $J_{8H-9H}$ = 7.4 Hz, but this does not make it possible to conclude on the relative stereochemistry of the dimer (**1**). The interatomic distances were also measured for the two stereoisomers in order to connect them with the nuclear Overhauser effect (NOE) correlations. Once again, the distance differences were not significant and did not allow for making a decision between the two relative configurations. Further studies involving circular dichroism would thus be required to solve the question concerning the absolute configurations of the two stereogenic centers on C8 and C9.

## 2.2. Biological Activity of Resveratrol and Labruscol

The inhibitory effects of the bioproduced resveratrol and labruscol (**1**) at various concentrations (0–200 µM) on the cell viability of the HT-144 melanoma cell line, considered to be an aggressive cancer cell line, for 24, 48 and 72 h were evaluated using an MTT (3-(4,5-dimethyl thiazol-2yl)-2,5-diphenyltetrazolium bromide) assay (Sigma-Aldrich, Saint-Quentin, France) [19] in the presence of fetal bovine serum (FBS). The data are presented in Figure 2A,B. At all tested concentrations and times, resveratrol displayed an inhibition of cancer line viability higher than the dimer, labruscol. Nonetheless, both compounds exerted almost 100% cell viability inhibition at the doses of 100 and 200 µM within 72 h of treatment. In the presence of FBS, the inhibiting concentration (IC$_{50}$) of resveratrol was around 25 µM and that of labruscol around 50 µM, i.e., twice that of resveratrol after 48 h of treatment (Figure 2A,B). Thus, resveratrol displays a higher inhibition of cancer cell viability than its dimers in the presence of FBS, confirming previous results [9].

Figure 2B shows that labruscol induced a higher inhibition of the cell viability of healthy human dermal fibroblasts as compared to resveratrol at all tested concentrations and treatment times. However, this inhibition was markedly lower than the one displayed by this compound on melanoma cells. The two compounds thus showed differential activities on healthy and cancer cell lines and this can lead to interesting applications in cancer therapy [20–22]. Without FBS, the determination of the IC$_{50}$ for resveratrol and labruscol led to very different results. With an IC$_{50}$ = 10 µM, labruscol was considerably more effective on cancer cell viability than resveratrol (IC$_{50}$ = 90 µM) (Figure 2C). These data thus confirm that the cytotoxic activity of resveratrol dimers such as ε-viniferin or labruscol is decreased in the presence of FBS, probably due to their interaction with serum proteins [9].

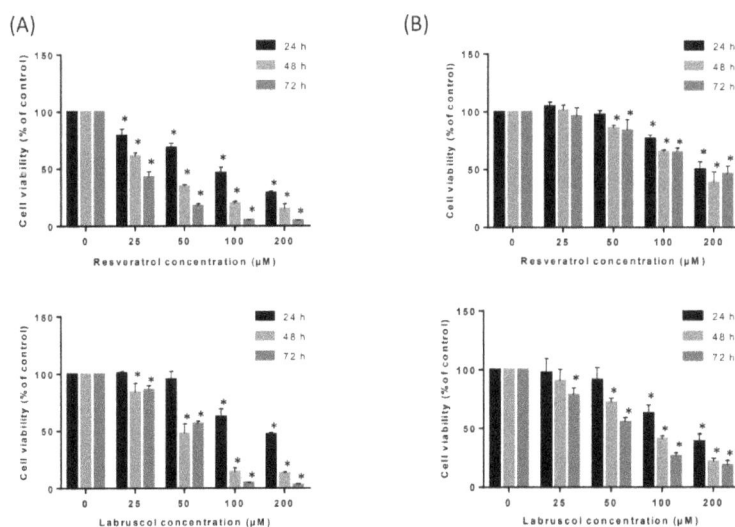

**Figure 2.** *Cont.*

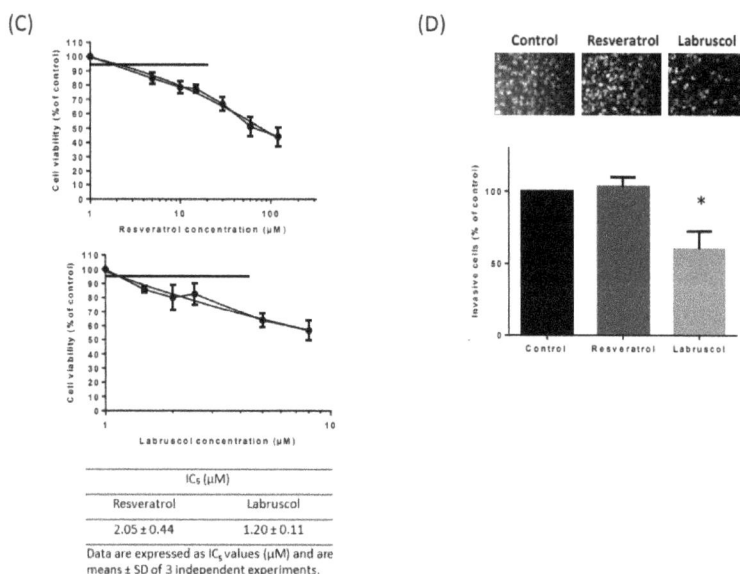

Figure 2. Effects of resveratrol and labruscol on cell viability and cell invasion in HT-144 and HDF cells. (**A**) HT-144 cells were treated with resveratrol and labruscol (0, 25, 50, 100 and 200 µM) in the presence of fetal bovine serum (FBS) for 24, 48 and 72 h before being subjected to an MTT (3-(4,5-dimethyl thiazol-2yl)-2,5-diphenyltetrazolium bromide) assay for cell viability determination. The values represented the means ± standard deviation (SD) of at least three independent experiments; (**B**) Human dermal fibroblast cells were treated with resveratrol and labruscol (0, 25, 50, 100 and 200 µM) for 24, 48 and 72 h before being subjected to an MTT assay for cell viability determination. The values represented the means ± SD of at least three independent experiments; (**C**) Determination of the inhibiting concentration (IC$_5$ values of resveratrol and labruscol. HT-144 cells were treated in a medium without FBS with resveratrol (0, 1.5, 10, 15, 30, 60 and 120 µM) and labruscol (0, 1.5, 2, 2.5, 5, 8 µM) for 24 h before being subjected to an MTT assay for IC$_5$ determination. The values represented the means ± SD of at least three independent experiments. Straight lines correspond to the IC$_5$; (**D**) Determination of cell invasion abilities. Cell invasion was measured using a Boyden chamber for 24 h with a matrigel® coating (25 µg/mL). HT-144 cells were treated in an FBS-free medium with vehicle (ethanol), resveratrol (2 µM) or labruscol (1.2 µM) for 24 h. The invasion abilities of HT-144 cells were quantified by counting the number of cells that invaded the underside of the transwell, as described in the Material and Methods section. The invasive activity of cancer cells was expressed as the mean number of cells that crossed the matrigel. The values represented the means ± SD of at least three independent experiments. * $p < 0.05$, compared with the vehicle group.

The effects of resveratrol and labruscol on tumor progression, including tumor migration and cell invasion, were tested in vitro on the HT-144 melanoma cell line. To this end we have identified the working concentrations of both resveratrol and labruscol in an FBS-free medium, at which cell viability inhibition does not exceed 5% (IC$_5$) (Figure 2C and Table). The MTT assay revealed that the IC$_5$ of resveratrol and labruscol were 2 and 1.2 µM, respectively. Concentrations of 1.2 and 2 µM were thus used to test the effect of both compounds on tumor progression. The migration assay was carried out by in vitro wound closure for 0, 12 and 24 h. Results showed that neither resveratrol nor labruscol had effects on tumor cell migration (data not presented). Interestingly, only labruscol was shown to induce a significant inhibition of 40% of cell invasion after 24 h of treatment, while resveratrol had no effect on cell invasion at the same concentration (Figure 2D). Although there are previous works reporting on the capacity of resveratrol to inhibit cell migration and invasion in other cell lines, this was only observed at concentrations higher than those used in this study (between 5 and 50 µM) [23–26].

The data presented here clearly show that resveratrol and the newly identified resveratrol dimer, labruscol, have interesting antiproliferative activities. Indeed, both compounds exert a marked inhibition of the cell viability of melanoma cells compared to healthy cells, although resveratrol displays a reduced impact on the cell viability of normal cells.

This tumor specificity has also been reported in the case of other phytostilbenes [6]. α-viniferin and *trans*-miyabenol C, two trimers of resveratrol, for example, were shown to exert a higher inhibiting activity on human colorectal carcinoma cells than on healthy colorectal lines [27]. The same activity was described for gnetin H, a resveratrol trimer in lung and breast carcinoma compared to healthy lines [28]. Due to the difficulty of obtaining resveratrol oligomers in sufficient amounts as pure compounds, there are only a few reports of other resveratrol dimers having similar inhibitory effects on the cell viability of cancer cells. These previously published results mainly concerned the effects of pallidol on human colorectal carcinoma cell lines (HCT1116, HT-29 and Caco-2) [27] and those of ε-viniferin on murine leukemia cell lines (P-388) [7], human oral squamous carcinoma cell lines (HL-60) [6], lymphoid and myeloid cell lines (U266, RPMI-8226, U937, K562, Jurkat) [8]. Pallidol and ε-viniferin were both reported to have inhibitory activities on the cell growth of two melanoma skin cancer cell lines, HT-144 and SKMEL-28 [9]. ε-viniferin also displays antiproliferative effects in human hepatocyte derived Hep G2 cells [4].

Labruscol and resveratrol (50 µM) were also shown to induce apoptosis in the HT-144 cancer cell line reaching 15% of apoptotic cells for both compounds within 72 h treatment (Figure 3).

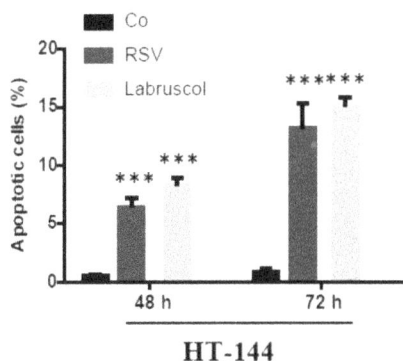

**Figure 3.** Apoptosis induction by resveratrol and labruscol in human skin melanoma cancer cells. Cells were treated during 48 h or 72 h with 50 µM resveratrol or labruscol. Apoptosis was assessed by staining (control cells or cells treated with resveratrol or labruscol) with 1 µg/mL Hoechst 33342. Co: control cells; RSV: resveratrol-treated cells; Labruscol: labruscol-treated cells. The values represented the means ± SD of at least 3 independent experiments. *** $p < 0.05$, compared with the control group.

This suggests that apoptosis induction in cancer cells could be a possible mechanism for the antiproliferative effects of labruscol, as reported in previous studies for resveratrol and related metabolites [29]. On the other hand, a transient senescence activity has already been described for resveratrol in human metastatic colon cancer cells [30]. As the chemical structure of labruscol is analogous to that of resveratrol (presence of a trans-resveratrol moiety), one can assume that labruscol can also inhibit tumor cell growth via a senescence induction pathway [30]. Further works are thus needed to characterize the mechanisms implied in the cytototoxic activities of labruscol reported in the present study.

In sum, the results obtained here with the aggressive cancer cell line HT-144 reveal the antiproliferative activity of a newly characterized resveratrol dimer, labruscol. Moreover, at the very low concentration of 1.2 µM (IC$_5$), this compound has shown a 40% inhibition of cancer cell invasion, a property not displayed by resveratrol. It thus seems that labruscol possesses complementary properties of resveratrol in particular regarding cell invasion. One can thus suggest that labruscol could be used in combination with resveratrol to improve its antiproliferative capacities. Other studies have indeed already shown a greater efficiency of mixtures of stilbenes compared to resveratrol alone [4,31,32]. As stilbene derivatives generally do not display high cytotoxic activity, the biological effects of labruscol could thus be assessed in the context of multi-drug resistance, where this compound alone or in combination can actively participate to cell resensitisation by therapeutic agents, helping in the reduction of their doses and their toxicities [29].

## 3. Materials and Methods

### 3.1. Chemicals, Reagents and Materials

Methanol (MeOH), ethyl acetate (EtOAc), and *n*-heptane (Hept) were purchased from Carlo ErbaReactifs SDS (Val de Reuil, France). Deuterated methanol (methanol-*d4*) was purchased from Sigma-Aldrich (Saint-Quentin, France). Deionized water was used to prepare all aqueous solutions.

### 3.2. Cultures in Bioreactor and Elicitation of Stilbene Production

Cell suspensions of *Vitis labrusca* L. var. Concord were cultured in a 14 L tank of a stirred bioreactor Bioflo 3000 (New Brunswick Scientific, Edison, New York, NY, USA) containing 10 L (final volume) of B5 medium [33]. The agitation (two marine turbines) was set to 50 rpm and the aeration rate maintained at 0.025 vvm. All other conditions were as previously described [9]. Stilbene production was induced by the use of two elicitors. The first elicitor was a β-cyclodextrin Kleptose® (Roquette, Lestrem, France). The second elicitor was methyljasmonate. These two elicitors were added to the cell cultures as previously reported [9].

### 3.3. Extraction of Total Stilbenes from the Culture Medium and CPC Purification of Labruscol

The whole culture medium (10 L) was filtered under reduced pressure and stored at −20 °C until use. A crude stilbene extract (3.1 g) was obtained from 5 L of the filtered culture medium by performing three successive extractions with ethyl acetate (3 × 2 L) in a separatory funnel followed by solvent elimination under vacuum at 40 °C.

Centrifugal partition chromatography (CPC) was carried out on a lab-scale FCPE300® column of 303 mL capacity (Rousselet Robatel Kromaton, Annonay, France) containing seven circular partition disks, engraved with a total of 231 oval partition twin-cells (~1 mL per twin-cell) and connected to a KNAUER Preparative 1800 V7115 pump (Berlin, Germany). The system was coupled to a UVD 170S detector set at 210, 254, 280, and 366 nm (Dionex, Sunnivale, CA, USA). Fractions were collected by a Pharmacia Superfrac collector (Uppsala, Sweden).

Labruscol was purified by using a gradient elution method as described previously [9]. Fractions were collected every minute, spotted on Merck thin layer chromatography (TLC) plates coated with silica gel 60 F254 and developed with chloroform/ethyl acetate/formic acid (6:4:1, *v/v/v*). After UV detection at 254 nm, the plates were sprayed with vanillin–sulfuric acid and heated to 100 °C for 5 min. Labruscol was detected as a pink stain with a retention factor of 0.67 in these TLC conditions.

### 3.4. Structural Elucidation of Labruscol

NMR analyses were performed in methanol-*d*$_4$ at 298 K on a Bruker Avance AVIII-600 spectrometer (Karlsruhe, Germany) equipped with a cryoprobe optimized for $^1$H detection and with cooled $^1$H, $^{13}$C and $^2$H coils and preamplifiers. 1D and 2D NMR spectra ($^1$H, $^{13}$C, HSQC, HMBC, NOESY and COSY) were recorded using standard Bruker pulse programs (Bruker, Karlsruhe, Germany). An aliquot of labruscol was also solubilized in MeOH and directly infused in a quadrupole time-of-flight hybrid mass spectrometer (QTOF micro®, Waters, Manchester, UK) equipped with an electrospray source. The mass range of the instrument was set at $m/z$ 100–1200 and scan duration was set at 2 s in the negative ion mode. The capillary voltage was 3000 V, the cone voltage was 35 V, and the temperature was 80 °C.

Conformational analysis to find the lowest energy conformers for the two possible relative stereochemistry *RR* or *RS* using the force field MM3* in octanol were done with Maestro module of the Schrödinger Suite, version 10.5.014 (Shrödinger Software, San Diego, CA, USA). No constraints were applied. Conformational search using mixed torsional/low-mode sampling was used. The number of separate conformers generated was 1000, with a maximum of 100 unique structures to be saved for each rotatable bond. A 21 kJ/mol energy cutoff was used to remove the higher energy conformers. A conformer was considered redundant and subsequently eliminated if its maximum atom deviation from an already-identified conformer was less than 0.5 Å. All conformers were subjected to further minimization using the Powell–Reeves conjugate gradient (PRCG) method for a maximum of 2500 steps, by using a convergence threshold of 0.05.

*3.5. Biological Tests*

3.5.1. Cell Cultures

The HT-144 cell line used in this study derived from human melanoma and was obtained from the American Tissue Culture Collection (ATCC). Cells were grown in a McCoy's 5a modified medium supplemented with 10% ($v/v$) FBS and 1% ($v/v$) antibiotic (penicillin, streptomycin) at 37 °C in a humidified atmosphere of 5% $CO_2$. Human dermal fibroblasts (HDF) were isolated from skin biopsies of healthy subjects. Their culture conditions were as described previously [9].

3.5.2. Experimental Treatments

The two cell types were seeded during 24 h before to be treated in triplicate wells with resveratrol and labruscol at different concentrations and treatment times. Compounds were dissolved in ethanol at a $5 \times 10^{-2}$ M final concentration and stored at $-20$ °C. Compounds were diluted in culture media, with or without FBS, to the desired final concentration. All control and treated cells received a maximal volume of 0.1% ($v/v$) of ethanol.

3.5.3. Cell Viability Assay

Cell viability was examined by MTT [3-(4,5-dimethyl thiazol-2yl)-2,5-diphenyltetrazolium bromide] assay (Sigma-Aldrich, Saint-Quentin, France) [19] as previously described [9]. The $IC_5$ for the bioproduced stilbenes, resveratrol and labruscol, was defined as the concentration producing 5% decrease in cell growth.

3.5.4. In Vitro Wound Closure

HT-144 cells ($2 \times 10^5$ cells/well) were plated in six-well plates for 24 h, wounded by scratching with a pipette tip, incubated in McCoy's 5a modified medium without FBS, and treated or not with resveratrol and labruscol at the same concentration, which did not exceed their $IC_5$, 1.2 µM for 0, 12 and 24 h. The cells were photographed, in three same fields by well, at 0, 12 and 24 h using a phase-contrast microscope (100×).

3.5.5. In Vitro Invasion Assays

The invasive potential of tumor cells was examined using modified Boyden chambers (6.5 mm diameter and 8 µM pore) (Greiner Bio-One, Les Ulis, France) according to the manufacturer's instructions. Briefly, HT-144 cells were suspended in a serum free McCoy's 5a modified medium and 100 µL of the cell suspension ($2 \times 10^4$ cells) were seeded onto the upper compartment of the

Transwell coated with 25 μg of Matrigel (Corning Life Sciences, Corning, NY, USA). Cells were treated, or not, 1 h after being seeded with resveratrol and labruscol at 1.2 μM respectively for 24 h. In the lower compartment, 800 μL of the McCoy's 5a modified medium containing 10% FBS were added 2 h after treatment and used as chemoattractant. After 24 h, cells were fixed with methanol, being the non-invading cells remaining on the upper side of the filter, scrapped off. Invading cells on the lower side of the filter were stained with Hoechst 33342 (Sigma-Aldrich, Saint-Quentin, France). Invading cells were observed with a fluorescence microscope and counted in five fields at 100× magnification. The invasive activity of cancer cells was expressed as the mean number of cells that crossed the Matrigel.

### 3.5.6. Apoptosis Identification

Apoptosis was identified by staining the nuclear chromatin of trypsinized cells (controls and resveratrol or labruscol-treated cells) with 1 μg/mL Hoechst 33342 (Sigma Aldrich, Saint Quentin, France) for 15 min at 37 °C. The percentage of apoptotic cells was determined by the analysis of 300 cells from randomly fields for each treatment.

### 3.6. Statistical Analysis

The data were expressed as the mean ± standard deviation (SD) of three independent experiments. Each experiment was performed in triplicate. The significance of differences was established with the Student's *t*-test.

**Acknowledgments:** This work is a part of the VSOP (Vitis Stilben Oligomers bioProduced) program which was supported by the Champagne-Ardenne and Bourgogne Regional Councils. The authors thank Vincenzo de Luca (Brock University, St Catharines, ON, Canada) for providing the Concord callus line.

**Author Contributions:** L.N. and J.H. designed the experiments; J.H., N.B., D.H., J.-M.N. and J.-H.R. designed labruscol purification and identification experiments; E.C. designed bio-production experiments; M.T., D.D., L.M. and C.C. designed the study; L.N., J.H. and P.J. analyzed the data; P.J., L.N., J-H.R. and J.H. wrote the paper.

**Conflicts of Interest:** The authors declare no conflict of interest.

### References

1. Rivière, C.; Pawlus, A.D.; Mérillon, J.-M. Natural stilbenoids: Distribution in the plant kingdom and chemotaxonomic interest in Vitaceae. *Nat. Prod. Rep.* **2012**, *29*, 1317–1333. [CrossRef] [PubMed]
2. Privat, C.; Telo, J.P.; Bernades-Genisson, V.; Vieira, A.; Souchard, J.-P.; Nepveu, F. Antioxidant properties of trans-ε-viniferin as compared to stilbene derivatives in aqueous and non aqueous media. *J. Agric. Food Chem.* **2002**, *50*, 1213–1217. [CrossRef] [PubMed]
3. Nassra, M.; Krisa, S.; Papastamoulis, Y.; Kapche, G.D.; Bisson, J.; André, C.; Konsman, J.-P.; Schmitter, J.-M.; Mérillon, J.-M.; Waffo-Téguo, P. Inhibitory activity of plant stilbenoids against nitric oxide production by lipopolysaccharide-activated microglia. *Planta Med.* **2013**, *79*, 966–970. [CrossRef] [PubMed]
4. Billard, C.; Izard, J.-C.; Roman, V.; Kern, C.; Mathiot, C.; Mentz, F.; Kolb, J.-P. Comparative antiproliferative and apoptotic effects of resveratrol, ε -viniferin and vine-shots derived polyphenols (vineatrols) on chronic B lymphatic leukemia cells and normal human lymphocytes. *Leuk. Lymphoma* **2002**, *43*, 1991–2002. [CrossRef] [PubMed]
5. Colin, D.; Gimazane, A.; Lizard, G.; Izard, J.-C.; Solary, E.; Latruffe, N.; Delmas, D. Effects of resveratrol analogs on cell cycle progression, cell cycle associated proteins and 5-fluoro-uracil sensitivity in human derived colon cancer cells. *Int. J. Cancer* **2009**, *124*, 2780–2788. [CrossRef] [PubMed]
6. Chowdhury, S.A.; Kishino, K.; Satoh, R.; Hashimoto, K.; Kikuchi, H.; Nishikawa, H.; Shirataki, Y.; Sakagami, H. Tumor-specificity and apoptosis-inducing activity of stilbenes and flavonoids. *Anticancer Res.* **2005**, *25*, 2055–2063. [PubMed]
7. Muhtadi, H.E.H.; Juliawaty, L.D.; Syah, Y.M.; Achmad, S.A.; Latip, J.; Ghisalberti, E.L. Cytotoxic resveratrol oligomers from the tree bark of *Dipterocarpus hasseltii. Fitoterapia* **2006**, *77*, 550–555. [CrossRef] [PubMed]

8. Barjot, C.; Tournaire, M.; Castagnino, C.; Vigor, C.; Vercauteren, J.; Rossi, J.-F. Evaluation of antitumor effects of two vine stalk oligomers of resveratrol on a panel of lymphoid and myeloid cell lines: Comparison with resveratrol. *Life Sci.* **2007**, *81*, 1565–1574. [CrossRef] [PubMed]

9. Nivelle, L.; Hubert, J.; Courot, E.; Jeandet, P.; Aziz, A.; Nuzillard, J.-M.; Renault, J.-H.; Clément, C.; Martiny, L.; Delmas, D.; Tarpin, M. Anti-cancer activity of resveratrol and derivatives produced by grapevine cell suspensions in a 14 L stirred bioreactor. *Molecules* **2017**, *22*, 474. [CrossRef] [PubMed]

10. Zghonda, N.; Yoshida, S.; Araki, M.; Kusunoki, M.; Mliki, A.; Ghorbel, A.; Miyazaki, H. Greater effectiveness of ε-viniferin in red wine than its monomer resveratrol for inhibiting vascular smooth muscle cell proliferation and migration. *Biosci. Biotechnol. Biochem.* **2011**, *75*, 1259–1267. [CrossRef] [PubMed]

11. Morales, M.; Bru, R.; García-Carmona, F.; Ros Barceló, A.; Pedreño, M.A. Effect of dimethyl-β-cyclodextrins on resveratrol metabolism in Gamay grapevine cell cultures before and after inoculation with shape *Xylophilus ampelinus*. *Plant Cell Tissue Org. Cult.* **1998**, *53*, 179–187. [CrossRef]

12. Bru, R.; Sellés, S.; Casado-Vela, J.; Belchí-Navarro, S.; Pedreño, M.A. Modified cyclodextrins are chemically defined glucan inducers of defense responses in grapevine cell cultures. *J. Agric. Food Chem.* **2006**, *54*, 65–71. [CrossRef] [PubMed]

13. Donnez, E.; Jeandet, P.; Clément, C.; Courot, E. Bioproduction of resveratrol and stilbene derivatives by plant cells and microorganisms. *Trends Biotechnol.* **2009**, *27*, 706–713. [CrossRef] [PubMed]

14. Jeandet, P.; Clément, C.; Courot, E. Resveratrol production at large scale using plant cell suspensions. *Eng. Life Sci.* **2014**, *14*, 622–632. [CrossRef]

15. Jeandet, P.; Clément, C.; Tisserant, L.-P.; Crouzet, J.; Courot, E. Use of grapevine cell cultures for the production of phytostilbenes of cosmetic interest. *C. R. Chim.* **2016**, *19*, 1062–1070. [CrossRef]

16. Friesen, J.-B.; McAlpine, J.-B.; Chen, S.-N.; Pauli, G.-F. Countercurrent separation of natural products: An update. *J. Nat. Prod.* **2015**, *78*, 1765–1796. [CrossRef] [PubMed]

17. Hubert, J.; Nuzillard, J.-M.; Purson, S.; Hamzaoui, M.; Borie, N.; Reynaud, R.; Renault, J.-H. Identification of natural metabolites in mixture: A pattern recognition strategy based on $^{13}$C-NMR. *Anal. Chem.* **2014**, *86*, 2955–2962. [CrossRef] [PubMed]

18. Tisserant, L.-P.; Hubert, J.; Lequart, M.; Borie, N.; Maurin, N.; Pilard, S.; Jeandet, P.; Aziz, A.; Renault, J.-H.; Nuzillard, J.-M.; et al. $^{13}$C-NMR and LC-MS profiling of stilbenes from elicited grapevine hairy root cultures. *J. Nat. Prod.* **2016**, *79*, 2846–2855. [CrossRef] [PubMed]

19. Morgan, D.M. Tetrazolium (MTT) assay for cellular viability and activity. *Methods Mol. Biol.* **1998**, *79*, 179–183. [PubMed]

20. Aziz, M.H.; Nihal, M.; Fu, V.X.; Jarrard, D.F.; Ahmad, N. Resveratrol-caused apoptosis of human prostate carcinoma LNCaP cells is mediated via modulation of phosphatidylinositol 3′-kinase/Akt pathway and Bcl-2 family proteins. *Mol. Cancer Ther.* **2006**, *5*, 1335–1341. [CrossRef] [PubMed]

21. Colin, D.; Limagne, E.; Jeanningros, S.; Jacquel, A.; Lizard, G.; Athias, A.; Gambert, P.; Hichami, A.; Latruffe, N.; Solary, E.; et al. Endocytosis of resveratrol via lipid rafts and activation of downstream signaling pathways in cancer cells. *Cancer Prev. Res.* **2011**, *4*, 1095–1106. [CrossRef] [PubMed]

22. Yeh, C.-B.; Hsieh, M.-J.; Lin, C.-W.; Chiou, H.-L.; Lin, P.-Y.; Chen, T.-Y.; Yang, S.-F. The antimetastatic effects of resveratrol on hepatocellular carcinoma through the downregulation of a metastasis-associated protease by SP-1 modulation. *PLoS ONE* **2013**, *8*, e56661. [CrossRef] [PubMed]

23. Tang, F.-Y.; Su, Y.-C.; Chen, N.-C.; Hsieh, H.-S.; Chen, K.-S. Resveratrol inhibits migration and invasion of human breast-cancer cells. *Mol. Nutr. Food Res.* **2008**, *52*, 683–691. [CrossRef] [PubMed]

24. Weng, C.-J.; Wu, C.-F.; Huang, H.-W.; Wu, C.-H.; Ho, C.-T.; Yen, G.-C. Evaluation of anti-invasion effect of resveratrol and related methoxy analogues on human hepatocarcinoma cells. *J. Agric. Food Chem.* **2010**, *58*, 2886–2894. [CrossRef] [PubMed]

25. Wang, H.; Zhang, H.; Tang, L.; Chen, H.; Wu, C.; Zhao, M.; Yang, Y.; Chen, X.; Liu, G. Resveratrol inhibits TGF-β1-induced epithelial-to-mesenchymal transition and suppresses lung cancer invasion and metastasis. *Toxicology* **2013**, *303*, 139–146. [CrossRef] [PubMed]

26. Ji, Q.; Liu, X.; Fu, X.; Zhang, L.; Sui, H.; Zhou, L.; Sun, J.; Cai, J.; Qin, J.; Ren, J.; et al. Resveratrol Inhibits Invasion and Metastasis of Colorectal Cancer Cells via MALAT1 Mediated Wnt/β-Catenin Signal Pathway. *PLoS ONE* **2013**, *8*, e78700. [CrossRef] [PubMed]

27. González-Sarrías, A.; Gromek, S.; Niesen, D.; Seeram, N.P.; Henry, G.E. Resveratrol oligomers isolated from *Carex* species inhibit growth of human colon tumorigenic cells mediated by cell cycle arrest. *J. Agric. Food Chem.* **2011**, *59*, 8632–8638. [CrossRef] [PubMed]

28. Gao, Y.; He, C.; Ran, R.; Zhang, D.; Li, D.; Xiao, P.-G.; Altman, E. The resveratrol oligomers, *cis-* and *trans-*gnetin H, from *Paeonia suffruticosa* seeds inhibit the growth of several human cancer cell lines. *J. Ethnopharmacol.* **2015**, *169*, 24–33. [CrossRef] [PubMed]

29. Aires, V.; Limagne, E.; Cotte, A.K.; Latruffe, N.; Ghiringhelli, F.; Delmas, D. Resveratrol metabolites inhibit human metastatic colon cancer cells progression and synergize with chemotherapeutic drugs to induce cell death. *Mol. Nutr. Food Res.* **2013**, *57*, 1170–1181. [CrossRef] [PubMed]

30. Colin, D.; Limagne, E.; Ragot, K.; Lizard, G.; Ghiringhelli, F.; Solary, E.; Chauffert, B.; Latruffe, N.; Delmas, D. Resveratrol transiently induces cell death and senescence in colon cancer cells. *Cell Death Dis.* **2014**, *5*, e1533. [CrossRef] [PubMed]

31. Colin, D.; Lancon, A.; Delmas, D.; Lizard, G.; Abrossinow, J.; Kahn, E.; Jannin, B.; Latruffe, N. Antiproliferative activities of resveratrol and related compounds in human hepatocyte derived HepG2 cells are associated with biochemical cell disturbance revealed by fluorescence analyses. *Biochimie* **2008**, *90*, 1674–1684. [CrossRef] [PubMed]

32. Giovannelli, L.; Innocenti, M.; Santamaria, A.R.; Bigagli, E.; Pasqua, G.; Mulinacci, N. Antitumoural activity of viniferin-enriched extracts from *Vitis vinifera* L. cell cultures. *Nat. Prod. Res.* **2014**, *28*, 2006–2016. [CrossRef] [PubMed]

33. Gamborg, O.L.; Miller, R.A.; Ojima, K. Nutrient requirements of suspension cultures of soybean root cells. *Exp. Cell Res.* **1968**, *50*, 151–158. [CrossRef]

**Sample Availability:** Samples of the compounds are not available from the authors.

*molecules*

MDPI

*Article*

# Resveratrol-Induced Changes in MicroRNA Expression in Primary Human Fibroblasts Harboring Carnitine-Palmitoyl Transferase-2 Gene Mutation, Leading to Fatty Acid Oxidation Deficiency

Virginie Aires [1,2], Dominique Delmas [1,2], Fatima Djouadi [3], Jean Bastin [3], Mustapha Cherkaoui-Malki [1] and Norbert Latruffe [1,*]

[1]   Laboratory BioPeroxIL, University of Bourgogne-Franche Comté, 6 Blvd Gabriel, 21000 Dijon, France; virginie.aires02@u-bourgogne.fr (V.A.); ddelmas@u-bourgogne.fr (D.D.); malki@u-bourgogne.fr (M.C.-M.)
[2]   INSERM UMR 866, Blvd Jeanne d'Arc, 21490 Dijon, France
[3]   INSERM U1124, Université Paris-Descartes, Rue des Saints-Pères, 75000 Paris, France; Fatima.djouadi@inserm.fr (F.D.); jean.bastin@inserm.fr (J.B.)
*   Correspondence: latruffe@u-bourgogne.fr; Tel.: +33-380-396-237

Received: 13 November 2017; Accepted: 16 December 2017; Published: 22 December 2017

**Abstract:** Carnitine palmitoyltransferase-2 (*CPT2*) is a mitochondrial enzyme involved in long-chain fatty acid entry into mitochondria for their β-oxidation and energy production. Two phenotypes are associated with the extremely reduced *CPT2* activity in genetically deficient patients: neonatal lethality or, in milder forms, myopathy. Resveratrol (RSV) is a phytophenol produced by grape plant in response to biotic or abiotic stresses that displays anti-oxidant properties, in particular through AP-1, NFκB, STAT-3, and COX pathways. Some beneficiary effects of RSV are due to its modulation of microRNA (miRNA) expression. RSV can enhance residual *CPT2* activities in human fibroblasts derived from *CPT2*-deficient patients and restores normal fatty acid oxidation rates likely through stimulation of mitochondrial biogenesis. Here, we report changes in miRNA expression linked to *CPT2*-deficiency, and we identify miRNAs whose expression changed following RSV treatment of control or *CPT2*-deficient fibroblasts isolated from patients. Our findings suggest that RSV consumption might exert beneficiary effects in patients with *CPT2*-deficiency.

**Keywords:** resveratrol; miRNA level; *CPT2*-deficient cells

## 1. Introduction

Resveratrol (RSV, trans-3,5,4′-trihydroxystilbene) is a phytoalexin produced by numerous plants in response to abiotic or biotic stress [1–3]. This polyphenol compound admittedly protects humans against various diseases (cardiovascular and inflammation-associated pathologies, infection, cancer, neurodegenerescence, aging, etc.) through the modulation of several signaling pathways, including those mediated by transcription factors AP-1 [4], NFκB, and STAT-3 [5] or the COX enzyme [6]. RSV has been shown to improve residual β-oxidation in primary human fibroblasts from *CTP2*-deficient patients by restoring normal fatty acid oxidation rates [7].

MicroRNAs are short, non-coding regulatory microRNAs present in plants, animals, and viruses. To date, more than 1500 miRNAs have been identified in humans. MiRNA primarily controls mRNA translation and stability. Due to their ability to regulate several hundred transcripts directly or indirectly through targeting components of key regulatory pathways, miRNAs behave as master regulators that impact all aspects of cell homeostasis and functions. Many miRNAs are considered either as tumor suppressor or onco-miRs, depending on the cellular context [8].

There is now a substantial amount of literature on miRNAs, including a few studies that have addressed the differential effects of polyphenols on miRNA expression [8–13]. In 2010, we demonstrated the regulatory effect of RSV on the expression of miRNAs involved in macrophage-associated inflammatory response [8] as well as on the expression of components of the TGFβ regulatory pathway in colon cancer cell lines [14]. Interestingly, in 2013, Milenkovic et al. [10] established that the expression of more than 100 miRNAs is modulated by polyphenols. These authors reported that various different polyphenols have both common and specific miRNA targets. Indeed, in mouse livers, over 137 miRNAs are modulated by phytophenols of the stilbenoids family (including resveratrol). While stilbenoids modulated the expression of 87 of these 137 miRNAs, 24 other miRNAs were targets of flavonoids, 6 of phenolic acids, and 20 of curcuminoids. This analysis further confirms that the different classes of polyphenols not only share common properties (as anti-oxidant for instance) but also have their own specific effects due to their unique chemical structure and reactivity and their differential effects on gene expression, especially miRNAs.

Presently, more than a hundred papers have confirmed that the effect of RSV in the prevention or treatment of various diseases, including prostate cancer [15], melanoma [16], breast cancer [17], lung tumors [18], white adipogenesis [19], liver steatosis [20], inflammation [21,22], neurodegenerative disorders [23], and osteoporosis [24], were mediated by miRNAs.

In the present study, we report the RSV-induced modulation of energy metabolism miRNA in human cells harboring mitochondrial fatty acid ß-oxidation-deficiency as a result of carnitine-palmitoyl transferase 2 (*CPT2*) gene mutations. *CPT2* is a mitochondrial inner membrane enzyme playing an essential role in the transfer of fatty acids from the cytosol to the mitochondrial matrix. In 2011, Bastin et al. [7] showed that RSV is able to enhance residual *CPT2* activity in human fibroblasts derived from patients harboring the muscular form of CTP2 deficiency and can restore near-normal fatty acids oxidation rates, opening potential clinical perspectives to successfully treat the *CPT2*-deficiency-associated myopathy. This stimulation was also observed with other analogues of the stilbene family [25]. In this study, we compared miRNA expression in control and *CPT2*-deficient primary human fibroblasts and analyzed RSV's impact on miRNA expression in both cell lines.

## 2. Results and Discussion

*2.1. Changes in miRNA Expression Associated with Mitochondrial CPT2-Deficiency in Primary Human Fibroblasts*

Table 1 shows that *CPT2*-deficiency was associated with significant changes in the levels of 51 miRNA in patient fibroblasts. More specifically, *CPT2*-deficient fibroblasts showed 13 upregulated miRNAs (with an 11-fold increase for *miR-301* in particular) and 38 downregulated miRNAs, including 3 miRNAs from the *miR-let-7* family. The multiplicity of miRNA target transcripts suggests that the fibroblast transcriptome might be widely affected by *CPT2*-deficiency.

**Table 1.** MiRNAs whose expression changed in human *CPT2*-deficient primary fibroblasts as compared with control primary human fibroblasts, as deduced from microRNA microarray analysis. Geometric mean of intensities <100 were considered as background and discarded. Changes were considered significant for $p < 0.05$.

| miRNAs | Fold Change | Increasing Parametric $p$ Value |
|---|---|---|
| miRNAs upregulated in *CPT2*-deficient fibroblasts: | | |
| 483 | 3.1 | $1.6 \times 10^{-6}$ |
| 301 | 11.43 | $4.1 \times 10^{-6}$ |
| 449b | 1.99 | $2.79 \times 10^{-6}$ |
| 206 | 3.38 | $9.39 \times 10^{-6}$ |
| 550-1 | 2.83 | 0.000171 |
| 539 | 2.04 | 0.0002213 |
| 661 | 2.79 | 0.0004408 |
| 371 | 2.65 | 0.0005968 |

Table 1. *Cont.*

| miRNAs | Fold Change | Increasing Parametric *p* Value |
|---|---|---|
| miRNAs upregulated in *CPT2*-deficient fibroblasts: | | |
| 10b | 2.75 | 0.0011091 |
| 9 | 4 4 | 0.0014253 |
| 550-2 | 2.1 | 0.0016581 |
| 651 | 2.87 | 0.0019172 |
| 196a-2 | 2.09 | 0.0019172 |
| miRNAs downregulated in *CPT2*-deficient fibroblasts: | | |
| let-7d | 0.16 | $<1 \times 10^{-7}$ |
| 211 | 0.14 | $4 \times -10^{-7}$ |
| let-7a3 | 0.22 | $1.2 \times 10^{-6}$ |
| 198 | 0.14 | $2.8 \times 10^{-6}$ |
| 141 | 0.28 | $4.6 \times 10^{-6}$ |
| 136 | 0.31 | $5.1 \times 10^{-6}$ |
| 203 | 0.24 | $5.7 \times 10^{-6}$ |
| 127 | 0.23 | $7.7 \times 10^{-6}$ |
| 181c | 0.26 | $1.85 \times 10^{-5}$ |
| 496 | 0.3 | $2.5 \times 10^{-5}$ |
| 126-5p | 0.14 | $3.64 \times 10^{-5}$ |
| 144 | 0.097 | $3.8 \times 10^{-5}$ |
| let-7g | 0.48 | $4.14 \times 10^{-5}$ |
| 181a2 | 0.44 | $4.42 \times 10^{-5}$ |
| 618 | 0.48 | $4.47 \times 10^{-5}$ |
| 41 | 0.15 | $4.5 \times 10^{-5}$ |
| 299-5p | 0.14 | $4.73 \times 10^{-5}$ |
| 1 | 0.41 | $4.88 \times -10^{-5}$ |
| 145 | 0.32 | $4.94 \times 10^{-5}$ |
| 25 | 0.26 | $6.37 \times 10^{-5}$ |
| 123 | 0.31 | $6.67 \times 10^{-5}$ |
| 200b | 0.27 | $8.33 \times 10^{-5}$ |
| 325 | 0.44 | $8.51 \times 10^{-5}$ |
| 593 | 0.42 | $9.19 \times 10^{-5}$ |
| 24-5p/189 | 0.14 | 0.0001524 |
| 125b2 | 0.1 | 0.0002071 |
| 123 | 0.25 | 0.0002196 |
| 154-5p | 0.3 | 0.0002281 |
| 184 | 0.43 | 0.0002499 |
| 199b | 0.47 | 0.0005099 |
| 22 | 0.25 | 0.0006033 |
| 363-3p | 0.37 | 0.0006076 |
| 338 | 0.24 | 0.0007282 |
| 146a | 0.42 | 0.0008154 |
| 212 | 0.28 | 0.0008813 |
| 196a-1 | 0.34 | 0.0008916 |
| 500 | 0.29 | 0.0013401 |
| 563 | 0.47 | 0.0016458 |

## 2.2. RSV-Induced Changes in miRNA Expression in Control and CPT2-Deficient Primary Human Fibroblasts

Table 2 shows changes in miRNA expression induced by RSV treatment in both control and *CPT2*-deficient fibroblasts. Twelve miRNAs were upregulated and 24 miRNAs downregulated in control fibroblasts, versus 8 miRNAs upregulated and 16 miRNAs downregulated in *CPT2*-deficient fibroblasts. Of note, *miR-566* and *miR-23a,b* were downregulated in both control and patient fibroblasts following RSV treatment, suggesting that these changes might result from RSV specific action on factors controlling transcription and/or maturation of these two miRNAs, irrespective of cell genotype or of energy metabolism deficiency. On the other hand, *miR-550-1,2* was upregulated and *miR-let7-a3* was downregulated in both untreated *CPT2*-deficient fibroblasts (Table 1) and RSV-treated control fibroblasts (Table 2). If one considers that RSV exerts beneficial effects on the cell, this suggests that the changes in expression of these two miRNAs might be advantageous in both cases. Finally, the expression of other miRNAs changed in both untreated and RSV-treated *CPT2*-deficient fibroblast. Thus, *miR-181a2,d*, *miR-let7d*, and *miR-146a* were downregulated in both untreated and RSV-treated *CPT2*-deficient fibroblast (Tables 1 and 2). This suggests that the upregulation of these three miRNAs might provide *CPT2*-deficient fibroblasts with an increased ability to survive with reduced catabolism of long chain fatty acids.

**Table 2.** RSV (75 μM) treatment impacts miRNA expression in control and in *CPT2*-deficient fibroblasts, compared with the corresponding DMSO-treated primary fibroblasts, as deduced from microRNA microarray analysis. Geometric mean of intensities <100 were considered as background and discarded. Changes were considered significant when $p < 0.05$.

| Control Fibroblasts | | | | | | CPT2-Deficient Fibroblasts | | | | | |
|---|---|---|---|---|---|---|---|---|---|---|---|
| Upregulation by RSV | | | Downregulation by RSV | | | Upregulation by RSV | | | Downregulation by RSV | | |
| miRNA | Fold Change | Increasing Parametric p Value | miRNA | Fold Change | Increasing Parametric p Value | miRNA | Fold Change | Increasing Parametric p Value | miRNA | Fold Change | Increasing Parametric p Value |
| 321 | 3.67 | 0.0003277 | 35 | 0.47 | 0.0011099 | 219 | 1.81 | 0.00028111 | 101-1/2 | 0.51 | 0.000758 |
| 594 | 3.33 | 0.000695 | 548a-1 | 0.28 | 0.0011964 | 299-5p | 1.94 | 0.00037058 | 181d | 0.45 | 0.00021178 |
| 550-2 | 2.65 | 0.0026216 | 566 | 0.49 | 0.0014556 | 193a | 1.96 | 0.00074255 | 16-1 | 0.48 | 0.00022522 |
| 565 | 2.87 | 0.0066109 | 620 | 0.49 | 0.0031365 | 199a1-5p | 1.8 | 0.035782 | 21 | 0.47 | 0.00023066 |
| 611 | 2.18 | 0.0100121 | 92b | 0.24 | 0.003498 | 548a1 | 2.3 | 0.041391 | 99a* | 0.27 | 0.00041291 |
| 483 | 2.29 | 0.0118661 | 378-5p | 0.41 | 0.0070053 | 337 | 1.89 | 0.0488272 | 20b | 0.47 | 0.00053956 |
| 335 | 2.69 | 0.0158687 | 579 | 0.19 | 0.020456 | | | | let-7d | 0.46 | 0.00071834 |
| 550-1 | 2.35 | 0.0182839 | 136 | 0.46 | 0.0206073 | | | | 17-5p | 0.43 | 0.0007255 |
| 449b-1 | 1.99 | 0.021485 | let-7f | 0.44 | 0.0220297 | | | | 146a | 0.27 | 0.0110431 |
| 661 | 2.45 | 0.0256413 | 211 | 0.24 | 0.0220881 | | | | 566 | 0.5 | 0.012436 |
| 326 | 3.58 | 0.0315627 | 376a-2 | 0.42 | 0.024334 | | | | 376b | 0.47 | 0.012361 |
| 196a-1 | 3.45 | 0.0369082 | 193a | 0.23 | 0.0255173 | | | | 26a | 0.46 | 0.0130816 |
| | | | 29a | 0.47 | 0.0265262 | | | | 103-1 | 0.35 | 0.0166397 |
| | | | 19b | 0.44 | 0.0294717 | | | | let-7c | 0.28 | 0.0192349 |
| | | | 141 | 0.26 | 0.029536 | | | | 423 | 0.14 | 0.0231317 |
| | | | 204 | 0.33 | 0.0295895 | | | | 23a | 0.18 | 0.0257826 |
| | | | 216 | 0.3 | 0.0311748 | | | | | | |
| | | | let-7a3 | 0.29 | 0.0323893 | | | | | | |
| | | | 618 | 0.44 | 0.0361713 | | | | | | |
| | | | 198 | 0.43 | 0.0375283 | | | | | | |
| | | | 22 | 0.39 | 0.0379847 | | | | | | |
| | | | 126-5p | 0.26 | 0.0463448 | | | | | | |
| | | | 23b | 0.49 | 0.0469568 | | | | | | |
| | | | 144 | 0.26 | 0.0496528 | | | | | | |

*2.3. Mirna Whose Expression Changed in CPT2-Deficient Primary Fibroblasts, Regardless of RSV Treatment, Target Pathways Involved in Fatty Acid Oxidation*

We have shown in previous studies that treatment by resveratrol similar to that used in the present study induced a dose-dependent increase in fatty acid oxidation in *CPT2*-deficient patients [25]. Furthermore, this dose of resveratrol was shown to correct not only *CPT2* deficiency, but also other fatty acid oxidation and respiratory chain deficiencies [7,26,27]. In all these experiments, the human fibroblasts did not exhibit growth changes or increased mortality. It was thus essential to use the same treatment conditions in order to investigate the changes in microRNA expression in response to resveratrol. We also showed previously that treating mouse RAW264.7 macrophages with a 10 to 100 µM range of resveratrol concentrations decreased JunB expression as well as AP-1 activity in a dose-dependent manner (Supplementary figure 3 in the manuscript of Tili et al. [4]). Thus, we believe that treatment by 75 µM resveratrol, despite representing a high concentration, should not have caused any bias in our microarrays analyses.

The marked beneficial effects of treatment on fatty acid oxidation in the patient fibroblasts could suggest that resveratrol directly regulates *CPT2* expression. However, the precise signaling pathway(s) by which RSV targets fatty acid oxidation in the context of *CPT2*-deficiency is still a matter of debate. Several polyphenols, including RSV have been shown to increase the activity and gene expression of SIRT1 accompanied by the increase in CPT1 mRNA encoding the rate-limiting enzyme of mitochondrial fatty acid oxidation [28]. SIRT1-dependent de-acetylation of PGC-1α leads to the transcriptional co-activation of nuclear and mitochondrial genes encoding for proteins promoting mitochondrial biogenesis, oxidative phosphorylation and energy production. On the other hand, SIRT3 mediates direct activation of proteins implicated in oxidative phosphorylation, tricarboxylic acid (TCA) cycle and fatty-acid oxidation, in addition to an indirect activation of PGC-1α and AMP-activated protein kinase (AMPK). SIRT1 is required for the activation of AMPK, which enhances energy-production through glucose transport, fatty acid oxidation, or mitochondrial biogenesis [29,30]. The action of resveratrol to correct *CPT2*-deficiency might therefore involve SIRT1, however, definite evidence based on silencing SIRT1 expression in patient fibroblasts is lacking. Altogether, there is a general consensus in the literature supporting RSV effects being mediated through an AMPK/SIRT1/PGC-1α pathway [31]. It has also been suggested that RSV effects might also occur through the estrogen receptor (ER), which RSV can bind and activate [23]. We therefore examined whether the 3′-untranslated regions of genes implicated in the two above pathways contain consensus target sites for those miRNAs whose expression changed either in untreated *CPT2*-deficient fibroblasts as compared with control fibroblasts (Table 1) or in *CPT2*-deficient fibroblasts treated with RSV as compared with DMSO-treated *CPT2*-deficient fibroblasts (Table 2).

The general mechanism associated with microRNA action involves the reduced expression of their target genes. Using the Targetscan software (www.targetscan.com), we found consensus target sequences for miRNAs that were upregulated, as well as for miRNAs that were downregulated, in untreated *CPT2*-deficient fibroblasts (Table 3). In particular, *miR-483*, the miRNA that increased with the lowest *P* value in Table 1, targeted seven genes of the above pathways. It was followed by *miR-449b*, *miR-371* (with often multiple target sequences in the same transcripts), and *miR-9* (6 genes targeted) and then by *miR-539* and *miR-301* (5 and 4 genes targeted, respectively). In contrast, except for *miR-181a2,c* (seven genes targeted), the miRNAs downregulated in untreated *CPT2*-deficient fibroblasts did not target more than 4 genes (*miR-211* and *miR-126a-5*) (Table 3), and, except *miR-211*, were not among the miRNAs that changed with the lowest *P* value in Table 1. Although it would be almost impossible to measure the relative effects of these miRNAs on the expression of each of these putative target genes, the above observation suggests that the effects of upregulated miRNAs as a whole might be greater than those of downregulated miRNAs, leading to reduced levels of expression of their respective target gene and, as a consequence, reduced fatty acid oxidation.

**Table 3.** Putative targets transcripts of miRNAs of Tables 1 and 2.

| Genes | Proteins | MiRNAs * |
|---|---|---|
| | Putative target transcripts of miRNAs upregulated in *CPT2*-deficient fibroblasts: | |
| *SIRT1* | SIRT1 | 2 × 449b/539/9/651/ |
| *STK11* | LKB1 | 483/ |
| *PRKAA1* | AMPK subunit | 301/449b/539/371/9/651/ |
| *PRKAA2* | AMPK subunit | 483/301/2 × 449b/206/4 × 371/3 × 10b/3 × 651/ |
| *PRKAB1* | AMPK subunit | 483/301/9/ |
| *PPARGC1A* | PGC-1 α | 2 × 301/539/196a-2/ |
| *ALDH7A1* | PDE | 2 × 483/449b/2 × 371 2 × 10b/651/ |
| *ESR1* | ER | 483/3 × 301/2 × 206/ 371/2 × 9/196a-2/ |
| *ESRRA* | ERR α | 449b/ |
| *NRF1* | NRF1 | 483/449b/2 × 539/3 × 371/9/ |
| *NFE2L2* | NRF2 | 651/ |
| *TFAM* | TFAM | 483/206/539/2 × 371/10b/9/651/ |
| | Putative target transcripts of miRNAs downregulated in *CPT2*-deficient fibroblasts: | |
| *SIRT1* | SIRT1 | 211/141/136/181a2,c/496/126-5p/ |
| *STK11* | LKB1 | - |
| *PRKAA1* | AMPK subunit | 496/126-5p/144/ |
| *PRKAA2* | AMPK subunit | let-7a3,d,g/2 × 141/203/3 × 181a2,c/4 × 126-5p/144/9/ |
| *PRKAB1* | AMPK subunit | 2 × 141/2 × 203/181a2,c/ |
| *PPARGC1A* | PGC-1 α | let-7a3,d,g/211/141/136/203/496/3 × 126-5p/144/ |
| *ALDH7A1* | PDE | 141/2 × 136/203/ |
| *ESR1* | ER | 211/136/203/181a2,c/496/ |
| *ESRRA* | ERR α | - |
| *NRF1* | NRF1 | 2 × 211/181a2,c/ |
| *NFE2L2* | NRF2 | 181a2,c/496/ |
| *TFAM* | TFAM | 211/4 × 141/2 × 136/2 × 203/181a2,c/496/4 × 126-5p/4 × 144/ |
| | Putative target transcripts of miRNAs upregulated after RSV treatment of *CPT2*-deficient fibroblasts: | |
| *SIRT1* | SIRT1 | 199a1-5p/ |
| *STK11* | LKB1 | 199a1-5p/ |
| *PRKAA1* | AMPK subunit | - |
| *PRKAA2* | AMPK subunit | 219/2 × 299/193a/199a1-5p/2 × 337/ |
| *PRKAB1* | AMPK subunit | 193a/ |
| *PPARGC1A* | PGC-1 α | 219/193a/2 × 199a1-5p/ |
| *ALDH7A1* | PDE | 199a1-5p/2 × 337/ |
| *ESR1* | ER | 219/299/2 × 193a/337/ |
| *ESRRA* | ERR α | - |
| *NRF1* | NRF1 | 199a1-5p/ |
| *NFE2L2* | NRF2 | 337/ |
| *TFAM* | TFAM | 299-5p/2 × 193a/199a1-5p/2 × 337/ |
| | Putative target transcripts of miRNAs downregulated after RSV treatment of *CPT2*-deficient fibroblasts: | |
| *SIRT1* | SIRT1 | 181d/23a/ |
| *STK11* | LKB1 | 20b/17-5p/ |
| *PRKAA1* | AMPK subunit | 2 × 101-1/2/16-1/21/26a/ |
| *PRKAA2* | AMPK subunit | 3 × 181d/21/20b/let-7c,d/17-5/2 × 146a/376b/26a/23a/ |
| *PRKAB1* | AMPK subunit | 181d/146a/ |
| *PPARGC1A* | PGC-1α | 101-1/2/let-7d/376b/26a/2 × 23a/ |
| *ALDH7A1* | PDE | 16-1/2 × 20b/2 × 17-5p/146a/ |
| *ESR1* | ER | 181d/21/3 × 20b/3 × 17-5p/146a/2 × 26a/2 × 103-1/23a/ |
| *ESRRA* | ERRα | 16-1/103-1/423/ |
| *NRF1* | NRF1 | 181d/2 × 21/ |
| *NFE2L2* | NRF2 | 181d/103-1/ |
| *TFAM* | TFAM | 181d/3 × 20b/3 × 17-5p/2 × 376b/26a/23a/ |

* MiRNAs are given in the same order as in Tables 1 and 2. The sign "x" indicates the number of putative target sequence for a given miRNA. Target transcripts were identified using the Targetscan software (www.targetscan.com). Numbers in front of miRNAs indicate that more than one consensus target site for this miRNA is present in the 3′-untranslated region of the transcript. For instance, *SIRT1* 3-untranslated region contains two consensus target sequences for *miR-449b*.

Consensus target sites in the 3′-untranslated regions of the same genes for miRNAs that were upregulated or downregulated following RSV treatment of *CPT2*-deficient fibroblasts (Table 2) show a similar distribution. While seven of the above genes are potential targets of *miR-199a1-5p* and 5 are potential targets of *miR-337* (both upregulated following RSV treatment). Table 3 also shows that seven genes of the above pathways are also putative targets of *miR-181d*, and five of them are also targets of *miR-20b*, *miR-17-5p*, *miR-26a*, and *miR-23a* (all of these miRNAs being downregulated following RSV treatment). Given that the 3′-untranslated regions of several of these genes contains more than one consensus target site for *miR-20b*, *miR-17-5p*, or other miRNAs, it is likely that RSV may change the expression of genes that encode factors implicated in these two pathways, and therefore fatty acid oxidation, through both miRNA-dependent and miRNA-independent mechanisms. Finally, as miRNAs that changed following RSV treatment of control fibroblasts are fairly different from those that changed following RSV treatment of *CPT2*-deficient fibroblasts (Table 2), it is possible that RSV effects on miRNA expression might depend on other factors—factors most likely implicated in modulating the activity of fatty acid oxidation in mitochondria.

In conclusion, this paper is the first to report changes in microRNA expression associated with *CPT2*-deficiency in human fibroblasts and sheds some new light on potential beneficial effects of RSV through modifying miRNA expression. In particular, it appears likely that changes in miRNA levels in *CPT2*-deficient cells might, at least in part, be involved in abnormal fatty acid oxidation. The emerging role of microRNAs in lipid metabolism has been emphasized in recent reviews [32] reporting that miRNAs are critical regulators of lipid synthesis, fatty acid ß-oxidation, and lipoprotein metabolism. Changes in the expression of crucial miRNAs can impact gene regulatory network, driving to metabolic syndrome and its related pathologies. This review introduced epigenetic and transcriptional regulation of miRNA expression, especially *miR-378* (controlling *FABP7*, *IGFBP3*, *PDCD4*, and *PPAR-α* mRNA expression) and *miR-21* (controlling *CRAT*, *MED13*, *ERRγ*, *GABP1*, and *IGF1α* mRNA expression). In this paper, we found that *miR-378* was downregulated by RSV in control fibroblasts and that *miR-21* was downregulated by RSV in *CPT2*-deficient fibroblasts. In addition, *miR-21* could putatively target *NRF1* mRNAs, which encode a transcription factor implicated in respiratory control (Table 3).

Further studies will be required to assess the impact of these changes in miRNA expression on RSV-induced stimulation of mitochondrial fatty acid oxidation in *CPT2*-deficient cells and to identify factors that mediate these RSV effects.

### 3. Materials and Methods

#### 3.1. Primary Human Fibroblasts and Cell Treatments

*CPT2*-deficient and control human skin fibroblasts used in this study have been described previously [5]. Point mutations and genotypes of the cells are the following: nucleotides changes, c.338C > T and c.371G > A, and consequently amino acid changes, S113L and R124Q. For cell treatment, a medium of Ham's F10 media containing glutamine, 12% fetal bovine serum, 100 U/mL penicillin, and 0.1 mg/mL streptomycin was removed and replaced with fresh medium containing either the vehicle, DMSO 0.1%, or 75 µM resveratrol (RSV). Cells were subsequently cultivated for 72 h before RNA extraction.

#### 3.2. RNA Extraction, Purification, and Micro RNAs Screening and Analysis

RNAs extracted with TRIzol (Invitrogen) were subsequently subjected to DNase digestion (Turbo-DNase-Ambion), as previously described [2]. MiRNA microarrays were analyzed as previously described [2]. Four independent repeats (i.e., cell cultures) were used for each group.

#### 3.3. RNA Labeling and Micro-Arrays

Five micrograms of total RNA were labeled by reverse transcription at 37 °C for 90 min using a biotin-labeled rand-octomer oligo primer. An RT reaction mix was further denatured in 0.5 N NaOH/1 mM EDTA at 65 °C for 15 min and neutralized by 1 M Tris HCl pH 7.6. Biotin signal was detected with an Alexa 647-Streptavidin conjugate. Chips were hybridized on Tecan HS 4800 hybridization station. Chips were pre-hybridized at 25 °C for 30 min in the buffer: 6× SSPE/30% formamide/1× Denhardt's solution. Chips were further hybridized with a labeled target in 6× SSPE/30% formamide at 25 °C for 18 h. Hybridization and post-hybridization washing was conducted in 0.75× TNT (Tris, sodium, Tween 20) at 37 °C for 40 min. The chips were stained by streptavidin–alexa647 (1:500) dilution in TNT for 30 min. Post-staining washing was conducted in 1× TNT FOR 40 min.

## 4. Conclusions

Taking account that RSV enhances residual *CPT2* activities in human fibroblasts derived from *CPT2*-deficient patients and restores normal fatty acid oxidation rates, we now report changes in miRNA expression linked to *CPT2*-deficiency, and we identify miRNAs whose expression changed following RSV treatment of control or *CPT2*-deficient fibroblasts isolated from patients. Our findings suggest that RSV consumption might exert beneficiary effects in patients with *CPT2*-deficiency through miRNAs expression modulation.

**Acknowledgments:** This work was supported by ANR (grant ANR-09-GENO-024-01). We thank Jean-Jacques Michaille from Université de Bourgogne-Franche Comté, Laboratoire Bio-PeroxIL, Dijon, France, for his review and advice.

**Author Contributions:** N.L. and D.D. conceived and designed the experiments; V.A. performed the experiments; V.A., D.D., M.C.-M., and N.L. analyzed the data; V.A., F.D., and J.B. contributed reagents/materials/analysis tools; N.L. wrote the paper.

**Conflicts of Interest:** The authors declare no conflict of interest.

## References

1. Latruffe, N.; Rifler, J.-P. Bioactive polyphenols from grape and wine. *Curr. Pharm. Des.* **2013**, *19*, 6053–6063. [CrossRef] [PubMed]
2. Jeandet, P.; Delaunois, B.; Conreux, A.; Donnez, D.; Nuzzo, V.; Cordelier, S.; Clément, C.; Courot, E. Biosynthesis, metabolism, molecular engineering, and biological functions of stilbene phytoalexins in plants. *BioFactors* **2010**, *36*, 331–341. [CrossRef] [PubMed]
3. Jeandet, P.; Hébrard, C.; Deville, M.A.; Cordelier, S.; Dorey, S.; Aziz, A.; Crouzet, J. Deciphering the role of phytoalexins in plant-microorganism interactions and human health. *Molecules* **2014**, *19*, 18033–18056. [CrossRef] [PubMed]
4. Tili, E.; Michaille, J.J.; Adair, B.; Alder, H.; Limagne, E.; Taccioli, C.; Ferracin, M.; Delmas, D.; Latruffe, N.; Croce, C.M. Resveratrol decreases the levels of miR-155 by upregulating miR-663, a microRNA targeting JunB and JunD. *Carcinogenesis* **2010**, *31*, 1561–1566. [CrossRef] [PubMed]
5. Limagne, E.; Lançon, A.; Delmas, D.; Cherkaoui-Malki, M.; Latruffe, N. Resveratrol Interferes with IL1-β-Induced Pro-Inflammatory Paracrine Interaction between Primary Chondrocytes and Macrophages. *Nutrients* **2016**, *8*, 280. [CrossRef] [PubMed]
6. Lin, H.Y.; Delmas, D.; Vang, O.; Hsieh, T.C.; Lin, S.; Cheng, G.Y.; Chiang, H.L.; Chen, C.E.; Tang, H.Y.; Crawford, D.R.; et al. Mechanisms of ceramide-induced COX-2-dependent apoptosis in human ovarian cancer OVCAR-3 cells partially overlapped with resveratrol. *J. Cell. Biochem.* **2013**, *114*, 1940–1954. [CrossRef] [PubMed]
7. Bastin, J.; Lopes-Costa, A.; Djouadi, F. Exposure to resveratrol triggers pharmacological correction of fatty acid utilization in human fatty acid oxidation-deficient fibroblasts. *Hum. Mol. Genet.* **2011**, *20*, 2048–2057. [CrossRef] [PubMed]
8. Karius, T.; Schnekenburger, M.; Dicato, M. MicroRNAs in cancer management and their modulation by dietary agents. *Biochem. Pharmacol.* **2012**, *83*, 1591–1601. [CrossRef] [PubMed]
9. Li, Y.; Kong, D.; Wang, Z.; Sarkar, F.H. Regulation of microRNAs by natural agents: An emerging field in chemoprevention and chemotherapy research. *Pharm. Res.* **2010**, *27*, 1027–1041. [CrossRef] [PubMed]
10. Milenkovic, D.; Jude, B.; Morand, C. MiRNA as molecular target of polyphenols underlying their biological effects. *Free Radic. Biol. Med.* **2013**, *64*, 40–51. [CrossRef] [PubMed]
11. Lançon, A.; Kaminski, J.; Tili, E.; Michaille, J.J.; Latruffe, N. Control of MicroRNA expression as a new way for resveratrol to deliver its beneficial effects. *J. Agric. Food Chem.* **2012**, *60*, 8783–8789. [CrossRef] [PubMed]
12. Lançon, A.; Michaille, J.J.; Latruffe, N. Effects of dietary phytophenols on the expression of microRNAs involved in mammalian cell homeostasis. *J. Sci. Food Agric.* **2013**, *93*, 3155–3164. [CrossRef] [PubMed]

13. Latruffe, N.; Lançon, A.; Frazzi, R.; Aires, V.; Delmas, D.; Michaille, J.J.; Djouadi, F.; Bastin, J.; Cherkaoui, J.; Malki, M. Exploring new ways of regulation by resveratrol involving miRNAs, with emphasis on inflammation. *Ann. N. Y. Acad. Sci.* **2015**, *1348*, 97–106. [CrossRef] [PubMed]

14. Tili, E.; Michaille, J.J.; Alder, H.; Volinia, S.; Delmas, D.; Latruffe, N.; Croce, C.M. Resveratrol modulates the levels of microRNAs targeting genes encoding tumor-suppressors and effectors of TGFβ signaling pathway in SW480 cells. *Biochem. Pharmacol.* **2010**, *80*, 2057–2065. [CrossRef] [PubMed]

15. Kumar, A.; Rimando, A.M.; Levenson, A.S. Resveratrol and pterostilbene as a microRNA-mediated chemopreventive and therapeutic strategy in prostate cancer. *Ann. N. Y. Acad. Sci.* **2017**. [CrossRef] [PubMed]

16. Wu, F.; Cui, L. Resveratrol suppresses melanoma by inhibiting NF-κB/miR-221 and inducing TFG expression. *Arch. Dermatol. Res.* **2017**. [CrossRef] [PubMed]

17. Venkatadri, R.; Muni, T.; Iyer, A.K.; Yakisich, J.S.; Azad, N. Role of apoptosis-related miRNAs in resveratrol-induced breast cancer cell death. *Cell Death Dis.* **2016**, *7*, e2104. [CrossRef] [PubMed]

18. Bai, T.; Dong, D.S.; Pei, L. Synergistic antitumor activity of resveratrol and miR-200c in human lung cancer. *Oncol. Rep.* **2014**, *31*, 2293–2297. [CrossRef] [PubMed]

19. Gracia, A.; Miranda, J.; Fernández-Quintela, A.; Eseberri, I.; Garcia-Lacarte, M.; Milagro, F.I.; Martínez, J.A.; Aguirre, L.; Portillo, M.P. Involvement of miR-539-5p in the inhibition of de novo lipogenesis induced by resveratrol in white adipose tissue. *Food Funct.* **2016**, *7*, 1680–1688. [CrossRef] [PubMed]

20. Gracia, A.; Fernández-Quintela, A.; Miranda, J.; Eseberri, I.; González, M.; Portillo, M.P. Are miRNA-103, miRNA-107 and miRNA-122 Involved in the Prevention of Liver Steatosis Induced by Resveratrol? *Nutrients* **2017**, *9*, 360. [CrossRef] [PubMed]

21. Ma, C.; Wang, Y.; Shen, A.; Cai, W. Resveratrol upregulates SOCS1 production by lipopolysaccharide-stimulated RAW264.7 macrophages by inhibiting miR-155. *Int. J. Mol. Med.* **2017**, *39*, 231–237. [CrossRef] [PubMed]

22. Tili, E.; Michaille, J.J. Promiscuous Effects of Some Phenolic Natural Products on Inflammation at Least in Part Arise from Their Ability to Modulate the Expression of Global Regulators, Namely microRNAs. *Molecules* **2016**, *21*, 1263. [CrossRef] [PubMed]

23. Wang, Z.H.; Zhang, J.L.; Duan, Y.L.; Zhang, Q.S.; Li, G.F.; Zheng, D.L. MicroRNA-214 participates in the neuroprotective effect of Resveratrol via inhibiting α-synuclein expression in MPTP-induced Parkinson's disease mouse. *Biomed. Pharmacother.* **2015**, *74*, 252–256. [CrossRef] [PubMed]

24. Guo, D.W.; Han, Y.X.; Cong, L.; Liang, D.; Tu, G.J. Resveratrol prevents osteoporosis in ovariectomized rats by regulating microRNA-338-3p. *Mol. Med. Rep.* **2015**, *12*, 2098–2106. [CrossRef] [PubMed]

25. Aires, V.; Delmas, D.; Le Bachelier, C.; Latruffe, N.; Schlemmer, D.; Benoit, J.F.; Djouadi, F.; Bastin, J. Stilbenes and resveratrol metabolites improve mitochondrial fatty acid oxidation defects in human fibroblasts. *Orphanet J. Rare Dis.* **2014**, *9*, 79. [CrossRef] [PubMed]

26. Lopes Costa, A.; Le Bachelier, C.; Mathieu, L.; Rotig, A.; Boneh, A.; De Lonlay, P.; Tarnopolsky, M.A.; Thorburn, D.R.; Bastin, J.; Djouadi, F. Beneficial effects of resveratrol on respiratory chain defects in patients' fibroblasts involve estrogen receptor and estrogen-related receptor alpha signaling. *Hum. Mol. Genet.* **2014**, *23*, 2106–2119. [CrossRef] [PubMed]

27. Mathieu, L.; Costa, A.L.; Le Bachelier, C.; Slama, A.; Lebre, A.S.; Taylor, R.W.; Bastin, J.; Djouadi, F. Resveratrol attenuates oxidative stress in mitochondrial Complex I deficiency: Involvement of SIRT3. *Free Radical Biol. Med.* **2016**, *96*, 190–198. [CrossRef] [PubMed]

28. Brenmoehl, J.; Hoeflich, A. Dual control of mitochondrial biogenesis by sirtuin 1 and sirtuin 3. *Mitochondrion* **2013**, *13*, 755–761. [CrossRef] [PubMed]

29. Rodriguez-Ramiro, I.; Vauzour, D.; Minihane, A.M. Polyphenols and non-alcoholic fatty liver disease: Impact and mechanisms. *Proc. Nutr. Soc.* **2016**, *75*, 47–60. [CrossRef] [PubMed]

30. Price, N.L.; Gomes, A.P.; Ling, A.J.; Duarte, F.V.; Martin-Montalvo, A.; North, B.J.; Agarwal, B.; Ye, L.; Ramadori, G.; Teodoro, J.S.; et al. SIRT1 is required for AMPK activation and the beneficial effects of resveratrol on mitochondrial function. *Cell Metab.* **2012**, *15*, 675–690. [CrossRef] [PubMed]

31.   Bastin, J.; Djouadi, F. Resveratrol and Myopathy. *Nutrients* **2016**, *8*, 254. [CrossRef] [PubMed]
32.   Yanga, Z.; Cappelloa, T.; Wang, L. Emerging role of microRNAs in lipid metabolism. *Acta Pharm. Sin. B* **2015**, *5*, 145–150. [CrossRef] [PubMed]

**Sample Availability:** Resveratrol commercially available.

![molecules] *molecules*

MDPI

Article

# The Effect of Resveratrol on Cell Viability in the Burkitt's Lymphoma Cell Line Ramos

Paola Jara [1], Johana Spies [1], Constanza Cárcamo [1], Yennyfer Arancibia [1], Gabriela Vargas [1], Carolina Martin [2], Mónica Salas [1], Carola Otth [3,4] and Angara Zambrano [1,4,*]

[1]   Instituto de Bioquímica y Microbiología, Facultad de Ciencias, Universidad Austral de Chile, Valdivia 5090000, Chile; paolad2864@gmail.com (P.J.); johana.spies@gmail.com (J.S.); constanza.carcamoz@gmail.com (C.C.); yennyfer.arancibia@gmail.com (Y.A.); gabs.vargas@gmail.com (G.V.); monsalas@gmail.com (M.S.)
[2]   Escuela de Tecnología Medica, Universidad Austral de Chile, Sede Puerto Montt 5480000, Chile; carolina.martin@uach.cl
[3]   Instituto de Microbiología Clínica, Facultad de Medicina, Universidad Austral de Chile, Valdivia 5090000, Chile; cotth@uach.cl
[4]   Center for Interdisciplinary Studies on the Nervous System (CISNe), Universidad Austral de Chile, Valdivia 5090000, Chile
*   Correspondence: angarahzambrano@gmail.com; Tel.: +56-63-222-1332

Received: 17 November 2017; Accepted: 13 December 2017; Published: 21 December 2017

**Abstract:** Resveratrol is a polyphenolic natural compound produced by a variety of crops. Currently, resveratrol is considered a multi-target anti-cancer agent with pleiotropic activity, including the ability to prevent the proliferation of malignant cells by inhibiting angiogenesis and curtailing invasive and metastatic factors in many cancer models. However, the molecular mechanisms mediating resveratrol-specific effects on lymphoma cells remain unknown. To begin tackling this question, we treated the Burkitt's lymphoma cell line Ramos with resveratrol and assessed cell survival and gene expression. Our results suggest that resveratrol shows a significant anti-proliferative and pro-apoptotic activity on Ramos cells, inducing the DNA damage response, DNA repairing, and modulating the expression of several genes that regulate the apoptotic process and their proliferative activity.

**Keywords:** resveratrol; cell viability; Ramos cells; DNA damage; DNA repair

## 1. Introduction

Resveratrol (3,5,4'-trihydroxystilbebe or RSV) is a polyphenolic natural product generated by a wide variety of crops, including grapes, peanuts, plums and berries, and some derived products such as red wine and fruit juice [1–3].

Accumulating evidence shows that RSV consumption may have many beneficial properties to human health, acting as an antioxidant, anti-aging, immunomodulator, anti-inflammatory and cardioprotective agent, reducing the risk of coronary artery disease. It has also been postulated to be a mimetic factor for the effects of caloric restriction on metabolism, including the enhancement of insulin sensitivity [4–7].

In addition, RSV has prompted a great interest in the biomedical industry mainly due to its anti-carcinogenic activity, whereby it can prevent the proliferation of cancer cells, inhibit angiogenesis and reduce invasive and metastatic factors [8–10]. Furthermore, some studies have shown that RSV inhibits tumor initiation, promotion, and progression [11].

The antiproliferative and proapoptotic activity of RSV has been reported in different human cancer cell lines that include colon, prostate, and breast cancer as well as leukemia [12–15]. Although there is abundant data regarding the chemopreventive role of RSV in many cancers, relatively little information exists on the antiproliferative activity of RSV in human lymphoma cells [16–18].

Although RSV is a promising multi-target anticancer agent with pleiotropic activities, its specific mechanisms of action remain unclear [19]. In general, RSV targets a great number of intracellular molecules implicated in apoptosis induction. For example, RSV induces cell death by altering proteins of the Bcl-2 family. Specifically, it is able to upregulate the expression of pro-apoptotic proteins such as Bak, Bax, NOXA and PUMA, while lowering the expression of anti-apoptotic members such as Bcl-2, Mcl-1 and Bcl-XL [20–22]. Moreover, RSV increases cellular apoptosis by modulating the extrinsic pathway that relies on binding of ligands such as FASL or TRAIL [6,23,24], whereas it also activates the intrinsic apoptotic pathway by inducing the mitochondrial release of cytochrome C, the generation of ROS, and the modulation of the p53 pathway [20,25–27].

Interestingly, RSV binds and activates the serine protein kinase ataxia telangiectasia mutated (ATM), inducing autophosphorylation and substrate phosphorylation [28]. ATM activates signaling checkpoints upon genotoxic stress, especially DNA double strand breaks. Indeed, the loss of ATM activity has been observed in various tumor types. RSV is able to induce extensive DNA damage and more specifically DNA double strand breaks in human colon carcinoma cells [29,30]. Also, RSV works on blocking topoisomerase (TOPO) activity. Indeed, RSV induced a delay in S-phase progression with the concomitant phosphorylation of the histone H2AX (H2A histone family, member X) [29,31]. Specifically, in B-cell lymphocytes, DNA double-strand breaks (DSBs) are generated in all developing lymphocytes, a process that is essential for normal lymphocyte development [32]. DSB activates an ATM-dependent signaling pathway that leads to phosphorylation and inactivation of the transcriptional coactivator CRTC2 [33]. CRTC2, in turn, regulates many target genes, some of which are involved in processes that modulate GC B-cell proliferation, self-renewal and also inhibit plasma cell differentiation [34].

In the case of lymphomas, Jazirehi et al., reported that RSV downregulates two anti-apoptotic proteins, Bcl-XL and Mcl-1, whereas it upregulates the pro-apoptotic proteins Bax and Apaf-1 in the Burkitt's lymphoma cell line Ramos [35]. RSV also promotes cell growth inhibition in NALM-6 cells [36] and blunts PI-3K signaling and glucose metabolism in germinal center-like LY1 and LY18 human diffuse large B-cell lymphomas (DLBCLs) [17]. However, the molecular mechanisms of action of RSV in lymphoma cells remain largely unknown.

In this study, we analyzed the effect of RSV on cell viability in the Burkitt's lymphoma cell line Ramos. We demonstrate that RSV has anti-proliferative and pro-apoptotic effects on these cells. We show that RSV induces the DNA damage response; activation of DNA repair; and modulates the expression of genes with key roles regulating the apoptotic process and the proliferative activity in this model cell line.

## 2. Results

*2.1. RSV Induces Decrease in Cell Viability in Ramos Cells*

Even though many beneficial properties of RSV have been reported, including anti-proliferative and pro-apoptotic activity in different human cancer cell lines, its effects in human lymphoma cells are poorly described.

To determine whether RSV has an antiproliferative effect on Ramos cells, we incubated this cell line with different concentrations of RSV for 24 h and 48 h, and determined viability using the MTT assay. Cell viability decreased in a concentration-dependent manner in response to RSV, by 25% with 50 μM RSV up to 60% using 150 μM RSV at 24 h, and by 30% with 20 μM RSV up to 60% using 150 μM RSV at 48 h (Figure 1A). Similar results were obtained independently using the Trypan Blue exclusion assay to determine cell survival, showing a maximum antiproliferative activity with 150 μM RSV at 24 h and 48 h of treatment (Figure 1B). These data demonstrate that RSV effectively reduces cell proliferation and viability in Ramos cells. For subsequent experiments we chose concentrations between 50 μM and 100 μM, which decrease cell viability by 50% after 24 h of treatment.

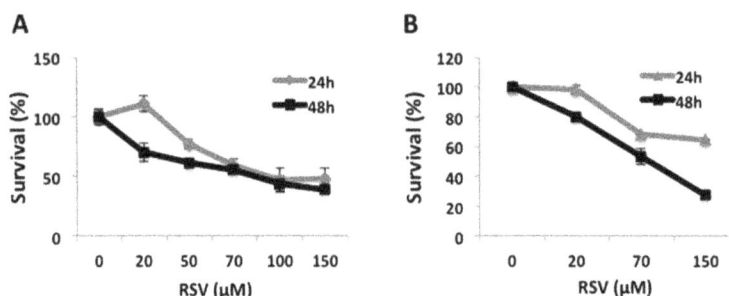

**Figure 1.** Effect of Resveratrol on cell viability in Ramos cells. Ramos cells were treated with resveratrol (RSV) at the indicated concentrations. (**A**) Cell viability was evaluated by MTT assay during 24 h and 48 h of treatment with resveratrol; (**B**) Cell viability was evaluated with the Trypan Blue exclusion assay during 24 h and 48 h of treatment with resveratrol. Data are presented as mean ± SD, for three independent experiments. * $p < 0.05$, ** $p < 0.01$, compared to control cells.

## 2.2. RSV Induce Apoptotic Cell Death in Ramos Cells

RSV exhibits many different mechanisms of action and apoptosis-related targets in various models. To determine whether RSV induces the apoptotic process in Ramos cells, we performed an immunoblotting assay against some important apoptotic markers, specifically, active-caspase 3 and fragmented PARP proteins after different RSV treatment. Antibodies against tubulin were used as loading control. Indeed, RSV triggered a significant increase in active-caspase 3 and cleaved-PARP using 70 μM and 100 μM for 24 h (Figure 2A,B). Treating Ramos cells with 70 μM RSV for different time periods also revealed a significant increase in active-caspase 3 and cleaved-PARP over 3h treatment (Figure 2C,D).

Interestingly, it has been reported that RSV is able to upregulate the expression of several pro-apoptotic mediators [21,22] therefore we treated Ramos cells with 70 μM RSV for 1 h and 3 h, and determined the mRNA levels for NOXA (Phorbol-12-myristate-13-acetate-induced protein 1), Fas (Tumor necrosis factor receptor superfamily member 6), and PUMA (p53 up-regulated modulator of apoptosis) by means of RT-qPCR. Our results shown that RSV induce a significant increase in the mRNA levels of NOXA and PUMA, but has no effect on the expression of Fas (Figure 2E). These results indicate that RSV can activate caspase 3 inducing the fragmentation of its downstream target, and upregulate the expression of a subset of genes know to be linked to apoptotic events.

## 2.3. RSV Induces DNA Damage and DNA Repair in Ramos Cells

Many reports suggest that RSV can induce extensive DNA damage, specifically DNA double strand breaks (DSB), in some tumor cells lines [29,30]. To determine whether RSV is able to induce DNA damage in Ramos cells, we treated cells with different concentrations of RSV for 24 h and tested the phosphorylation levels of ATM (ataxia-telangiectasia mutated kinase) and BRCA1 (breast cancer type 1 susceptibility protein), two proteins associated with the activation of the DNA damage response. Our results show that treatment with RSV over 50 μM induces a significant increase in p-ATM and p-BRCA1 (Figure 3A,B). Also, treating Ramos cells with 70 μM RSV for different time periods revealed a significant increase in ATM and BRCA1 phosphorylation over 3 h treatment (Figure 3C,D). We also treated Ramos cells with two different concentrations of RSV (50 μM and 100 μM) to detect the presence of γ-H2AX. Our results show that treatment with RSV induces a significant increase in γ-H2AX using both concentrations (Figure 3E,F). These data suggest that the induction of DNA damage might be one of the molecular mechanisms involved in the loss of cell viability caused by RSV in Ramos cells.

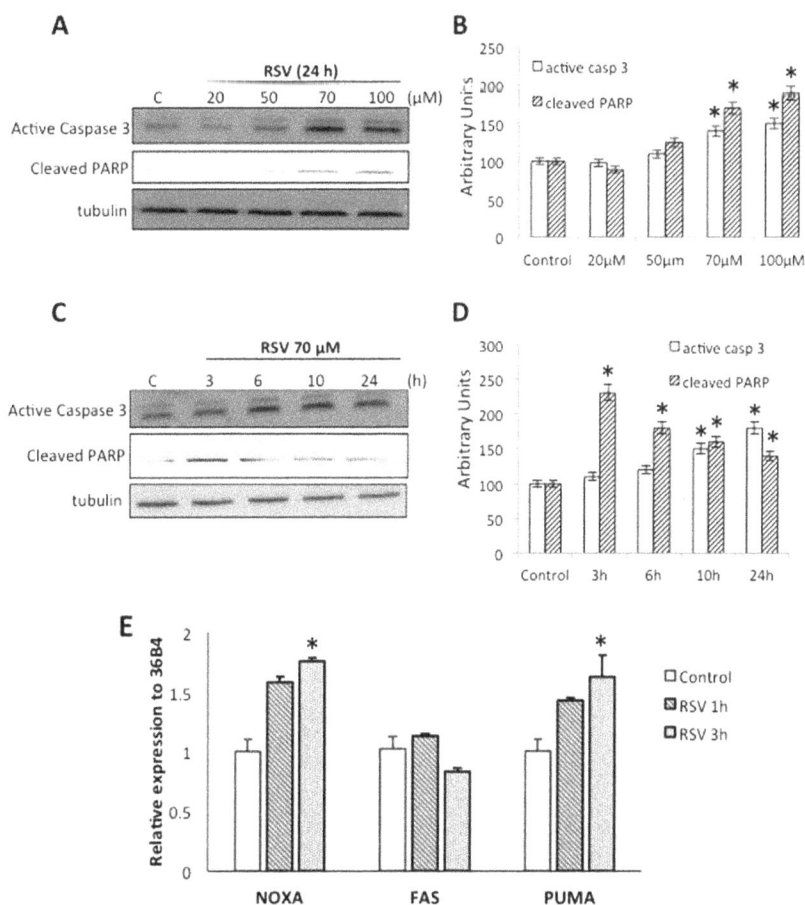

**Figure 2.** Effect of Resveratrol on apoptosis markers. (**A**) Western blot analysis of proteins from cells treated with different concentrations of resveratrol (RSV) for 24 h using anti-active caspase 3 and anti-cleaved PARP (poly ADP-ribose polymerase) antibodies. Antibodies against total tubulin were used as loading control. (**B**) Quantification of the proteins shown in panel A is represented as a column plot with error bars. (**C**) Western blot analysis of proteins from cells with 70 µM RSV for different time periods using anti-active caspase3 and anti-cleaved PARP antibodies. Antibodies against total tubulin were used as loading control. (**D**) Quantification of the proteins shown in panel C is represented as a column plot with error bars. (**E**) The mRNA levels of apoptotic-related genes were measured by quantitative reverse transcriptase polymerase chain reaction (NOXA; Phorbol-12-myristate-13-acetate-induced protein 1, FAS; Tumor necrosis factor receptor superfamily member 6 and PUMA; p53 up-regulated modulator of apoptosis). Cells were treated with 70 µM RSV for 1 h or 3 h. Results are representative of three independent experiments, * $p < 0.05$ compared to control cells.

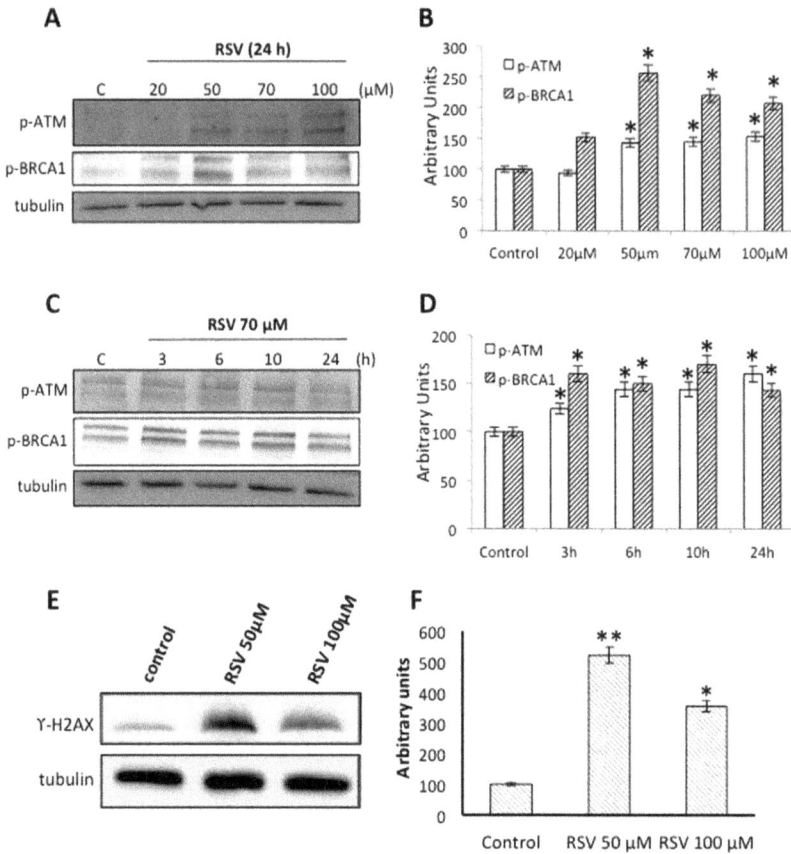

**Figure 3.** Effect of Resveratrol on the activation of the DNA damage response. (**A**) Western blot analysis of proteins from cells treated with different concentrations of resveratrol (RSV) for 24 h using anti-p-ATM (ataxia-telangiectasia mutated kinase) and anti-p-BRCA1 (breast cancer type 1 susceptibility protein) antibodies. Antibodies against total tubulin were used as loading control. (**B**) Quantification of the proteins shown in panel A is represented as a column plot with error bars. (**C**) Western blot analysis of proteins from cells treated with 70 µM RSV for different time periods using anti-p-ATM and anti-p-BRCA1 antibodies. Antibodies against total tubulin were used as loading control. (**D**) Quantification of the proteins shown in panel C is represented as a column plot with error bars. (**E**) Western blot analysis of proteins from cells treated with 50 µM or 100 µM RSV for 24 h using anti-γ-H2AX antibodies. Antibodies against total tubulin were used as loading control. (**F**) Quantification of the proteins shown in panel E is represented as a column plot with error bars. Results are representative of three independent experiments, $* p < 0.05$ compared to control cells.

Non-homologous end-joining (NHEJ) and homology-directed repair (HDR) are the main pathways in all organisms for repairing DBS [37,38]. To determine whether RSV is able to induce the repair mechanisms in Ramos cells, we treated cells with different concentrations of RSV for 24 h and tested the protein levels of Rad50, Mre11 and p-p95/NBS1, proteins that represent the primary DSB sensor. Our results show that treatment with RSV induces a significant increase in all these proteins, suggesting that RSV induces their expression (Figure 3A,B). We also treated cells with different concentrations of RSV for 24 h and tested the protein levels of DNA-PKcs and KU80, essential proteins in the initiation of NHEJ repair pathway. Our results show that treatment with RSV induces a significant increase in both proteins, suggesting that RSV induces the activation of this pathway (Figure 4).

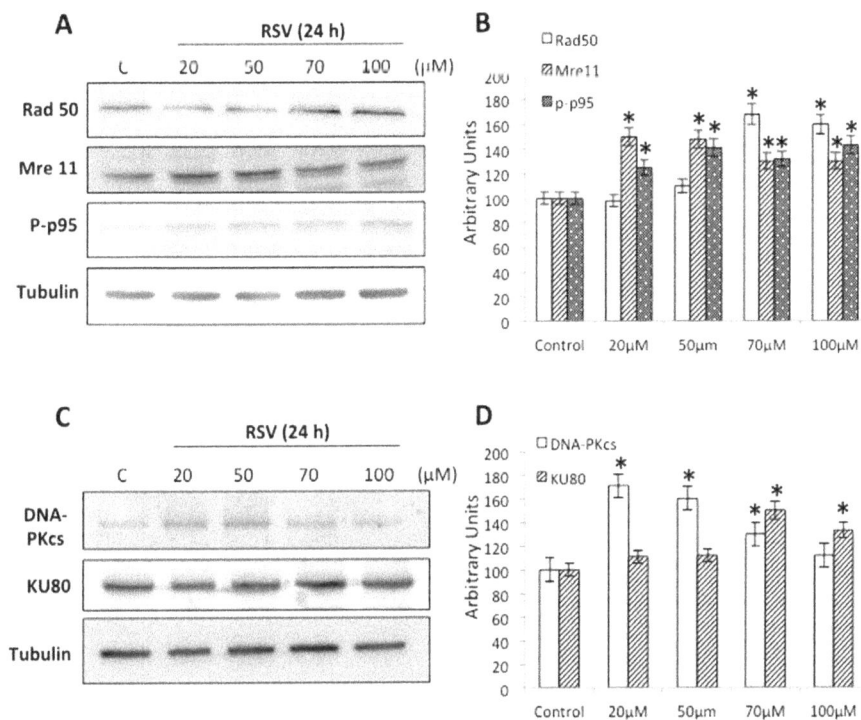

**Figure 4.** Effect of Resveratrol on the activation of double-strand break repair pathways. (**A**) Western blot analysis of proteins from cells treated with different concentrations of RSV for 24 h using anti-Rad50, anti-Mre11 and anti-p-p95/NBS1 antibodies. Antibodies against total tubulin were used as loading control. (**B**) Quantification of the proteins shown in panel A is represented as a column plot with error bars. (**C**) Western blot analysis of proteins from cells treated with different concentrations of RSV for 24 h using anti-DNA-PKcs and anti-KU80 antibodies. Antibodies against total tubulin were used as loading control. (**D**) Quantification of the proteins shown in panel C is represented as a column plot with error bars. Results are representative of three independent experiments, * $p < 0.05$ compared to control cells.

## 2.4. RSV Regulates Gene Expression Related to Proliferation and B Cell Differentiation

DNA double-strand breaks (DSB) are generated in all developing lymphocytes, and it is an essential event for normal lymphocyte development [32]. This process is led by the inactivation of the transcriptional coactivator CRTC2 [33,34], which controls many direct target genes, some of them involved in processes that regulate GC B-cell proliferation, self-renewal, and inhibit plasma cell differentiation.

To determine whether RSV regulates B cell differentiation by reducing the expression of genes involved in B-cell proliferation, we measured three transcripts that are widely studied as reporters of this process: TCL-1 (T-cell leukemia/lymphoma protein 1A), Bach2 and Myc. We treated cells with 50 μM or 100 μM RSV for 24 h and then analyzed mRNA expression using RT-qPCR. We observed a significant decrease in the mRNA levels of TCL-1, Myc and Bach2 (Figure 5). Analogous results were obtained with Etoposide (Eto), a positive control for DNA damage that is capable of downregulating CRTC2-target genes [33].

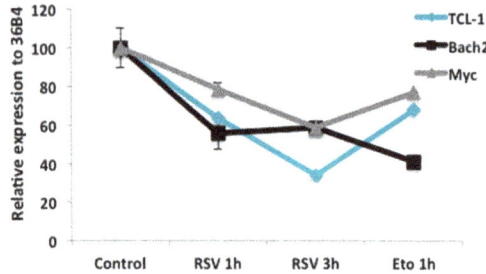

**Figure 5.** Effect of Resveratrol on the expression of proliferative genes essential in lymphomagenesis. The mRNA levels of proliferative were measured by quantitative reverse transcriptase polymerase chain reaction. Cells were treated with 100 μM resveratrol (RSV) for 1 h or 3 h. The graph represents the relative expression of TCL-1 (T-cell leukemia/lymphoma protein 1A), Bach2 and Myc genes. Results are representative of three independent experiments, * $p < 0.05$.

## 3. Discussion

Resveratrol is a multi-target anti-cancer agent with pleiotropic activity. In this study, we assessed the effects of RSV on cell death in the Burkitt's lymphoma cell line Ramos. We show that, as in other model cell lines, RSV has anti-proliferative and pro-apoptotic activities in these cells. Furthermore, our results demonstrate that RSV induces the DNA damage response, DNA repairing and modulates the expression of genes with key roles in the apoptotic process and the proliferative activity of Ramos cells.

Resveratrol is a polyphenolic natural product generated by a wide variety of plants. It has prompted great interest in the biomedical community mainly due to its anti-carcinogenic properties. Resveratrol prevents the proliferation of cancer cells by inducing inhibition of tumor initiation, promotion, and progression [10,11]. Although there is accumulating evidence on the chemopreventive role of RSV in many cancers, further data in relation to human lymphoma cells is still lacking [12–16,39].

Here we show that RSV reduces proliferation and cell viability in Ramos cells (Figure 1), concomitant with the induction of caspase-3 and PARP fragmentation (Figure 2), which are normally associate with apoptotic cell death. These results are consistent with previous studies demonstrating that RSV promotes cell growth inhibition in NALM-6 cells [36]; induces cytotoxicity in human Burkitt's lymphoma, Raji, and Daudi cell lines [40]; and also induces cell-cycle arrest in germinal center-like LY1 and LY18 human diffuse large B-cell lymphomas (DLBCLs) [15,17]. Also, in mantle cell lymphomas (MCL), specifically in jeko-1 cell line, resveratrol induces apoptosis modulating several key molecules involve in cell cycle and apoptosis [41].

RSV exhibits different mechanisms of action implicated in cell cycle control and apoptosis induction. For example, it interacts directly with the human GLUT1 hexose transporter thus inhibiting the transport of hexoses [42]. In addition, RSV treatment results in decreased glycolytic flux, with a parallel reduction in the expression of several mRNAs encoding rate-limiting glycolytic enzymes [39]. Also, RSV induces cell death by altering the expression Bcl-2 family proteins [20], upregulating the expression of Bax, Bak, PUMA, and NOXA, whereas decreasing the anti-apoptotic members Bcl-2, Mcl-1, and Bcl-XL [21,22]. We observed similar results in Ramos cells, where RSV upregulated the expression of genes associated with apoptotic events (PUMA and NOXA). We also analyzed the expression of Fas receptor, where, in contrast to previous studies in ALCL cell lines, where RSV increased cellular apoptosis by enhancing Fas/CD95 expression in a dose-dependent manner [23,24], we did not find major changes in the expression of Fas receptors in Ramos cells.

Another relevant molecular pathway induced by RSV is DNA damage. Our results show a robust activation of the DNA damage response, with increased phosphorylation of the damage sensor ATM (Figure 3A,B), increased p-BRCA1 (Figure 3C,D), and the concomitant phosphorylation of γ-H2AX (Figure 3E,F). ATM activates cell cycle checkpoint signaling upon genotoxic stress, especially DNA double strand breaks. Interestingly, loss of ATM activity is a hallmark of various tumor types and RSV can bind to ATM increasing autophosphorylation and substrate phosphorylation [28,43]. BRCA1 is an essential tumor suppressor involved in DSB repair, preserving genome stability. Ours results are consistent with published evidence suggesting that RSV induces extensive DNA damage, more specifically, DNA double strand breaks in human colon carcinoma cells [29,30]. Interestingly, a well-documented anticancer mechanism of RSV is through the inhibition of topoisomerase (TOPO) activity. Indeed, RSV induced a delay in S-phase progression with the concomitant phosphorylation of histone H2AX [29,31], which is fully consistent with our results. A principal effect of the DNA damage response (DDR) is to maintain genomic stability inducing DNA repair. NHEJ and HDR seem to be the major pathways trigged by DDR in eukaryotic cells. The MRN complex formed by MRE11/RAD50/NBS1 proteins is essential for DNA end resection during HDR repair [38]; our results show a clear increased in the levels of these proteins induced by RSV (Figure 4A,B). At the same time, our results show a clear increased in the levels of DNA-PKcs and KU80 induced by RSV, proteins involved in the initiation of NHEJ pathway [44]. These results demonstrate that DDR induced by RSV in able to promote both NHEJ and HDR repair pathway to maintain genome stability. Importantly, to use different drugs that generate DSBs is extremely beneficial in cancer chemotherapy, increasing the impact of RSV as a complementary therapy.

DNA double-strand breaks are essential for the normal development and differentiation of lymphocytes [32], which is also apparent in Ramos cells as a critical step to induce plasma cell differentiation [33,34]. It has been demonstrated that DNA damage regulates signaling pathways that lower the expression of many genes associated with proliferation and survival. For example, TCL-1 [45], Bach2 [46], and Myc [47] are all downregulated during differentiation in these cells. Our results show decreased mRNA levels for TCL-1, Myc, and Bach2, comparable with the effect of Etoposide, a TOPO2 inhibitor that induce DSBs in Ramos cells [33].

In summary, the results presented here suggest that RSV is a natural molecule with a significant anti-proliferative and apoptotic activity on Ramos cells, inducing the DNA damage response and modulating the expression of several genes that regulate the apoptotic process and the proliferative activity in this model lymphoma cell line.

## 4. Materials and Methods

### 4.1. Cell Culture

Ramos cells, a B lymphocyte cell line from Burkitt's Lymphoma (ATCC CRL-1596), was grown in RPMI-1640 (Hyclone, Logan, UT, USA) containing 10% fetal bovine serum (FBS, Hyclone, Logan, UT, USA), 50 U/mL penicillin, 50 mg/mL streptomycin, and 2 mM L-glutamine, 1mM Sodium Pyruvate, and essential amino acids (from Hyclone, Logan, UT, USA), at 37 °C in humidified 5% $CO_2$ atmosphere.

### 4.2. Cell Viability Assays

Ramos cells were seeded in 96-well plates and treated with different concentrations of resveratrol (Sigma Chemical, St. Louis, MO, USA), we used DMSO as vehicle. After 24 h incubation, mitochondrial activity was measured by the modified 3-[4,5-dimethylthiazol 2-yl]-2,5 diphenyltetrazolium bromide (MTT) assay [48]. Cells were incubated for 4 h at 37 °C with MTT (10 μL of 5 mg/mL MTT solution per well, Sigma-Aldrich, St. Louis, MO, USA). The reaction was stopped with the addition of cell lysis buffer (50% dimethylformamide and 20% SDS, pH 7.4). ΔA values at 550–650 nm were determined using an automatic microtiter plate reader (Metertech Σ960) and the results were expressed as a percentage of control. Cell viability was also assayed by Trypan Blue exclusion and the results were expressed as a percentage of control. The RSV concentrations used in this work are in accordance with previous reports [42].

*4.3. Western Blot Analysis*

Cells were cultured and treated with resveratrol for different time periods and different concentrations. Next, cells were lysed in RIPA buffer (50 mM Tris, pH 7.5, 150 mM NaCl, 5 mM EDTA, 1% NP-40, 0.5% sodium deoxycholate, 0.1% SDS, 100 mg/mL PMSF, 2 mg/mL aprotinin, 2 mM leupeptin, and 1 mg/mL pepstatin) and protein concentration was determined using the Bradford assay. Protein extracts were resolved by SDS–PAGE (60 mg per lane) on a 10% polyacrylamide gel and transferred into immobilon membranes (Millipore, Bedford, MA, USA). After blocking with 5% skimmed milk, membranes were incubated with 1:1000 dilutions of primary antibodies. We used antibodies against p-ATM, p-BRCA1 and γ-H2AX (Cell Signaling Technology, Inc., Danvers, MA, USA) as markers of DNA damage response; active-caspase 3 and cleaved PARP (Cell Signaling Technology, Inc., Danvers, MA, USA) to determine apoptosis marker; and finally, Rad50, Mre11, p-p95/NBS1, DNA-PKcs, and KU80 (Cell Signaling Technology, Inc, Danvers, MA, USA) as markers of DNA repair; Tubulin (Calbiochem, Darmstadt, Germany) was used as loading control.

*4.4. Quantitative Real-Time Reverse Transcriptase-Polymerase Chain Reaction (RT-qPCR)*

Total RNA was isolated from cells using Trizol reagent (Life Technologies, Waltham, MA, USA) following the manufacturer's instructions. Total RNA was subjected to RT-PCR. 1–5 μg of total RNA was used to synthetize first-strand cDNAs with the iScript kit (BIO-RAD, CA, USA). Quantitative RT-PCR (RT-qPCR) analysis was performed as described previously [49]. Expression was normalized to a 36B4 mRNA control sequence. Oligonucleotide primers for real-time RT-PCR are: TCL-1 FW 5′-CGATACCGATCCTCAGACTCCAGTT-3′, RV 5′-AAAGGAGACAGGTGCTGCCAAG-3′, Myc FW 5′-AGCGACTCTGAGGAGGAACAAGAAGAT-3′, RV 5′-TTGGCAGCAGGATAGTCCTTCCG-3′, Bach2 FW 5′-CTTGCCTGAGGAGGTCACAGC-3′, RV 5′-AGCATCCTTCCGGCACACAAA-3′, 36B4 FW 5′-TGGCAGCATCTACAACCCTGAAGT-3′, RV 5′-TGGGTAGCCAATCTGAAGACAGACA-3′.

*4.5. Statistical Analysis*

Data are presented as mean ± SE of the values from the number of experiments performed in triplicate as indicated in the corresponding figures. Data were analyzed for statistically significant differences ($p < 0.05$) using the Student's t-test.

**Acknowledgments:** We thank Pablo Mardones and Francisco J. Morera for critical reading of the manuscript. This work was supported by grants FONDECYT REGULAR 1141067 (to A.Z.), and 1150574 (to C.O.); and DID UACh S-2016-48 (to M.S.).

**Author Contributions:** M.S., C.O. and A.Z. conceived and designed the experiment; P.J., J.S., Y.A., G.V., C.M., C.C. and A.Z. performed experiments and acquired the data; C.C., M.S., C.O. and A.Z. analyzed and interpreted the data; M.S. and A.Z. drafted, edited and revised manuscript.

**Conflicts of Interest:** The authors declare no conflict of interest. The founding sponsors had no role in the design of the study; in the collection, analyses, or interpretation of data; in the writing of the manuscript, and in the decision to publish the results.

**References**

1. Jeandet, P.; Bessis, R.; Maume, B.F.; Meunier, P.; Peyron, D.; Trollat, P. Effect of Enological Practices on the Resveratrol Isomer Content of Wine. *J. Agric. Food Chem.* **1995**, *43*, 316–319. [CrossRef]
2. Jeandet, P.; Delaunois, B.; Conreux, A.; Donnez, D.; Nuzzo, V.; Cordelier, S.; Clement, C.; Courot, E. Biosynthesis, metabolism, molecular engineering and biological functions of stilbene phytoalexins in plants. *Biofactors* **2010**, *36*, 331–341. [CrossRef] [PubMed]

3. Jeandet, P.; Hebrard, C.; Deville, M.A.; Cordelier, S.; Dorey, S.; Aziz, A.; Crouzet, J. Deciphering the Role of Phytoalexins in Plant-Microorganism Interactions and Human Health. *Molecules* **2014**, *19*, 18033–18056. [CrossRef] [PubMed]

4. Agarwal, B.; Baur, J.A. Resveratrol and life extension. *Ann. N. Y. Acad. Sci.* **2011**, *1215*, 138–143. [CrossRef] [PubMed]

5. Wood, J.G.; Rogina, B.; Lavu, S.; Howitz, K.; Helfand, S.L.; Tatar, M.; Sinclair, D. Sirtuin activators mimic caloric restriction and delay ageing in metazoans. *Nature* **2004**, *431*. [CrossRef]

6. Athar, M.; Back, J.H.; Tang, X.; Kim, K.H.; Kopelovich, L.; Bickers, D.R.; Kim, A.L. Resveratrol: A review of preclinical studies for human cancer prevention. *Toxicol. Appl. Pharmacol.* **2007**, *224*, 274–283. [CrossRef] [PubMed]

7. Fauconneau, B.; WaffoTeguo, P.; Huguet, F.; Barrier, L.; Decendit, A.; Merillon, J.M. Comparative study of radical scavenger and antioxidant properties of phenolic compounds from Vitis vinifera cell cultures using in vitro tests. *Life Sci.* **1997**, *61*, 2103–2110. [CrossRef]

8. Juan, M.E.; Alfaras, I.; Planas, J.M. Colorectal cancer chemoprevention by trans-resveratrol. *Pharmacol. Res.* **2012**, *65*, 584–591. [CrossRef] [PubMed]

9. Varoni, E.M.; Lo Faro, A.F.; Sharifi-Rad, J.; Iriti, M. Anticancer Molecular Mechanisms of Resveratrol. *Front. Nutr.* **2016**, *3*. [CrossRef] [PubMed]

10. Tang, F.Y.; Su, Y.C.; Chen, N.C.; Hsieh, H.S.; Chen, K.S. Resveratrol inhibits migration and invasion of human breast-cancer cells. *Mol. Nutr. Food Res.* **2008**, *52*, 683–691. [CrossRef] [PubMed]

11. Aziz, M.H.; Kumar, R.; Ahmad, N. Cancer chemoprevention by resveratrol: In vitro and in vivo studies and the underlying mechanisms. *Int. J. Oncol.* **2003**, *23*, 17–28. [CrossRef] [PubMed]

12. Lee, Y.J.; Lee, Y.J.; Im, J.H.; Won, S.Y.; Kim, Y.B.; Cho, M.K.; Nam, H.S.; Choi, Y.J.; Lee, S.H. Synergistic anti-cancer effects of resveratrol and chemotherapeutic agent clofarabine against human malignant mesothelioma MSTO-211H cells. *Food Chem. Toxicol.* **2013**, *52*, 61–68. [CrossRef] [PubMed]

13. Can, G.; Cakir, Z.; Kartal, M.; Gunduz, U.; Baran, Y. Apoptotic Effects of Resveratrol, a Grape Polyphenol, on Imatinib-Sensitive and Resistant K562 Chronic Myeloid Leukemia Cells. *Anticancer Res.* **2012**, *32*, 2673–2678. [PubMed]

14. Iguchi, K.; Toyama, T.; Ito, T.; Shakui, T.; Usui, S.; Oyama, M.; Iinuma, M.; Hirano, K. Antiandrogenic Activity of Resveratrol Analogs in Prostate Cancer LNCaP Cells. *J. Androl.* **2012**, *33*, 1208–1215. [CrossRef] [PubMed]

15. Hussain, A.R.; Uddin, S.; Bu, R.; Khan, O.S.; Ahmed, S.O.; Ahmed, M.; Al-Kuraya, K.S. Resveratrol suppresses constitutive activation of AKT via generation of ROS and induces apoptosis in diffuse large B cell lymphoma cell lines. *PLoS ONE* **2011**, *6*. [CrossRef] [PubMed]

16. Shimizu, T.; Nakazato, T.; Xian, M.J.; Sayawa, M.; Ikeda, Y.; Kizaki, M. Resveratrol induces apoptosis of human malignant B cells by activation of caspase-3 and p38 MAP kinase pathways. *Biochem. Pharmacol.* **2006**, *71*, 742–750. [CrossRef] [PubMed]

17. Faber, A.C.; Chiles, T.C. Resveratrol induces apoptosis in transformed follicular lymphoma OCI-LY8 cells: Evidence for a novel mechanism involving inhibition of BCL6 signaling. *Int. J. Oncol.* **2006**, *29*, 1561–1566. [CrossRef] [PubMed]

18. Guha, P.; Dey, A.; Sen, R.; Chatterjee, M.; Chattopadhyay, S.; Bandyopadhyay, S.K. Intracellular GSH Depletion Triggered Mitochondrial Bax Translocation to Accomplish Resveratrol-Induced Apoptosis in the U937 Cell Line. *J. Pharmacol. Exp. Ther.* **2011**, *336*, 206–214. [CrossRef] [PubMed]

19. Khan, O.S.; Bhat, A.A.; Krishnankutty, R.; Mohammad, R.M.; Uddin, S. Therapeutic Potential of Resveratrol in Lymphoid Malignancies. *Nutr. Cancer* **2016**, *68*, 365–373. [CrossRef] [PubMed]

20. Delmas, D.; Solary, E.; Latruffe, N. Resveratrol, a phytochemical inducer of multiple cell death pathways: Apoptosis, autophagy and mitotic catastrophe. *Curr. Med. Chem.* **2011**, *18*, 1100–11021. [CrossRef] [PubMed]

21. Shankar, S.; Singh, G.; Srivastava, R.K. Chemoprevention by resveratrol: Molecular mechanisms and therapeutic potential. *Front. Biosci.* **2007**, *12*, 4839–4854. [CrossRef] [PubMed]

22. Park, J.W.; Choi, Y.J.; Suh, S.I.; Baek, W.K.; Suh, M.H.; Jin, I.N.; Min, D.S.; Woo, J.H.; Chang, J.S.; Passaniti, A.; et al. Bcl-2 overexpression attenuates resveratrol-induced apoptosis in U937 cells by inhibition of caspase-3 activity. *Carcinogenesis* **2001**, *22*, 1633–1639. [CrossRef] [PubMed]

23. Clement, M.V.; Hirpara, J.L.; Chawdhury, S.H.; Pervaiz, S. Chemopreventive agent resveratrol, a natural product derived from grapes, triggers CD95 signaling-dependent apoptosis in human tumor cells. *Blood* **1998**, *92*, 996–1002. [PubMed]

24.  Krammer, P.H. CD95's deadly mission in the immune system. *Nature* **2000**, *407*, 789–795. [CrossRef] [PubMed]

25.  Benitez, D.A.; Hermoso, M.A.; Pozo-Guisado, E.; Fernandez-Salguero, P.M.; Castellon, E.A. Regulation of cell survival by resveratrol involves inhibition of NF kappa B-regulated gene expression in prostate cancer cells. *Prostate* **2009**, *69*, 1045–1054. [CrossRef] [PubMed]

26.  Nonn, L.; Duong, D.; Peehl, D.M. Chemopreventive anti-inflammatory activities of curcumin and other phytochemicals mediated by MAP kinase phosphatase-5 in prostate cells. *Carcinogenesis* **2007**, *28*, 1188–1196. [CrossRef] [PubMed]

27.  Hu, Y.; Sun, C.Y.; Huang, J.; Hong, L.; Zhang, L.; Chu, Z.B. Antimyeloma effects of resveratrol through inhibition of angiogenesis. *Chin. Med. J.* **2007**, *120*, 1672–1677. [PubMed]

28.  Lee, J.H.; Guo, Z.; Myler, L.R.; Zheng, S.; Paull, T.T. Direct activation of ATM by resveratrol under oxidizing conditions. *PLoS ONE* **2014**, *9*. [CrossRef] [PubMed]

29.  Demoulin, B.; Hermant, M.; Castrogiovanni, C.; Staudt, C.; Dumont, P. Resveratrol induces DNA damage in colon cancer cells by poisoning topoisomerase II and activates the ATM kinase to trigger p53-dependent apoptosis. *Toxicol. In Vitro* **2015**, *29*, 1156–1165. [CrossRef] [PubMed]

30.  Schroeter, A.; Marko, D. Resveratrol modulates the topoisomerase inhibitory potential of doxorubicin in human colon carcinoma cells. *Molecules* **2014**, *19*, 20054–20072. [CrossRef] [PubMed]

31.  Basso, E.; Fiore, M.; Leone, S.; Degrassi, F.; Cozzi, R. Effects of resveratrol on topoisomerase II-alpha activity: Induction of micronuclei and inhibition of chromosome segregation in CHO-K1 cells. *Mutagenesis* **2013**, *28*, 243–248. [CrossRef] [PubMed]

32.  Bredemeyer, A.L.; Helmink, B.A.; Innes, C.L.; Calderon, B.; McGinnis, L.M.; Mahowald, G.K.; Gapud, E.J.; Walker, L.M.; Collins, J.B.; Weaver, B.K.; et al. DNA double-strand breaks activate a multi-functional genetic program in developing lymphocytes. *Nature* **2008**, *456*, 819–823. [CrossRef] [PubMed]

33.  Sherman, M.H.; Kuraishy, A.I.; Deshpande, C.; Hong, J.S.; Cacalano, N.A.; Gatti, R.A.; Manis, J.P.; Damore, M.A.; Pellegrini, M.; Teitell, M.A. AID-induced genotoxic stress promotes B cell differentiation in the germinal center via ATM and LKB1 signaling. *Mol. Cell* **2010**, *39*, 873–885. [CrossRef] [PubMed]

34.  Walsh, N.C.; Teitell, M. B-cell differentiation stimulated by physiologic DNA double strand breaks. *Cell Cycle* **2011**, *10*, 176–177. [CrossRef] [PubMed]

35.  Jazirehi, A.R.; Bonavida, B. Resveratrol modifies the expression of apoptotic regulatory proteins and sensitizes non-Hodgkin's lymphoma and multiple myeloma cell lines to paclitaxel-induced apoptosis. *Mol. Cancer Ther.* **2004**, *3*, 71–84. [PubMed]

36.  Ghorbani, A.; Zand, H.; Jeddi-Tehrani, M.; Koohdani, F.; Shidfar, F.; Keshavarz, S.A. PTEN over-expression by resveratrol in acute lymphoblastic leukemia cells along with suppression of AKT/PKB and ERK1/2 in genotoxic stress. *J. Nat. Med.* **2015**, *69*, 507–512. [CrossRef] [PubMed]

37.  Shibata, A.; Jeggo, P.A. DNA double-strand break repair in a cellular context. *Clin. Oncol. (R. Coll. Radiol.)* **2014**, *26*, 243–249. [CrossRef] [PubMed]

38.  Aparicio, T.; Baer, R.; Gautier, J. DNA double-strand break repair pathway choice and cancer. *DNA Repair* **2014**, *19*, 169–175. [CrossRef] [PubMed]

39.  Faber, A.C.; Dufort, F.J.; Blair, D.; Wagner, D.; Roberts, M.F.; Chiles, T.C. Inhibition of phosphatidylinositol 3-kinase-mediated glucose metabolism coincides with resveratrol-induced cell cycle arrest in human diffuse large B-cell lymphomas. *Biochem. Pharmacol.* **2006**, *72*, 1246–1256. [CrossRef] [PubMed]

40.  Yan, Y.; Gao, Y.Y.; Liu, B.Q.; Niu, X.F.; Zhuang, Y.; Wang, H.Q. Resveratrol-induced cytotoxicity in human Burkitt's lymphoma cells is coupled to the unfolded protein response. *BMC Cancer* **2010**, *10*. [CrossRef] [PubMed]

41.  Cecconi, D.; Zamo, A.; Parisi, A.; Bianchi, E.; Parolini, C.; Timperio, A.M.; Zolla, L.; Chilosi, M. Induction of apoptosis in Jeko-1 mantle cell lymphoma cell line by resveratrol: A proteomic analysis. *J. Proteome Res.* **2008**, *7*, 2670–2680. [CrossRef] [PubMed]

42.  Salas, M.; Obando, P.; Ojeda, L.; Ojeda, P.; Perez, A.; Vargas-Uribe, M.; Rivas, C.I.; Vera, J.C.; Reyes, A.M. Resolution of the direct interaction with and inhibition of the human GLUT1 hexose transporter by resveratrol from its effect on glucose accumulation. *Am. J. Physiol. Cell. Physiol.* **2013**, *305*, C90–C99. [CrossRef] [PubMed]

43.  Lapenna, S.; Giordano, A. Cell cycle kinases as therapeutic targets for cancer. *Nat. Rev. Drug Discov.* **2009**, *8*, 547–566. [CrossRef] [PubMed]

44.  Neal, J.A.; Meek, K. Choosing the right path: Does DNA-PK help make the decision? *Mutat. Res.* **2011**, *711*, 73–86. [CrossRef] [PubMed]
45.  Shen, R.R.; Ferguson, D.O.; Renard, M.; Hoyer, K.K.; Kim, U.; Hao, X.; Alt, F.W.; Roeder, R.G.; Morse, H.C., 3rd; Teitell, M.A. Dysregulated TCL1 requires the germinal center and genome instability for mature B-cell transformation. *Blood* **2006**, *108*, 1991–1998. [CrossRef] [PubMed]
46.  Swaminathan, S.; Huang, C.; Geng, H.; Chen, Z.; Harvey, R.; Kang, H.; Ng, C.; Titz, B.; Hurtz, C.; Sadiyah, M.F.; et al. BACH2 mediates negative selection and p53-dependent tumor suppression at the pre-B cell receptor checkpoint. *Nat. Med.* **2013**, *19*, 1014–1022. [CrossRef] [PubMed]
47.  Linden, M.A.; Kirchhof, N.; Carlson, C.S.; Van Ness, B.G. Targeted overexpression of an activated N-ras gene results in B-cell and plasma cell lymphoproliferation and cooperates with c-myc to induce fatal B-cell neoplasia. *Exp. Hematol.* **2012**, *40*, 216–227. [CrossRef] [PubMed]
48.  Mosmann, T. Rapid colorimetric assay for cellular growth and survival: Application to proliferation and cytotoxicity assays. *J. Immunol. Methods* **1983**, *65*, 55–63. [CrossRef]
49.  Kuraishy, A.I.; French, S.W.; Sherman, M.; Herling, M.; Jones, D.; Wall, R.; Teitell, M.A. TORC2 regulates germinal center repression of the TCL1 oncoprotein to promote B cell development and inhibit transformation. *Proc. Natl. Acad. Sci. USA* **2007**, *104*, 10175–10180. [CrossRef] [PubMed]

**Sample Availability:** Samples of the compound resveratrol (RSV), and the antibodies are available from the authors.

MDPI AG

St. Alban-Anlage 66

4052 Basel, Switzerland

Tel. +41 61 683 77 34

Fax +41 61 302 89 18

http://www.mdpi.com

*Molecules* Editorial Office

E-mail: molecules@mdpi.com

http://www.mdpi.com/journal/molecules